CONFERENCE PROCEEDINGS

The People's Liberation Army in the Information Age

Edited by

James C. Mulvenon
and
Richard H. Yang

Supported by the Center for Asia-Pacific Policy and the U.S. Air Force

NATIONAL SECURITY RESEARCH DIVISION

Project AIR FORCE

RAND

PREFACE

This volume is the product of a conference, jointly sponsored by the RAND Center for Asia-Pacific Policy (CAPP) and the Taiwan-based Chinese Council of Advanced Policy Studies (CAPS), held in San Diego, California, from 9–12 July 1998. The meeting brought together Chinese military experts to discuss a subject too long ignored: the non-hardware side of the People's Liberation Army's (PLA's) modernization. The result is a comprehensive examination of the critical "software" side of China's military modernization, covering topics as diverse as civil-military relations, professionalism, logistics, training, doctrine, systems integration, and force structure.

Whereas financial and logistical support for the conference was supplied by CAPS and CAPP, funding for the publication of this volume was provided by RAND's Project AIR FORCE Strategy and Doctrine Program, under the leadership of Dr. Zalmay Khalilzad. This program is in the third year of a comprehensive study of issues related to Chinese military and security affairs for the United States Air Force; the project is entitled "Chinese Defense Modernization and Its Implications for the U.S. Air Force." It focuses on the fundamental question of how U.S. policy should deal with China, a rising power that could have the capability, in the not too distant future, of challenging the U.S. position in East Asia and its military, political, and economic access to that dynamic and important region. It then addresses the implications for the Air Force, in the areas of shaping the environment, deterrence, and war fighting. To achieve these objectives, RAND is building a "Center of Excellence" for the study of China and the PLA.

PROJECT AIR FORCE

Project AIR FORCE, a division of RAND, is the Air Force federally funded research and development center (FFRDC) for studies and analyses. It provides the Air Force with independent analyses of policy alternatives affecting the development, employment, combat readiness, and support of current and future aerospace forces. Research is performed in three programs: Strategy and Doctrine, Force Modernization and Employment, and Resource Management and System Acquisition.

CONTENTS

ACKNOWLEDGMENTS

This conference would not have been possible without the generous financial support of the Chinese Council of Advanced Policy Studies, whose tireless advocacy for study of the People's Liberation Army (PLA) is well known within the community of scholars but deserves much wider recognition. Special thanks to Andrew and Yi-Su Yang for their consistently excellent conference management.

They always create the perfect environment for genuine progress toward a better understanding of the often opaque Chinese military system.

Everyone in the field looks forward to their annual meeting, both as a gathering of friends and a community of scholarship.

Special thanks also to Dr. Zalmay Khalilzad, director of the Project AIR FORCE Strategy and Doctrine Program, for his financial support of the publication of this conference volume.

1. INTRODUCTION

James Mulvenon
Richard H. Yang

It is now generally held by military analysts that the dramatic victories by the United States in the 1991 Gulf War were as much a product of advanced American technology as they were a result of competence in the less sexy but perhaps more important areas of logistics, training, education, systems integration, and information-based C4ISR (command, control, communications, computers, intelligence, surveillance, and reconnaissance). Most observers believe that critical U.S. advances in these latter areas acted as critical "force multipliers," significantly enhancing the lethal combat effectiveness of coalition forces arrayed against Saddam Hussein. The payoff for years of expensive investment in such human capital and information technologies was the overwhelming dominance over opposition forces and the minimization of U.S. casualties.

For its part, China since 1978 has begun to make the necessary changes in its weapons inventory to be considered a modern military force. Purchases from abroad, such as fourth-generation SU-27 fighters, *Sovremenny*-class destroyers, SA-10 air defense batteries, and *Kilo*-class submarines, have joined indigenous equipment to form the outlines of a future, high-tech military organization. These advances have been carefully documented by numerous outside observers. Less frequently analyzed, however, have been Chinese attempts to reform and modernize their logistics networks, C4I infrastructure, systems integration capabilities, doctrinal frameworks, and force structure. From the conference papers assembled here, it appears that the Chinese are making halting but positive advances in these areas, although they clearly have a long way to go before they can claim to have developed a military on par with that of the United States or many of its allies.

The volume begins with a chapter by David Shambaugh that grew out of his presentation on the "State of the PLA Field" at the conference on the final day. After the conference had adjourned, the resulting paper was judged to be of sufficient merit to warrant inclusion in this volume, even though it does not directly address the topic at hand. Instead, it is a sweeping view of the last thirty years of Chinese military studies. Shambaugh's chapter examines both the producers and consumers of Chinese military analysis, and concludes that the field has increased in both quantity and quality in recent years. He evaluates the data used by PLA analysts, asking the critical questions: what do we know and what don't we know? Finally, he lays out a list of challenges for the foreseeable future, emphasizing the need for the

People's Liberation Army (PLA) field to reach out to other disciplines within Asian studies and political science.

Ellis Joffe's chapter builds on a monograph originally published for the Chinese Council of Advanced Policy Studies (CAPS). To update this effort, the author has added an epilogue to include the 15th Party Congress in October 1997 and the National People's Congress in March 1998. He argues that civil-military relations have entered a new mode, characterized by unprecedented potential for the wielding of political power by the military, although this is tempered by professionalization and countervailing political factors. This new potential has manifested itself in the military's new capacity to influence high politics and policies, not routine involvement in political or administrative affairs. Joffe traces the origins of this change to three important trends in the Chinese system: (1) the transformation of the role of the paramount leader under Jiang Zemin; (2) the military's new leverage in narrowly defined policymaking arenas, in particular foreign policy and internal PLA decisionmaking; and (3) a growing party-army separation marked by conflicting interests. The author concludes, however, by pointing out the PLA's new potential power has not been exercised and in fact has led to a decline in interventions, thanks to Jiang Zemin's clear dependence on the PLA and generational change throughout the system. Joffe cautions that Jiang will retain the military's support only as long as he delivers economic growth and social stability.

Andrew Yang and Milton Liao of CAPS offer a deeply detailed analysis of the origins and evolution of Rapid Reaction Forces (RRFs) in the PLA. This chapter examines the PLA's establishment of RRFs within the existing force structure, and assesses the force's stated purpose of fighting regional wars under high-tech conditions and suppressing internal disturbances. The authors' analysis focuses on the various RRF force structures, weapon systems, and logistics capabilities, as well as training and exercises conducted from 1996–1997. Yang and Liao conclude that the RRFs provide the PLA with new operational possibilities, but also create substantial challenges for the existing system, especially in the areas of force coordination, logistics support, and C3I (command, control, communications, and intelligence).

Harlan Jencks' chapter examines the restructuring effort under way in the Chinese defense-industrial complex, particularly since the announcements made at the National People's Congress in March 1998. The author describes and attempts to evaluate the status of that reorganization after its first months, with special reference to the defense scientific, technical, and industrial sectors. He points out that there is considerable confusion even among senior Chinese officials as to which organizations they currently represent or will represent in the near future, as well as which higher-level organizations they are currently answerable to. Jencks concludes that the true outlines of the restructuring will not be clear for several years, especially given ample evidence of discontent and resistance among Chinese officials and institutions. Indeed, he argues there is a significant possibility that the restructuring will never be completed at all.

Kenneth Allen's chapter analyzes the long, uneven, but positive journey that the Chinese Air Force has undertaken to modernize its obsolescent force of 1950s and

1960s vintage combat aircraft. The analysis updates the research that originally appeared in Kenneth Allen et al., *China's Air Force Enters the 21st Century*, RAND, MR-580-AF, 1995. To modernize rapidly, the author argues, the People's Liberation Army Air Force (PLAAF) and China's aviation industry have had to seek foreign assistance for entire weapon systems, subsystems, and technical support. He asserts that they have used this foreign assistance to successfully develop and produce indigenous weapon systems, form a rapid reaction force, train with new combat tactics, computerize some of their command centers, and revise logistics and maintenance support capabilities. At the same time, Allen points out that the PLAAF continues to be plagued by a set of serious deficiencies, most of which will need to be addressed to achieve clear breakthroughs. The author concludes by raising two critical questions: Will the PLAAF be able to carry out these reforms, especially in a joint force arena? Will it be able to support its weapons systems acquired in full or in part from foreign sources?

Using the Army War College model for analyzing strategy, David Finkelstein's chapter offers a notional national military strategy for China. It asks the question, "If China published a national military strategy as does the Pentagon, what would it look like?" To answer this question, the author examines China's national security objectives writ large, derives national military objectives, and then discusses the ways in which the PLA is working to achieve those objectives. The author places the PLA's current reforms within a comprehensive, strategic context, and derives a methodology to explain the future paths the PLA will take. This chapter therefore offers a road map for explaining the rationale for reform of the Chinese military for some time to come.

Nan Li's chapter attempts to flesh out the specific components of the PLA's new campaign doctrine. The author introduces a number of new concepts from the Chinese military literature, including "war zone campaign," "elite forces and sharp arms," "gaining initiative by striking first," and "fighting a quick battle to force a quick resolution." He argues that these doctrinal principles are designed to transform the PLA's technological inferiority into local and momentary superiority in future "local war under high-tech conditions." At the same time, he cautions that more careful analysis is necessary to determine to what extent these theoretical changes may translate into actual changes in PLA practice.

"The PLA and Information Warfare," by James Mulvenon, introduces us to a fertile new area in Chinese military doctrine. Mining tens of books and hundreds of articles by PLA authors, Mulvenon presents key Chinese definitions of IW terminology, compares them with their counterparts in the U.S. literature, and charts the evolution of Chinese IW strategy before and after the 1991 Gulf War. He identifies critical theorists and centers involved in the development of IW doctrine in China, as well as the key principles of the People's Republic of China (PRC) information warfare writings, assessing to what extent the literature is derivative of Western scholarship. He concludes that there is a "uniquely Chinese IW strategy," in particular the emphasis on "preemption," although the vast majority of writings are plagaristic renderings of U.S. authors. Finally, Mulvenon presents an unsettling IW scenario involving Taiwan, in which the PLA carries out computer network attacks against civilian and military targets in the United States, with the goal of delaying American military deployment to the Taiwan theater, while simultaneously bringing Taiwan to the negotiating table through a campaign of ballistic missile attacks, fifth column sabotage, and IW attacks against critical infrastructure.

John Frankenstein's chapter on Chinese defense industries provides a comprehensive overview of the experiences of the CDIC (Chinese defense-industrial complex) since the onset of the reform era. The author examines the contradictions of the defense industrial system, identifying the many failures and relatively few successes. Frankenstein then probes the origins of these failures, analyzing the multiple organizational, financial, and bureaucratic problems of China's defense production base. He evaluates the various strategies used by the industry to break out of its downward cycle, especially conversion to civilian production, and concludes that China has achieved little success with either a dual-use spin-off or spin-on approach. In the few cases where military factories have become profitable by switching to civilian output, they paradoxically have become more resistant to continuing their less-profitable military production. The author concludes that the primary motivation behind defense industrial policies in the reform era has been the maintenance of social stability, which argues for a defense industry marked by full employment and inefficiency. Reforms introduced at the March 1998 National People's Congress suggest a new willingness to consider a more Darwinian attitude toward the industry, but economic downturn may foil these attempts, leaving China's defense-industrial base to continue to muddle through its problems.

The chapter offered by RADM Eric McVadon (ret.) examines a crucial element of military modernization: systems integration. The author argues that the ability of the PLA to carry out systems integration provides insights into the future of Chinese military modernization that are possibly more important than examinations of technology acquisition and assimilation. As an example, he points to the PLA Strategic Rocket Forces, which illustrate how the PLA can succeed domestically by avoiding requirements for systems integration. In addition, he analyzes the newest major combatant ships of the PLA Navy and explains why China must look to foreign sources for the essential integration of complex systems. The concept of systems integration is defined, and five levels of systems integration are explained, with discussion of the state of PLA systems integration achievement at each level. The aspirations of the PLA with respect to systems integration are described as well as the views of American technical specialists of Chinese success (or lack thereof). McVadon concludes that the PLA may be losing ground technologically because other countries are achieving systems integration rapidly. While some in the PLA think Western systems integration will produce exploitable vulnerabilities, McVadon asserts that the ability to effectively conduct such exploitation is almost certainly beyond the reach of the PLA in the foreseeable future. At the same time, he cautions that some of the areas in which the PLA may attain limited success in systems integration could indeed be troublesome. These narrow advances could be significant in the delicate balance of forces with Taiwan in certain warfare areas, and could also cause consternation for U.S. forces in other threat areas.

Mulvenon and Bickford's chapter examines the role of PLA enterprises in the Chinese telecommunications industry up to July 1998 and outlines a number of issues related to these enterprises. First, the authors examine the origins of PLA involvement in the telecommunications industry, arguing that the military's exploitation of both its internal communications infrastructure and priority access to civilian infrastructure was a natural consequence of the military's participation in commercial activities. Second, they outline the extent of PLA activities in this area in the 1980s and 1990s, examining two case studies: the China Telecom–Great Wall code division multiple access (CDMA) project and PLA unit involvement in the radio-paging industry. Third, Mulvenon and Bickford assess the implications of these joint venture telecommunications projects for technology transfer to the PLA's C4I modernization. The authors conclude by arguing that while the PLA's potential for growth in this industry is significant, the PLA's recent decision to divest its economic interests indicates that the civilian sector is likely to dominate the telecommunications industry at the PLA's expense in the medium term.

Dennis Blasko's chapter envisions the outlines of a future PLA force structure, based on the trends of the past twenty years and recent statements in a 1998 defense White Paper. He forecasts that the PLA will be reduced to perhaps 1.5 million, as PLA civilians and business operations are stripped of their active duty roles. The resulting force will still lean toward the ground forces, but the percentage share of personnel in the naval and air forces will increase, as will the numbers of reserve and PAP (armed police force) soldiers. For the foreseeable future, the author predicts that units will have a mix of high-, medium-, and low-technology weapons and equipment, and will strive to find new ways to maximize the use of their existing equipment to defeat a high-technology enemy. The numbers and types of logistics and technical units will increase throughout the force to maintain and support the PLA's modern equipment. The Chinese defense industries, in his view, will be able to produce limited amounts of modern weapons for the PLA, but most truly advanced weapons will be of foreign origin and relatively few in number. Rapid deployment of conventional forces will be enhanced through acquisition of transport ships and aircraft as well as by unit consolidation near points of embarkation. He forecasts that naval and air forces will acquire more offensive capabilities and the ability to operate farther from the Chinese land mass, but an operational aircraft carrier capability will not enter the force until at least the end of the second decade of the 21st century. Cruise missile, ballistic missile, and nuclear forces will be improved gradually and incrementally and will remain the key to China's deterrent force. In his opinion, changes in the command and control structure will contribute to better integration of forces and capabilities. Several regional headquarters will be eliminated, resulting in five "theater-like" headquarters. A few smaller headquarters will be formed for the Army, Special Operations Forces, and Space Forces. Tactical units will be restructured during a period of experimentation. Blasko concludes that the PLA must do a better job of explaining these changes to the outside world, lest China's neighbors and the United States misinterpret these reforms as proof of a more aggressive posture.

In sum, the chapters present a picture of a PLA that has learned many of the right lessons about the importance of "software" to successful combat performance, but is enjoying more variable success in implementing the necessary changes throughout the system. It is our contention that the Chinese military's progress on these more-difficult problems will ultimately determine whether the PLA will transform itself into a peer competitor with the United States in the 21st century or remain merely a regional military power.

2. PLA STUDIES TODAY: A MATURING FIELD

David Shambaugh[1]

It is useful in all areas of scholarly endeavor to periodically stand back and reflect upon the state of a field. This essay was inspired by listening to the discussion of the other contributions to this volume as they were presented at the San Diego conference. What follows is *not* intended as either a critical survey of the existing secondary literature or a comprehensive assessment of the issues that intellectually occupy the field (as previously undertaken by Jonathan Pollack, Harlan Jencks, and June Teufel Dreyer[2]), nor is it an effort in primary research. With these caveats in mind, let me address several themes.

PRODUCERS AND CONSUMERS

Who are the contributors to PLA studies today, how do they differ from previous generations, and what are the distinguishing features of this community? This discussion also assesses the changing "consumers" of the research published by the PLA studies community.

There is little doubt that the field has increased in both quantity and quality in recent years. However, it is not easy to calculate the field's parameters and size. Unlike a country club, with a roll and dues-paying membership, inclusion in the field of PLA studies is not so easily counted. One simple criterion—publication—may reveal only a partial list. For example, some of the keenest analysts—like the legendary Ellis Melvin of Tamaroa, Illinois—labor away in obscurity and never publish or attend conferences. This can, of course, also be said of the large cadre of government and

[1]David Shambaugh is Professor of Political Science and International Affairs, and Director of the China Policy Program, at The George Washington University. He is also nonresident Senior Fellow in the Foreign Policy Studies Program at the Brookings Institution, and is active in a number of other organizations. He was previously editor of *The China Quarterly* and taught at the School of Oriental and African Studies of the University of London. He has published extensively on Chinese domestic politics, foreign relations, and security affairs.

The author is grateful to those who offered useful comments and suggestions on an earlier draft of this chapter: Ellis Joffe, Lucian Pye, John Frankenstein, David Finkelstein, Dennis Blasko, Harlan Jencks, Michel Oksenberg, and K. P. Ng.

[2]Jonathan D. Pollack, "The Study of Chinese Military Politics: Toward a Framework for Analysis," in Catherine M. Kelleher (ed.), *Political-Military Systems: Comparative Perspectives*, Beverly Hills: Sage, 1974; June Teufel Dreyer, "State of the Field Report: Research on the Chinese Military," *AccessAsia Review*, Vol. 1, No. 1, Summer 1997, pp. 5–30; and Harlan W. Jencks, "Watching China's Military: A Personal View," *Problems of Communism*, May–June 1986, pp. 71–78.

intelligence analysts whose work never sees the light of day. Some, like John Lewis, publish a lot and have been major contributors to the field over a number of years, but elect not to attend conferences and have limited interaction with others in the field. Others, like Mel Gurtov, suddenly reemerge after years of silence to publish significant books and articles.[3] Another subgroup, which includes William Whitson, Doug Lovejoy, Tom Roberts, Eden Woon, Michael T. Byrnes, and Richard Latham, have left for other professional endeavors.[4] Some, like Al Wilhelm or Chas Freeman, are fountains of knowledge about, and frequently interact with, the PLA, yet they rarely publish in the field. Some academics who contributed in the 1970s—Harvey Nelsen, Harry Gelber, and Tom Robinson come to mind—remain intellectually engaged in the issues but have stopped publishing in the field.[5] Some who contributed major studies of the PLA prior to and during the Cultural Revolution— like Alexander George and John Gittings—remained active authors but no longer wrote about the PLA.[6] Others—like Benjamin Ostrov, Paul Folta, Mark Ryan, and Cheng Hsiao-shih—contributed important monographs, never to be heard from again.[7] Some scholars of domestic Chinese politics occasionally venture into the field and publish noteworthy articles but tend not to be regularly engaged in research on the PLA. Examples include Wang Shaoguang, Wei Li, Jeremy Paltiel, and Avery Goldstein.[8] Similarly, there are well-known scholars of Chinese foreign policy—Allen Whiting, Richard Solomon, Harry Harding, John Garver, Robert Ross, Gerald Segal, Tom Christiansen, A. Doak Barnett, and others—who have contributed major case studies in the national security sphere and on China's crisis behavior, but who are not generally thought of as active PLA watchers.

Thus, it is not a simple task to determine the composition of the PLA studies community today. One cannot judge membership based on publications alone. This

---

[3]Mel Gurtov and Byong-Moo Hwang, *China's Security: The New Roles of the Military*, Boulder, CO: Lynne Rienner, 1998; and Mel Gurtov, "Swords into Market Shares: China's Conversion of Military Industry to Civilian Production," *The China Quarterly*, June 1993, pp. 231–241. Twenty-seven years ago, Gurtov and Hwang published *China Under Threat: The Politics of Strategy and Diplomacy*, Baltimore: Johns Hopkins, 1971.

[4]See, for example, William Whitson, *The Chinese High Command*, New York: Praeger, 1973; Charles D. Lovejoy, Jr. and Bruce W. Watson (eds.), *China's Military Reforms*, Boulder, CO.: Westview, 1986; and Thomas C. Roberts, *The Chinese People's Militia and the Doctrine of People's War*, Washington, D.C.: National Defense University Press, 1983. Richard Latham's contributions over the years are too numerous to list.

[5]See Harvey Nelsen, *The Chinese Military System*, Boulder, CO.: Westview, 1977, 1981; Harry G. Gelber, *Technology, Defense, and External Relations in China, 1975–1978*, Boulder, CO: Westview, 1979.

[6]Alexander George, *The Chinese Communist Army in Action*, New York: Columbia, 1967; John Gittings, *The Role of the Chinese Army*, London: Oxford University Press, 1967; and Gerald Corr, *The Chinese Red Army*, London: Osprey Publishers, 1974.

[7]See Benjamin C. Ostrov, *Conquering Resources: The Growth and Decline of the PLA's Science and Technology Commission for National Defense*, Armonk, NY: M. E. Sharpe, 1991; Paul Humes Folta, *From Swords to Plowshares: Defense Industry Reform in the PRC*, Boulder, CO: Westview, 1992); Mark A. Ryan, *Chinese Attitudes Towards Nuclear Weapons*, Armonk, NY: M. E. Sharpe, 1991; and Cheng Hsiao-shih, *Party-Military Relations in the PRC and Taiwan*, Boulder, CO.: Westview, 1990.

[8]Wang Shaoguang, "Estimating China's Defense Expenditure: Some Evidence From Chinese Sources," *The China Quarterly*, September 1996, pp. 889–911; Wei Li, "The Security Service for Chinese Central Leaders," *The China Quarterly*, September 1995, pp. 814–827; Jeremy Paltiel, "PLA Allegiance on Parade: Civil-Military Relations in Transition," *ibid.*, pp. 784–800; Avery Goldstein, "Great Expectations: Interpreting China's Arrival," *International Security*, Vol. 22, No. 3, Winter 1997–1998, pp. 36–73.

is certainly the case with the large cohort of PLA specialists in the United States and foreign governments and militaries. Inside the U.S. government and military, for example, there may be as many as 100 analysts devoted full-time to studying the PLA in the Department of Defense, Defense Intelligence Agency (DIA), Central Intelligence Agency, National Security Agency, National Reconnaissance Office, U.S. Pacific Command (PACOM), and other offices. On rare occasion these individuals participate in academic conferences, but they never contribute papers or publish in academic journals (even under pseudonyms). Their participation in conferences is to be welcomed, and they make important contributions on these occasions, yet their failure to publish in the unclassified world is regrettable. Apparently, the Byzantine bureaucratic procedures necessary to gain clearance to publish serve as a significant deterrent, although it is not necessarily the case that these individuals would be so inclined if the bureaucratic strictures were not such an impediment. This has not always been the case; in years past such analysts were regular contributors to journals such as *The China Quarterly*. William Whitson's landmark edited volume *The Military and Political Power in China in the 1970s* contained a large number of chapters written by intelligence analysts.[9] The recent contributions by several DIA analysts to the 1997 Staunton Hill conference volume may be suggestive of renewed contributions to public discourse.[10]

Overseas, not surprisingly, the government and military on Taiwan devote substantial human resources to analyzing the PLA. This information does not tend to reach the public domain, in either English or Chinese, but fruitful discussions can be had with such experts on the island. The governments and militaries in Japan, South Korea, Australia, India, and Russia also contain small cohorts of PLA watchers, and there is a smattering of academic and private-sector PLA experts in these countries. In Europe, PLA specialists in government are minimal and outside of government they are virtually non-existent (three come to mind in Germany, two in France, none in Britain, one in Scandinavia, and none in the Mediterranean countries or Central Europe). Special mention should be made of Israel, where two leading members of the field reside: Ellis Joffe and Yitzak Shichor.

So who comprises the PLA studies community today? Institutionally, the community of active Western PLA specialists resident in academe in the United States and abroad remains woefully small: Ellis Joffe, Yitzhak Shichor, June Dreyer, John Frankenstein, Iain Johnston, myself, and the group of younger scholars mentioned below. The dearth of university-based PLA specialists is abundantly evident when one examines the list of contributors to this or other conference volumes in the field. This is lamentable, but is unlikely to change for the simple fact that there are few incentives for a China scholar to take up the PLA as a subject area of primary research: there are no university jobs in comparative politics, international relations, or security studies that are specifically designated for a PLA specialist; there are few knowledgeable professors to train such students; there exist extremely limited fieldwork opportunities in China; and few academic journals welcome article

---

[9]William Whitson (ed..), *The Military and Political Power in the 1990s*, NY: Praeger, 1972.

[10]See special issue of the *China Strategic Review*, Vol. 3, No. 1, Spring 1998.

submissions in this field (although this is improving in recent years with *Survival, International Security, Security Studies, Asian Survey, The China Quarterly,* and *The China Journal* all publishing increasing numbers of articles on Chinese security and defense matters[11]). These are all real disincentives to write a doctoral dissertation on the PLA or work in the field. Indeed, it is a field without a comfortable home in the social sciences. In political science, it does not comfortably fit into either comparative politics or international relations (IR); its closest "home" is in the tenuous subfield in IR of security studies (a field which itself suffers from a chronic identity crisis). As is discussed further below, China political scientists have long shunned PLA studies—a bizarre peculiarity considering the pivotal role the military has played in the political life of the PRC.[12] Yet, those in academe who have ventured into the PLA studies field have largely done so by way of the study of Chinese domestic politics.

Of late, there are encouraging signs of a new generation of young scholars entering the field who have received their doctorates in recent years: James Mulvenon from UCLA, Thomas Bickford from California-Berkeley, Evan Feigenbaum from Stanford; Andrew Scobell and Fang Zhu from Columbia, Huang Jing from Harvard, Nan Li from Johns Hopkins, and You Ji from the Australian National University.[13] Although Mulvenon has opted for a career outside the Academy, it is hoped the others will continue to research and write on the PLA in the post-doctoral period. This new generation is armed with solid social science training, a sound base in China area studies, and good Chinese language skills. Several of these individuals come from China, and bring obvious linguistic and research assets with them. Some are children of PLA officers. It can be hoped that these individuals will invigorate PLA studies as PRC émigrés have done for the study of Chinese politics, political economy, foreign policy, and diplomatic history. They are to be welcomed into our ranks, although, as foreign nationals, their presence at conferences poses distinct problems for the participation of government experts.

If the PLA academic studies community remains small, who comprise the majority? As the contributors to this volume make clear, the bulk of PLA specialists today work

---

[11]Submissions of manuscripts to these journals have increased severalfold in recent years.

[12]One can survey many of the major monographs on Chinese politics over the years— from Franz Schurmann's *Ideology and Organization in Communist China* to Roderick MacFarquhar's edited *Chinese Politics*—and find scant attention paid to the PLA either as an actor in the political process or as an institution worthy of attention in its own right. I encourage colleagues in comparative politics and foreign policy studies to integrate the military into their work. Whether at a central, regional, or local level, the military is an important actor in politics, commerce, security, and governance. I have attempted elsewhere to "bring the soldier back in" to our understanding of the pre–Cultural Revolution period. See David Shambaugh, "The Building of the Civil-Military State in China 1949–1965: Bringing the Soldier Back In," in Timothy Cheek and Tony Saich, (eds.), *The Construction of State Socialism in China, 1949–1965,* Armonk, NY: M. E. Sharpe, 1997.

[13]James Mulvenon, "Soldiers of Fortune: The Rise of the Military-Business Complex in the Chinese People's Liberation Army, 1978–98," Dissertation, University of California, Los Angeles, 1998; Nan Li, "Bureaucratic Behavior, Praetorian Behavior, and Civil-Military Relations: Deng Xiaoping's China (1978–89) and Gorbachev's Soviet Union," Dissertation, Johns Hopkins University, 1994; Fang Zhu, *Gun Barrel Politics: Party-Army Relations in Mao's China,* Boulder, CO: Westview Press, 1998; Thomas J. Bickford, "Marching into the Abyss: The Changing Role of the People's Liberation Army in Chinese Politics," Dissertation, University of California, Berkeley, 1995; Andrew Scobell, "Civil-Military Relations in the People's Republic of China in Comparative Perspective," Dissertation, Columbia University, 1995.

in private sector research institutes, corporations, and consultancies—e.g.: Jonathan Pollack, Michael Swaine, and James Mulvenon of RAND; Bates Gill of The Brookings Institution; Rick Fisher of The Heritage Foundation; James Lilley and Arthur Waldron of the American Enterprise Institute; Michael McDevitt and David Finkelstein of the Center for Naval Analysis; Monte Bullard of the Center for Non-Proliferation Studies at the Monterey Institute of International Studies; Dennis Blasko of International Trade and Technology Associates; Wendy Frieman of Science Applications International Corporation; Ken Allen of The Stimson Center; and Tai Ming Cheung of Kroll Associates. These individuals spend much of their time, but by no means all, researching the PLA, but their research agendas are largely driven by the needs of their corporate or government sponsors and clients. They do not generally have the freedom or the time to structure their own research agendas, and their output takes different forms from traditional academic work, i.e., generally shorter, policy-oriented studies instead of books or detailed case studies (RAND reports being the notable exception to this rule). Other members of the field, such as Paul Godwin, Bud Cole, Ron Montaperto, Larry Wortzel, Harlan Jencks, and Shirley Kan, work in U.S. government institutions. Their publications do not appear to be influenced by their employers, and each has made significant contributions to the field over the years, but inevitably their research agendas are somewhat influenced by the needs of their sponsors. Others, like John Corbett, Karl Eikenberry, Mark Stokes, Roy Kamphausen, and others, are currently in active military or government service. Others, like John Caldwell and Ed O'Dowd, are on assignments that have temporarily taken them out of the field. Some, such as Eric McVadon, have retired from active service and continue as active members of the field as individual consultants. Finally, there are some, like Michael Pillsbury, who straddle the worlds of government and think tanks.

Many of these individuals have previously served as either military attachés in China and/or in the Department of Defense. Several were Foreign Area Officers (FAOs) in the U.S. Army or other services. This distinguished group has made an important mark on the field of PLA studies. They have brought obvious expertise and a "feel" for understanding the PLA that most academics lack. They have benefited from first-hand exposure to the highest reaches of the PLA command structure down to basic units and training facilities. There is just no replacement for this experience. But the real value-added of this group is that they have spent their entire adult lives as military officers, and concomitantly bring a comprehension of the weapons systems, technologies, training regimens, and operations and maintenance routines for militaries that academics are hard-pressed to understand. The addition of this cohort to the PLA studies community is relatively recent (although there have always been some in the field), and it has provided an incredible boost to the field. Curiously, though, to date the field has not benefited from the addition of former PLA officers, soldiers, or defectors.

Before leaving the subject of producers for the consumers of the PLA studies field, mention must be made of the sense of "community" and collegiality that is shared among many members in the field. Fortunately, this is not a field beset by backbiting, factionalism, enlarged egos, petty jealousies, or the like. This is not to suggest that there are never heated debates or sharp differences of opinion, as there are. A year or so ago the field seemed sharply divided over the "China threat" question, while a few years earlier the nature of civil-military relations was a point of contention. But, overall, it is a remarkably congenial community, both personally and intellectually. This compatibility is forged annually at one or more of the annual conferences that have become staples for the field: those sponsored by the Council on Advanced Policy Studies (Taipei) and its foreign partners (RAND since 1996), the American Enterprise Institute, Army War College, and the Staunton Hill gatherings. These venues have done much not only to forge a sense of community, but, perhaps more important, they have advanced the field intellectually by commissioning carefully considered sets of papers and engaging in frank discussion. Each conference has resulted in a published volume, which has pushed the field forward. Given the dearth of individual books in the field (Ellis Joffe's 1987 *The Chinese Army After Mao* was one of the last), these collective volumes have served an instrumental purpose in advancing the field. Many fields advance in a more indirect way— through publication of journal articles, scholarly monographs, and paper or poster presentations at annual meetings of the discipline. Our field is most fortunate to have had these three venues in recent years to push the study of the field forward, and to forge friendships and conviviality. The thematic nature and planning that go into these conferences have provided analytical coherence to the field, have filled in many important gaps in our knowledge, and have focused the discourse in the field.

Thus, the PLA studies field has enjoyed substantial growth over the years, as compared with either the pre–Cultural Revolution period, when the community could be counted on two hands (e.g., Ralph Powell, Hal Ford, William Whitson, Alice Langley Hsieh, Samuel Griffith, Ray Bradbury, Ellis Joffe, John Gittings), or the 1970s and 1980s. As is discussed below, the research foci have evolved, as have the available data. Also, the field is no longer as insular as in the past, as there is a much more diverse "market" for the field's scholarship.

If the "producers" in the field have changed over the years, so too have the "consumers." For much of its history the PLA studies field was very insular. Those in the field would basically write for and talk only with each other. The field was cut off from mainstream contemporary China studies, from East Asian security studies, and from international security studies more broadly. Moreover, there was little public interest in the subject and only a few journalists (such as Drew Middleton of *The New York Times*) paid it heed. Those who followed the PLA in the government and intelligence communities had little interaction with those outside.

All of this has changed in recent years. During the last decade, those in other subfields of contemporary China studies have begun to show greater interest in the PLA, as the PLA has become recognized as an increasingly important actor in domestic politics, the economy, and foreign affairs. As the PLA began to modernize its forces and flex its muscles in the Taiwan Strait and South China Sea, the East Asian security studies community began to pay greater heed to our work. International nongovernment organizations (NGOs) such as the International Institute of Strategic Studies and Stockholm International Peace Research Institute all began to pay increasing attention to the PLA, and this has been reflected in their publications. Government-affiliated research institutes, such as the National Institute for Defense Studies in Japan, the Stiftung für Wissenschafft ünd Politik in

Germany, the Institute for Defense and Security Analysis in India, the Institute for Defense Studies in Singapore, the Korea Institute of Defense Analyses, and others, have also begun to devote greater resources and staff to studying the PLA. The international media has definitely increased the attention paid to the PLA, although articles in the mainstream press are still few and far between. In fact, one should not overstate the attention paid to the PLA or writings by PLA experts, as our field still remains a remarkably insular one. When events thrust the PLA into the public spotlight, then we are suddenly called upon for instant analysis. Unfortunately, when this occurs, we are often forced to oversimplify complex issues, organizations, doctrines, weapons, and so on for journalists who only wish to give their readers bare-bones analysis. Of course, journalists from magazines such as *Jane's Defense Weekly, Defense News,* and *Aviation Week and Space Technology* are of a different nature, as their questions are far more nuanced, their articles are considerably more detailed, and their readership is far more knowledgeable.

Other new consumers are to be found in the U.S. Congress and foreign parliaments, multinational corporations, and foreign policy élites. As China's arms sales and commercial activities abroad have gained notoriety, legislative attention has increased. As China has recovered from its post-Tiananmen isolation and renormalized its foreign relations, military-to-military exchanges with the PLA have been added to the diplomatic agenda, and there has been a concomitant need among diplomats and foreign policy élites outside of government to know more about the PLA. "Track II" exchanges with the PLA are becoming an increasingly important channel of contact and information for government and nongovernmental specialists.

With this increased consumer base, it is not surprising that some producers try to tailor their analyses to certain consumers. There is little doubt that certain think tank researchers, independent analysts, and congressional staff allow ideological, partisan, and political incentives to color their analyses and policy prescriptions. Certain consumers are also selective in what they "hear" and wish to hear from analysts, frequently ignoring important details, nuance, or simply distorting empirical facts. Rumor generation and embellishment are not uncommon among some PLA watchers. In a field where hard data are relatively scarce and PLA transparency is low, there are often incentives to speculate, create a "fact" out of a report or rumor, not look for independent confirmation of evidence, or otherwise be less than rigorous in analysis. These tendencies are definitely the exception rather than the rule in our field, but we all need to be as empirical as possible in our work.

DATA

All fields are prisoners of their data. To a certain extent, the PLA studies field has mirrored other parts of contemporary China studies. A perusal of footnotes in the first generation of PLA studies during the 1950s and 1960s reveals an almost singular reliance on *Liberation Army Daily* and other print and broadcast media, often reported in the survey of China mainland press. On rare occasion, original documentation was made available to the scholarly community, often as the result of intelligence operations by the nationalists on Taiwan, the Central Intelligence Agency (CIA), or British secret service. Such was the case with the so-called "secret military papers" captured in 1961 in Tibet.[14] Human subject interviews did not figure prominently for PLA scholars of the first generation, and they were certainly not as important as for specialists in domestic Chinese politics, who based many books on interviewing refugees in Hong Kong. Alexander George was one to take advantage, by systematically interviewing nonrepatriated prisoners of war (POWs) from the Korean War.[15] Like other subject areas, the Cultural Revolution produced a flood of new data for PLA watchers. Important insights were gained into party-army relations, commanders and commissars, the defense industrial sector, force structure, and individual units. More recently, largely as a result of the publication of the phenomenal volume, *The PLA in the Cultural Revolution (Wenhua da geming de renmin jiefangjun)*, have we become more fully aware of how the armed forces were affected during these tortuous years.[16] The 1970s yielded little new by way of documentary data for PLA studies, but for the first time it became possible to interact with the Chinese military. This came as a result of two processes: the establishment of diplomatic relations between China and Western countries and China's shopping for arms and defense technology in the West. Then strategic cooperation against the former Soviet Union also brought the PLA in closer contact with Western political and military leaders.

The great boom in documentary data for PLA studies did not come until the 1980s and 1990s. As in other areas of Chinese studies, there has been a publishing explosion in materials on the PLA in the last two decades. Certain events, such as the 70th anniversary of the PLA in 1997, stimulate outpourings of materials, but there has been a marked increase in many spheres in recent years. Several thousand volumes have been published in this decade alone, the vast majority of which are not "restricted internal circulation" (*neibu* or *junnei faxing*) and are available for purchase by foreign researchers. Visits to the PLA publishing house outlets in Beijing and Nanjing, and the Academy of Military Sciences press outlet (the NDU press is on campus and generally inaccessible), should be regular stops for all in the field. I will not attempt to catalogue or discuss such materials here, as I have done so in a preliminary way elsewhere[17] and it lies beyond the purpose of this essay, but suffice it to say that there is hardly an area of Chinese military affairs that is not covered in these books: doctrine, logistics, political work, battle and campaign histories, force

---

[14]The "secret papers" were, in fact, the low-classification PLA *Work Bulletin* (*Gongzuo Tongxun*), although at the time they were a treasure trove of primary data. Twenty-nine issues from 1961 were released by the Department of State to the public in 1963. The translated texts were subsequently published as J. Chester Cheng (ed.), *The Politics of the Chinese Red Army*, Stanford, CA: Hoover Institution Press, 1966. Also see the special issue of *The China Quarterly*, No. 18, 1964.

[15]Alexander L. George, *The Chinese Communist Army in Action*, op. cit.

[16]Li Ke and Chi Shengzhang, '*Wenhua da geming' zhong de renmin jiefangjun*, Beijing: Zhonggong dangshi ziliao chubanshe, 1989. This book was subsequently banned and the authors (researchers at the Academy of Military Sciences and National Defense University) were arrested.

[17]David Shambaugh, "A Bibliographical Essay on New Sources for the Study of China's Foreign Relations and National Security," in Thomas W. Robinson and David Shambaugh (eds.), *Chinese Foreign Policy*, Oxford: Clarendon Press, 1994.

structure and management, command and control, each of the services, high-tech warfare, regional and international security, foreign militaries, and so on. In addition, there are over 100 military journals published—although the vast majority of these are *neibu*. Michael Pillsbury's edited Chinese views of future warfare is a good illustration, albeit a small sampling, of what is discussed these days in the pages of Chinese military journals.[18]

The problem in PLA studies today is not that we lack data, but rather three interrelated problems:

- Gaining access to the data

- Insufficient exploitation

- Poor bibliographic control.

Unfortunately, woefully few members of the field collect and use primary data from China. Data are not that hard to get hold of if one only visits the publishing outlets noted above, as well as the *Xinhua Shudian* network of bookstores. Journals are another story, as generally one must have access to a library on a residential basis. Even open (*gongkai*) journals such as *China Military Science* (*Zhongguo junshi kexue*) and *National Defense University Journal* (*Guofang daxue xuebao*) are hard to come by.[19] Finding complete runs of back issues is even more difficult. As there is just a handful of PLA journals that can be subscribed to outside of China, one really must trawl for materials in the PRC. The Universities Services Center (USC) on the campus of the Chinese University of Hong Kong and Taiwan's Institute of International Relations (IIR) libraries also house substantial collections of materials on the PLA. The latter includes comprehensive press clippings files from PRC and PLA media and specialized publications. The IISS in London and SIPRI in Stockholm contain similar clippings files drawn from English language publications. Our major Chinese studies library collections in the United States and other countries are doing a poor job of collecting these materials. Surveys of the best—the Fairbank Center Library at Harvard and the Sinological Institutes in Heidelberg and Leiden, respectively—reveal that these facilities are acquiring only a fraction of what is possible.[20] The Library of Congress is even worse.[21] The exception appears to be the Center for Chinese Studies library at the University of California-Berkeley, which appears to have the best collection of PLA materials outside of China, the USC in Hong Kong, or the IIR in Taipei.

Second, even if one gains access to these materials, having the time to carefully read, digest, and use them in research is no small challenge for nonnatives. While the total

---

[18]Michael Pillsbury (ed.), *Chinese Views of Future Warfare*, Washington, D.C.: National Defense University Press, 1997.

[19]I have been able to secure nearly complete runs of these journals for the last decade.

[20]The Sinological Seminar library at the University of Heidelberg has the only continuous paper run of *Jiefangjun bao* (*Liberation Army Daily*) in the Western world.

[21]I am presently undertaking a survey of the contemporary China collection in the Library of Congress, commissioned by the Librarian of Congress, with the intention of devoting greater resources and attention to building this into an unrivaled collection.

volume is but a fraction of what our colleagues in Chinese economic or political studies must cope with, it is still a daunting task to digest dozens of books in Chinese. The U.S. government (Foreign Broadcast Information Service [FBIS]) could make a tremendous contribution not only to our field but also to U.S. national security through the sustained translation of such materials. An investment of $500,000 in translating carefully selected books and articles would result in a tremendous "multiplier effect" for analysts inside and outside of government. Obviously, one must possess very good Chinese language skills and knowledge of the specialized vocabulary to exploit the available materials (presuming they are accessible). Unfortunately, only a handful of scholars in the field (including government) possess such linguistic prowess. Thus, both lack of access and poor exploitation have hampered the field to date.

Finally, those who do delve into the primary materials face the task of gaining bibliographic control. Do we cross-check our data from one source to the next? How is one to know which is more authoritative? Indeed, in many cases, how is one to know the identity or institutional location of the author? Only a careful monitoring of materials can distinguish new information. For example, just in the last year or so, PLA publishing sources have begun to publish books and articles detailing the weapons capabilities in the military's arsenal and several volumes on command and control issues. How do we know what is truly new information (a proper approach would entail cross-checking with a wide variety of Western sources)? It is important that analysts and scholars be careful not to confuse ambition with capability, as much PLA writing explores doctrines and technologies that the PLA aspires to, but does not currently possess.

Aside from published sources, mention should be made of human sources. Unfortunately, our field has not benefited very much from defectors or émigrés, a category that had provided much information on the Soviet and East European militaries.[22] As in other areas of Chinese studies, interviews have come to play an important role in scholarly research since the 1980s. These have been particularly significant in discussions of the PLA's perspectives on regional security, but also on a wide variety of aspects of the PLA: doctrine,[23] political work,[24] party-army

---

[22]Xue Litai's and Hua Di's collaborations with John Lewis are an exception (and the latter case resulted, in part, in Hua Di's arrest upon his return to China).

[23]David Shambaugh, "The Insecurity of Security: The PLA's Evolving Doctrine and Threat Perceptions Towards 2000," *Journal of Northeast Asian Studies*, Spring 1994.

[24]David Shambaugh, "The Soldier and the State in China: The Political Work System in People's Liberation Army," *The China Quarterly*, September 1991.

relations,[25] national security policymaking,[26] logistics,[27] the budget,[28] commercial activity,[29] and other areas.

As in other aspects of contemporary China studies, PLA specialists need to utilize multiple sources of data and research strategies. Simply scouring the English language press and publications in defense publications is far from sufficient. Nor is use of FBIS and BBC/SWB translations adequate, as they fail to pick up books altogether and capture only a fraction of the periodical and newspaper literature published in China. The Hong Kong press can often provide important information but must be treated with great care. These three sources need to be used in tandem with a wide range of primary source Chinese language material and, where possible, interviews. Getting interviews with the PLA is, however, much easier said than done. We must admit that we have very little contact with real "operators," officers, and soldiers. If one has the chance to visit bases, they are always showcase units. When one does have the opportunity for discussions on bases, they are usually carefully scripted. The cohort with whom interaction is most frequent are the "barbarian handlers" and intelligence collectors from the Second Department of the General Staff, military attachés posted abroad, and specialists in international security at the Institute of Strategic Studies of the National Defense University and the Foreign Military Studies division of the Academy of Military Sciences. One is often left with the impression that these individuals know more about foreign militaries than they do about their own! There is a resulting disconnect as our PLA specialists are meeting with their American specialists. Nonetheless, meetings can be arranged with strategists and other experts in these institutions for serious discussion of PLA issues. Of course, military attachés have the best access, but even that is carefully controlled and scripted.[30] Slowly, officers are beginning to go abroad as visiting scholars or for conferences, thus opening further channels of information. Track II dialogues with the PLA are also beginning to yield important insights. Gradually, the PLA's door is being pried open, access is improving, and transparency is growing. There is a variety of reasons why it will continue to open very slowly and it will be a painstaking and frequently frustrating endeavor to "engage" the PLA, but the trendline is in the right direction.[31]

---

[25]Michael D. Swaine, *The Military and Political Succession in China*, RAND, R-4254-AF, 1992.

[26]Michael D. Swaine, *The Role of the Chinese Military in National Security Policymaking*, RAND, MR-782-OSD, 1998.

[27]Tai Ming Cheung, "Reforming the Dragon's Tail: Military Logistics in the Era of High-Technology Warfare and Market Economics," in James Lilley and David Shambaugh (eds.), *China's Military Faces the Future*, Armonk, NY and Washington, DC: M. E. Sharpe and AEI Press, 1999.

[28]David Shambaugh, "World Military Expenditure: China," *SIPRI Yearbook*, Oxford: Oxford University Press, 1995; and David Shambaugh, "Wealth in Search of Power: The Chinese Defense Budget and Budgeting Process," paper presented at IISS/CAPS conference, Hong Kong, 1996.

[29]Tai Ming Cheung has a forthcoming book.

[30]See the observations in Larry M. Wortzel, "The U.S.-China Military Relationship in the 21st Century," paper presented at the Eighth Annual Conference on the PLA, Wye Plantation, Maryland, 1998.

[31]See the discussion in David Shambaugh, *Enhancing Sino-American Military Relations*, Washington, D.C.: Sigur Center for Asian Studies, 1998.

WHAT DO AND DON'T WE KNOW?

We should be modest in our self-assessments as a field, as there is still a lot that remains unknown and many sources that remain untapped. Yet there is also much to be satisfied with, as the general knowledge and information base among specialists on the PLA are impressive. The contributions to this volume are indicative of the state of the field today, and it must be judged to be quite informed and sophisticated. Other conference volumes and published monographs sustain this judgment. The current crop of doctoral dissertations, notably those by Mulvenon and Feigenbaum, are of excellent quality and show what can be done with newly available sources. The field has progressed substantially in recent years, and it can truly be said to have matured.

In the last decade, the field has produced extensive and informed studies of all services (ground, air, naval, and nuclear), civil-military relations, the officer corps, doctrine and strategy, defense policy decisionmaking, the national security environment and threat perceptions, military finance and budgeting, the military-industrial complex, arms sales and purchases, a wide variety of weapons systems and capabilities, as well as "software" issues (such as those contained in this volume). Even the People's Armed Police and the PLA's role in arms control have been examined. The field has matured through hard work by individuals with a passion for their subject and a strong will to know more. We actually know quite a lot about the PLA, far more, I would submit, than the PLA is aware of. If the PLA had a better sense of what is known *in the public domain and based on unclassified sources*, and if they read our publications, they may have less incentive to try to hide that which is already common knowledge among specialists abroad.

Of course, our knowledge base is far from sufficient and there is much more work to be done. Moreover, we are "shooting at a moving target" that is evolving and changing rapidly. What needs to be done? In my view, a careful "institutional mapping" of key organizations is still needed. We really do not have a clear sense of how the general departments, group armies, and regional commands are organized and function. We need to know more about how training and tactics are evolving, and how the services are adapting to the new doctrine of "limited war under high-technology conditions." Despite being the most heavily studied subject in the field, civil-military relations still remains a black hole, with woefully inadequate data, which forces us to speculate often beyond what hard evidence can sustain. Nascent signs of increased state control over the armed forces, with concomitant autonomy from the party, need to be monitored carefully. With wholesale turnover in the officer corps and in the High Command taking place in recent years, we are confronted with a completely new generation about whom we are poorly informed. Much more information needs to be gained about the socialization, training, and interrelationships in the upper echelons of this new generation of officers and military leaders. We may have a relatively clear understanding of the PLA's perceptual worldview and strategic outlook, but that is not the same as understanding intentions and how a military will act in a given crisis. Hence, we need to know how the PLA prepares for certain contingencies involving Korea, Japan, Taiwan, India, the United States, and possible conflict in the South China Sea. We know that the military-industrial complex is in dire straits on the whole, but we have a poor sense of the linkage between research and development, exactly why certain systems never make it to or past the prototype stage, and a whole range of issues related to systems integration and assimilation of technology. Indeed, we need to gain a clearer sense of the interaction of technology, procurement, doctrine, and strategy. Does doctrine drive procurement decisions, or do technological impediments constrain doctrine and strategy? How are decisions made on what to buy abroad, and what problems of assimilation is the PLA experiencing in the systems it has bought from Russia and other sources? Is strategy driven by threat perception and possible contingencies, or does a wider set of variables affect the calculus? How does doctrine affect strategy, and vice versa? What is the PLA's "calculus of deterrence" (to borrow Allen Whiting's original phrase), and what is its warfighting doctrine? What role, if any, do tactical nuclear weapons and other forms of weapons of mass destruction (WMD) play in warfighting doctrine and exercises? What will the force structure look like in five years' time, after the downsizing, equipment retirements, and demobilizations? Is there a clear vision on where the PLA seeks to be five, ten, fifteen, and twenty years into the future? These questions are just the tip of the iceberg. There is much that we need to know.

WHAT ELSE IS NEEDED?

The following are some random suggestions of needs in the field:

- More input from the government intelligence and military communities and cross-fertilization with nongovernment scholars and analysts

- More and continued input from retired military, particularly those with first-hand experience in China

- More input from, and interaction with, PLA "scholars" and officers in China

- More Chinese émigrés to enter the field

- More doctoral dissertations and young scholars encouraged into the field

- More comprehensive assessments of the PLA that begin to put back together the pieces of the puzzle that have been disaggregated in recent years

- More attempts to theorize about developments

- More thought on how to "engage" the PLA and shape a coherent long-term strategy for coexisting with the PLA.

As noted above, the PLA studies field also tends to operate in a vacuum. It needs to be much better integrated with other subfields:

- Wed PLA studies to the broader study of Chinese bureaucratic politics and institutional culture

- Wed PLA studies to broader theories of technological innovation, change, and diffusion

- Wed PLA studies to the study of the military during the republican period (1911–1949) and in some areas to the imperial period, and engage in dialogue with historians of these periods

- Wed PLA studies to comparative military studies and the study of civil-military relations, particularly in East Asian and communist systems, where there are multiple instructive comparisons to be made

- Wed PLA studies better to debates in post–Cold War strategic studies, e.g., the impact of globalization, technology and information diffusion, economic security, and a variety of aspects of non-conventional security

- Wed PLA studies to development theory, particularly in the area of technology absorption, specialization, and role differentiation.

I stress that these are just some random suggestions, but their thrust is that the field is too insular and can benefit from some comparative perspectives.

CONCLUSION

The contributions to this volume are testimony to the maturation of our field. Each is significant empirical contribution in its own right, but collectively they illustrate many of the themes noted above. Notwithstanding the areas for improvement noted above, the field of PLA studies has grown and matured well in recent years. It is also like an extended family with its share of eminent grandfather figures, overworked middle-aged professionals, and young up-and-comers. But it is a dynamic field, growing rapidly intellectually and analytically. We may complain about the lack of PLA transparency (rightfully so), but we can no longer grumble about insufficient data and research materials. But we need to learn how to better exploit these new materials and gain bibliographic control over them. We also need to collectively reflect on our past scholarship and analysis, to ascertain where we were right and where we were wrong and why. This, appropriately, will be the theme of next year's conference.

In closing, it may be worth recalling the note on which Harlan Jencks concluded a similar stock-taking a dozen years ago.[32] Citing Harvey Nelsen, Jencks recalls that the field had consensus and a good grasp of civil-military relations, PLA hardware, the PLA's strategic priorities, resource allocation, and long-term modernization policy. I would estimate that these are still the areas where the field has the greatest confidence and consensus (albeit not absolute). Nelsen further observed that the field was still obsessed with structure and function, as it has been since the 1960s, with spartan understanding of decisionmaking and policy implementation methods. I would submit that institutional mapping is always necessary, as organizations must be thought of as evolving organisms that need to be carefully tracked. PLA decisionmaking and implementation are, by necessity, exercises in educated guesswork. The "black box" remains black in the case of the PLA. But these are two

---

[32] Harlan W. Jencks, "Watching China's Military: A Personal View," *op. cit.,* p. 78.

areas where we have learned a great deal in political science over the last decade, and it might be instructive to more carefully study the possible adaptation of what we have learned on the civilian side to the study of Chinese military policymaking and implementation. Nelsen concluded with two other observations that are still worthy of contemplation: that, ultimately, PLA modernization would be adverse to U.S. strategic interests and that, second, there was little that any country can do to affect PLA modernization as China had built up a plethora of defense contacts and access to international resources. The former may still be true (although the jury is still out), although the latter definitely changed as a result of Tiananmen. The PLA remains under a near-total Western embargo on defense technology, weaponry, and spare parts, and there is little sign this will change in the foreseeable future. This makes our analytical task easier, although it may not make it any easier to deal with the PLA.

3. THE MILITARY AND CHINA'S NEW POLITICS: TRENDS AND COUNTER-TRENDS

Ellis Joffe[1]

The political development of the Chinese army in recent years has followed several trends, which sometimes run in different directions. One set of trends has led to a substantial increase of the PLA's influence. Already apparent as Deng faded from the scene, these trends have been accentuated after his death, and will shape the PLA's behaviour on the political scene in coming years. Other trends, however, have qualified this increase. The result is uncertainty about the state of the PLA in Chinese politics.

This uncertainty is compounded by a widespread tendency to focus on one trend only, which produces a partial and misleading picture. For example, an emphasis on the new political influence of military leaders tends to ignore the trend's limitations and portrays the leaders as dominating Chinese politics and impelling China toward external conflicts. Or, an emphasis on the PLA as a political and economic force tends to downplay the impressive upgrading of professional capabilities that has occurred in recent years.

The PLA's political position, in short, is shaped by trends and counter-trends, and an attempt to draw up a balanced assessment must look in both directions. This is not easy, since available information is almost entirely circumstantial and inferential. Therefore, the conclusions, and the reasoning that leads to them, have to be speculative in part.

INSIDE THE NEW POLITICS

Since the twilight of the Deng period, the PLA has been undergoing a transformation in all major areas of its activities. One of the most important has been its role in politics, which has an impact on all other areas. This has been the result of changes which have occurred on the political scene, and which have caused the PLA to veer from past patterns of involvement and move in new directions. The extent of the shift has yet to be determined, but it is already apparent that in the transition to the post-

---

[1]Ellis Joffe is professor of Chinese Studies and International Relations at the Hebrew University of Jerusalem. He is the author of two books and many articles and book chapters on the Chinese military.

Deng period, the PLA has acquired a potential for the exercise of unprecedented political influence.

In the early stages of the transition this potential has been realized only in part. Even so, the military has moved closer to the center of decisionmaking on certain issues, and, as the Hong Kong handover demonstrated, is publicly playing a newly prominent role in upholding the most valued national asset of all—Chinese sovereignty. More important, conditions have been created for the further transformation of this potential into actual influence, a prospect which may turn the PLA into a more powerful force in future Chinese politics, and which, as a result, has enhanced its political position in the present.

In order to assess this position, it is necessary to examine the changes that brought it about. Although the military may have capitalized on these changes, it did not initiate them. No one did. They have been the result of the succession process which began several years ago, and the adaptation of Chinese politics and institutions to the new times. Their impact on the military has been on three related levels. The first, and by far the most important, has been on the stature of China's paramount leader and his standing in the PLA. The second, on the procedural setup for reaching major national decisions and the military's part in it. And the third, on the institutional relationship between the party and the army.

THE PARAMOUNT LEADER AND THE PLA

Despite the vast changes that it has undergone, China is still ruled by an authoritarian one-party regime. In such a regime, politics are shaped first of all by the personality and position of the paramount party leader. What distinguishes the politics of the post-Deng period from the preceding four decades is the stature of Jiang Zemin and his relations with the PLA. Since this distinction applies not only to Jiang but also, most probably, to whoever succeeds him, it can be viewed as a major feature of China's new politics. In order to illuminate it, it will be useful to compare Jiang with Mao and Deng—as paramount leaders, and as supreme commanders of the PLA.[2]

These two aspects are critically connected, and this connection, more than any other factor, defines the political influence of the military, for several reasons. First, the political strength or weakness of the paramount leader determines to what extent he is susceptible to pressure or faces threats from rival groups in the leadership. Second, his standing in the military determines to what extent he can rely on it for support against such rivals. Third, a combination of the first two determines the extent to which the military can make demands on the leader and his capacity to handle them.

Three categories of relations can possibly derive from this connection. In the first, the position of the paramount leader is unquestionably strong both politically and in the military. In this case, the political influence of the military as an independent

---

[2]On Mao and Deng as paramount leaders, see Andrew J. Nathan and Robert S. Ross, *The Great Wall and the Empty Fortress: China's Search for Security*, New York: W. W. Norton and Company, 1996, pp. 124–129.

factor is minimal. In the second, the stature of the leader on the political scene is not entirely secure, but he has unswerving backing from the military. In this case, his political vulnerability is offset by this support, as is his need to meet specific demands from the military. In the third situation, the leader is weak politically and in the military. In this case, he is susceptible to political pressure, but cannot rely on unqualified support from the military. To gain such support he has to curry favor with them, which puts the military in a powerful position. How then does Jiang Zemin measure up against Mao and Deng from the vantage point of this connection?

Mao fell into the first category without any reservation. His stature as China's supreme leader was unique. It was accepted like a "mandate of heaven" by all other leaders in their public behavior, and his personal position was unassailable. It derived from a combination of several elements, though their intensity varied over time: his brilliant record as the victorious leader of the revolution; the great success of the new rulers under his leadership in unifying the country politically and rebuilding it economically; and the ties that he had formed over the years. To these must be added Mao's distinction as a social visionary, his skills as a political street-fighter, his charisma, and the respect accorded to the supreme leader in Chinese political culture.

As a result, Mao's personal authority was unparalleled. This enabled him to disregard institutional boundaries, and, like an emperor, to rule everything. He could—and did—intervene in whatever area he chose. Whether his intervention took the form of approving decisions, initiating major policies, or forcing a change of direction, Mao's top colleagues accepted his imperious role—from a mixture of fear, faith, and fidelity.

The most prominent area in which no boundaries existed for Mao was the PLA. And with good reason. As founding father of the Red Army and its leader throughout the revolutionary years, Mao regarded himself as eminently qualified and justified to intervene in military affairs in later years as well, and always acted as an active commander-in-chief of the PLA. Although after 1949 the party and army developed as separate hierarchies, at the apex there was no distinction between Mao's roles as China's paramount political and military leader.

This situation was fully accepted by PLA leaders. To them Mao was not only nominal chairman of the party's Central Military Commission, but China's *de facto* supreme military authority. Their attitude derived from the same qualities that undergirded Mao's political stature, which were augmented by particular ones that bonded Mao and the PLA: his record of successful military leadership, and the connections that he had forged with PLA commanders. The result was a special relationship that was a hallmark of the Chinese communist regime.

This relationship had several implications for the political position of the PLA. The first was that the support of the military leadership for Mao and his policies was given as a matter of course, even when the leadership disagreed with him on specific issues. This meant that he could count on military backing in leadership circles in the pursuit of his policies or in personal struggles, and all players in the system knew that he could. A second implication was that Mao could use the PLA as his personal power base in elite struggles, confident that his personal authority would ensure its positive response, even if the situation required deeper involvement, including the use of troops. This was demonstrated by the participation of the military in a political conflict against an increasingly recalcitrant party in the prelude to the Cultural Revolution, and by its massive embroilment in subsequent nationwide struggles.

The third implication was that because of Mao's personal authority, the PLA had little leverage to sway him or to dissent from his orders. This does not mean that its chiefs were meek yes-men; indeed, in inner councils they voiced their opposing views. Their resistance to Maoist anti-professionalism campaigns, especially during Peng Dehuai's tenure as PLA chief, is a case in point. However, from the little that has been revealed by disputes that became public, it appears that when Mao made up his mind, he prevailed. Because of his stature, military leaders did not have the power, nor probably the will, to defy him. At any rate, Mao's predominant position was never threatened by opposition from the military. This was demonstrated in several instances.

The first was during the deliberations that preceded China's intervention in the Korean War. Mao stood for sending Chinese troops against U.S. forces, but was opposed by military leaders. Nonetheless, he decided on this fateful step despite their dissent, and they accepted the decision.[3]

Another instance occurred at the Lushan Plenum in 1959, when Mao rejected the criticism of Defense Minister Peng Dehuai, and got the party leaders to remove Peng from his senior posts. Although Peng explicitly criticized Mao for the economic effects of the Great Leap Forward, the confrontation between them was also fuelled by his dissatisfaction with the campaign against military professionalism, which Mao had launched at the same time.[4] Since this dissatisfaction was widespread among professional officers, Peng's dismissal also signaled Mao's rejection of their views. The acquiescence of the officers, who made no move whatsoever against Mao, was a convincing indication of their unquestioned subordination to Mao's commanding position.

The third instance was triggered by the escalation of American involvement in Vietnam in 1965. The Chinese leadership disagreed on the gravity of the threat and on the response. Most concerned was Luo Ruiqing, then chief of staff, who presumably reflected the views of the professional military, and who evidently considered the possibility of an American attack on China far more serious that did the political leaders. To meet it, Luo proposed a linear defense by the PLA, instead of the Maoist strategy of strategic retreat, which would have compelled a reconciliation with the Soviet Union to get vital military aid. However, his view was rejected and he

---

[3]Chen Jian, *China's Road to the Korean War: The Making of the Sino-American Confrontation*, New York: Columbia University Press, 1994, especially Part Three.

[4]Ellis Joffe, *Between Two Plenums: China's Intraleadership Conflict, 1959–62*, The University of Michigan, Center for Chinese Studies, Michigan Papers in Chinese Studies, No. 22, 1975, pp. 8–22.

was dismissed.[5] As in the case of Peng Dehuai, the military made no move to support their chief.

The only time military commanders seemingly violated Mao's orders and got away with it was during the Cultural Revolution chaos, when PLA units suppressed the Red Guards they had been told to support. However, this was not outright insubordination. PLA commanders either had received impossible orders, which instructed them to support revolutionary activity and to maintain order at the same time, or received no clear-cut instructions at all. Whatever the case, they chose to support stability, which meant cracking down on riotous Red Guards. However, they did this while proclaiming allegiance to Mao and the Center, and made no effort to use the power that had devolved to them to defy the Center openly. The only exception was the Wuhan commander in 1967, but he was removed by the Center and not backed by his fellow commanders.[6]

The conclusion is that Mao towered over the PLA, whose leaders may have functioned as an interest group on professional military issues, but remained subordinate to Mao's final authority. To the extent that it had gained political influence as an institution—as during the Cultural Revolution and its aftermath—it was by supporting Mao and subject to his leadership. Under Mao, the PLA did not become an independent political player.

Neither did the PLA become a political force under Deng Xiaoping. This was despite the differences between Deng's stature as paramount leader and Mao's. Deng fits the second category—a leader whose political authority was less than complete, but who had strong backing in the military.

As a political leader Deng drew on qualities that resembled Mao's in some respects, but were sharply dissimilar in others. Like Mao, Deng had a record of achievements as a revolutionary leader, a vast network of connections, and a successful party career after the revolution. However, unlike Mao, Deng was a conciliator and arbiter rather than a despot, and his ruthlessness was tempered by collegiality and affability. More important, unlike Mao, he did not completely overshadow all veteran leaders and was not looked upon—and did not behave—as their superior. His return to supreme power in 1978, after the second of two Mao-induced periods in the political wilderness, obviously pointed to his preeminent status, but not to his unquestioned superiority. He still had to prove himself and, unlike Mao, his personal authority also depended very much on the success of his policies.

In this area lies the most striking difference between Mao and Deng as paramount leaders. Mao was eminently successful as a revolutionary leader, and during the first few post-revolutionary years, but after that he started to make monumental mistakes—starting with the Great Leap Forward and culminating in the Cultural Revolution. Although his personal stature was strong enough to withstand these

---

[5]Harry Harding and Melvin Gurtov, *The Purge of Lo Jui-ch'ing: The Politics of Chinese Strategic Planning*, RAND, R-548-PR, February 1971.

[6]Harvey Nelson, *The Chinese Military System: An Organizational Study of the Chinese People's Liberation Army*, Boulder, CO.: Westview Press, 1977, pp. 27–43.

catastrophes, Mao's political power declined as a result. On the other hand, Deng's personal stature as he entered the final phase of his career was not unshakable, but it grew and gathered strength as his daring policies reaped remarkable successes. Mao's policies eroded his stature, Deng's raised it.

In this process, Deng's own special relationship with the PLA played a crucial part. This relationship was forged during the Sino-Japanese War, when Deng served as the respected political commissar of what later became the PLA's Second Field Army. After the establishment of the communist regime, he became a key figure in the administration of the southwest region that had been established by that army, working closely with military leaders who would later move to the capital with him.[7]

Deng's PLA years gave him the necessary credentials to qualify as a military leader and to obtain permanent entry into the inner councils of the military hierarchy. More important, they enabled him to form the personal connections with senior military leaders which provided the basis for his invaluable support in the PLA. Although he moved into civilian posts in later years, Deng continued to maintain close ties to the PLA—as vice-chairman of the National Defense Council before the Cultural Revolution, as PLA chief-of-staff in the second half of the 1970s, and as chairman of the Central Military Commission until 1989—and was always looked upon by PLA commanders as one of their own.[8]

This bond accounts for the protection he received from PLA commanders when he was hunted by Red Guards in the mid-1960s; for the backing he got from PLA commanders when he made his bid for power in the late 1970s; and for their reluctant readiness to intervene in the Tiananmen crisis of 1989 and to fire on the demonstrators. It also accounts for the steady support of PLA leaders for Deng's reforms, which he could take for granted as an invaluable asset in his unyielding efforts to implement the new policies.

Deng's standing in the PLA was likewise vital in toning down the sectoral demands of its commanders. Even though Deng placed military upgrading as one of the nation's "four modernizations," it came last on the list, which meant that, despite the PLA's dire needs, precedence would be given to economic development. As a result, the military budget was more or less frozen for the decade of the 1980s, and declined by about 10 percent as a share of national expenditure. Although there was some compensation in the form of massive manpower cuts, efficiency campaigns, and involvement in profit-making enterprises (as yet limited), allocations to the PLA were far from adequate for its daily requirements, let alone for weapons upgrading. This situation could have been fertile ground for PLA alienation from Deng. But although there were periodic complaints about the low level of military financing, PLA leaders did not make any move that could weaken Deng, such as publicly withdrawing its support or aligning with rival leaders. They would not do this to one of their own.

---

[7]David S. G. Goodman, *Deng Xiaoping and the Chinese Revolution: A Political Biography*, London: Routledge, 1994, pp. 37–48.

[8]*Ibid.*, pp. 117–125.

Jiang Zemin's position, when he was thrust to power in 1989, was completely different. Although it has improved over the years, it is still incomparable to that of his predecessors. Jiang falls squarely into the third category—a leader over whose political and military standing there are big question marks.

When Jiang became paramount leader, his personal qualities were hardly notable. His greatest disadvantage was that he lacked the three most important sources of personal authority on which Mao and Deng could draw: personality, past achievements, and a network of connections. As a member of the post-revolutionary generation, he inevitably missed out on the years of struggle during which illustrious records were created and long-lasting loyalties formed. According to most observers in China, his presence did not inspire much awe or respect, either.

Jiang's political career had been relatively short, Before ascending to the top, he had been mayor of Shanghai, and prior to that a nondescript technical administrator. As mayor, he was reportedly known as a "flower pot"—ornamental but ineffective. This was widely believed to have been the chief reason for his appointment: various factions agreed on it because Jiang had not made serious enemies in his inconspicuous path to the peak.[9] Unlike Mao, he had not articulated a grand vision and, unlike Deng, he had no original blueprint for China's development. When this void was added to his unproven leadership abilities, his personal authority appeared shaky at best. For this reason, support for Jiang did not come naturally from members of China's ruling elite. It had to be granted, and Jiang was decisively dependent on their readiness to accept him as leader. This dependence has been reduced somewhat over the years, but it is still critical. To remain in power, he needs the support of the principal personalities who head the vast hierarchies which make up the Chinese power structure. The most important are the armed forces.

Their importance stems from the basic fact that military backing is indispensable to the survival of a paramount leader in China. The most obvious reason is that they command the forces which may be used to oust him. However, this is hardly an ever-present concern for Jiang, since the possibility of a coup is remote in the Chinese scheme of things. What is not so remote is that he may have to call upon the armed forces for support in less extreme situations. For example, they may be required to intervene in a leadership conflict, as they had on the eve of the Cultural Revolution. They may be needed to maintain public order in a time of turmoil, as they had been during the Cultural Revolution. They may be used to remove a rival leadership group, as they were in the case of the "gang of four." They may be employed to put down major anti-government demonstrations, as they were in the Tiananmen crisis. They may be used to suppress a rebellious provincial figure, a possibility that is not entirely unrealistic, given the growing power of China's local leaders.

Aside from specific crisis situations, Jiang has to be confident that the armed forces are on his side in his political dealings and the pursuit of his policies, and that they

---

[9]Tai Ming Cheung, *An Eye on China: Jiang Zemin*, Hong Kong: Kim Eng Securities, November 1994, pp. 4–5.

will come to his aid if the need arises. Without such confidence, his position will be unstable, and even if he is able to plod along, he will not be able to rule effectively.

Military support has to be manifested in various ways. The critical point is that all the players in the political system have to be fully aware of its existence. This the military can do by frequently praising the leader and his policies in statements by military chiefs, and in the organs of the PLA; by prominently showing him respect—according him the central place in viewing military maneuvers is one example; by educating the troops accordingly, and making such efforts public; by avoiding any action that might undermine his position, such as criticizing his policies; and most important, by refraining from showing any support for a rival leader.

Under Mao and Deng, such support was axiomatic; under Jiang it is not. It was axiomatic because they had enjoyed the utmost respect and complete confidence of the military. In contrast, Jiang does not. His command of the armed forces derives first of all from his chairmanship of the Central Military Commission, but unlike Mao and Deng, he clearly lacks the personal qualifications to hold this position. Even by his own admission, Jiang was not fit to assume the top post in the Chinese military establishment.[10] The reason was glaring: he had no military record, no military experience, no particular knowledge of military affairs, and no connections in the armed forces. He came to the PLA position only by virtue of his elevation to paramount leader, and Deng reportedly had put pressure on the veteran PLA chiefs to give Jiang the essential backing.[11] This backing did not come naturally and has remained conditional.

From this fact derive two essential features of Jiang's rule and relations with the military, as they had shaped up in the first few years of his rule. First, because he cannot rely on the automatic backing of the military, Jiang's political position cannot be secure, since other players know that in a major crisis it may not be forthcoming. Second, he has to gain such backing by making concessions to PLA leaders. This puts them in a position to exert unusual political influence. However, these features have been losing some of their initial force. As a result, the influence of the military is limited by several factors.

JIANG ZEMIN'S SOURCES OF STRENGTH

Although not as obvious as those highlighting Jiang's shortcomings, these factors partially offset the weaknesses stemming from his personality and past experience. They extend both to his political position and to his standing in the military.

The first factor is institutional. As paramount leader, Jiang occupies pivotal positions—since 1992 he has been general secretary of the Chinese Communist Party, chairman of the Central Military Commission, and president of the People's Republic of China. The first two mean that he stands at the apex of the two most

---

[10] David Shambaugh, "China's Commander-In-Chief: Jiang Zemin and the PLA," in C. Dennison Lane et al. (eds.), *Chinese Military Modernization*, Washington, DC: The AEI Press, 1996, p. 216.
[11] *Ibid.*

important hierarchies in the Chinese political system, while from the third he gets luster and international exposure.

These posts give Jiang substantial power. Despite the supreme importance of personal authority in Chinese political culture, official position is also highly significant. This is because traditional culture emphasizes the centrality of hierarchy and subordination to power-holders in line with their standing in it. Those at the top command much symbolic prestige, which in itself is a source of power. Jiang has capitalized on his formal posts in an effort to bolster his prestige by maintaining high public visibility that spotlights his role as national leader—for example, in the ceremonies surrounding the Hong Kong handover.

Symbols aside, Jiang's official posts have given him the invaluable advantage of placing trusted officials in key posts. The loyalties and obligations stemming from this advantage have always been a core component of power in China's authoritarian political system, which is still moved above all by personal ties between the leader and his supporters. They are absolutely crucial to a leader like Jiang, whose personal authority is not upheld by charismatic leadership traits or connections that reach far into the past. Personal traits cannot be created, but connections can.

Jiang began to create connections shortly after ascending to the top post. Most of Jiang's appointees are old colleagues from Shanghai, prompting observers to term them the "Shanghai clique." Here are some examples. One of its leading members is Zeng Qinghong, Jiang's personal assistant of many years, who, as head of the party Central Committee's Central Office, manages the work of the party's most important committees and supervises its vast bureaucracy. Another is Zhang Quanjing, who controls the party's all-powerful Organization Department. Through these aides Jiang has gained a vital foothold at the core of the party apparatus—the inner wheel that turns all the other wheels in China's ruling mechanism. Two additional members are Ding Guan'gen, head of the party's Central Propaganda Department, and Shao Huaze, chief editor of the party's chief organ, the People's Daily. They—together with minister of Television and Radio, Sun Jiazheng, and director of the PLA's Liberation Army Daily, Sun Zhongtong—have ensured that Jiang's activities receive widespread laudatory coverage in the media.[12]

More important, Jiang has made every effort to build up his own power base in the military by steadily appointing trusted officers to important posts. Unlike the sprawling party and government organizations, which consist of numerous bureaucratic domains and are difficult to control, the unified structure of the military guarantees that the insertion of reliable officers in key posts goes a long way in magnifying Jiang's influence.

As a novice overshadowed by veteran officers, Jiang could not do much at the beginning of his administration. However, his opportunity came after the removal of the two dominant figures in the PLA. Masterminded by Deng in 1992, this removal

---

[12]Tai Ming Cheung, *An Eye on China: China Under Jiang Zemin*, Hong Kong: Kim Eng Securities, April 1997, pp. 3–8.

forced out the veteran party leader Yang Shangkun, vice-chairman of the Central Military Commission, and his younger half-brother, Yang Baibing, secretary-general of the Commission and director of the PLA's General Political Department. Yang Shangkun had been the point man of the party "elders" in the Tiananmen crisis, and Yang Baibing was their hatchet man in purging the PLA after the massacre. Together they had a tight grip on movement of personnel in the armed forces. Despite this, and apparently because of it, they lost their positions when Deng and his colleagues reportedly decided that they had become overly powerful and ambitious.[13]

At the same time, Admiral Liu Huaqinq and General Zhang Zhen were brought out of semi-retirement by Deng and appointed to the two top posts in the military as vice-chairmen of the Central Military Commission. As veteran leaders enjoying great prestige in the PLA, their task was apparently to ease Jiang's way into the top military echelons in the face of suspicious commanders, and to oversee its transition to a new generation of commanders. Their relationship with Jiang could hardly have been an easy one. While towering over him in military stature, they have treated Jiang publicly with the respect due their senior in his capacity as chairman of the Central Military Commission, and have supported him without reservation. However, as chairman, Jiang presides over meetings of China's foremost military leaders, and has probably had to defer to them on matters concerning the PLA and beyond.

As long as the two old veterans remain in office, Jiang's standing in the military will be overshadowed by their presence. For this reason, he has reportedly tried, albeit unsuccessfully, to ease Liu and Zhang into retirement.[14] If he succeeds in attaining this at the 15th party congress in the autumn of 1995, his position in the PLA will be strengthened considerably. This will result not from Jiang's sudden transformation into a military leader, but from the ascension to top posts of military leaders who belong to his generation, and who will not be able to pull revolutionary rank on him. Furthermore, they will have been appointed by Jiang and will owe their allegiance to him, or at least be in his debt.

Presumably with this in mind, Jiang has shown utmost concern for overseeing personnel shifts in the PLA. He began doing this immediately after the downfall of the Yangs. First to fall were generals who had been associated with them. These changes were presumed to have had Deng's blessing, since working closely with Jiang in the housecleaning was General Wang Ruilin, Deng's military secretary and confidante (who was later promoted by Jiang to the rank of full general).

In the next stage, he installed new leaders in key positions throughout the PLA. At the uppermost level, in October 1995 he moved two allies into the top echelon of the Central Military Commission as vice-chairmen: defense minister General Chi Haotian and former PLA chief-of-staff General Zhang Wannian. Chi and Zhang are obviously set to take over from Liu Huaqing and Zhang Zhen as the two top PLA commanders when the old veterans retire. Further down the hierarchy, under the

---

[13]Shambaugh, pp. 221–226.

[14]*New York Times*, February 23, 1997.

popular slogan of advancing leaders "younger in average age and more professional," Jiang has removed or retired senior officers who owed no allegiance to him and replaced them with younger ones. These have included the directors of the three General Departments in the PLA Headquarters—General Staff (whose head is the chief-of-staff), Logistics, and Political Departments—commanders of Military Regions, Military Districts, and Group Armies, commandants of military academies, commanders of the People's Armed Police and the Central Guards Bureau (charged with the protection of the central leadership), commander of the Navy, the chief of the People's Armed Police, and the director of the Commission for Science, Technology and Industry for National Defense.[15]

Such sweeping changes could not have been carried out without the close cooperation of Liu, Zhang, and other senior commanders, and not all the new appointees could have been Jiang's personal choice. In fact, given Jiang's short familiarity with the inner workings of the PLA, it is a safe bet that most shifts were primarily the work of senior professional commanders, especially Zhang Zhen. Still, it is unlikely that Jiang would have sanctioned the advancement or retention of senior officers not favorably disposed towards him. Whatever the role of senior commanders in the reshuffles, the ultimate responsibility, and credit, for the appointments and promotions lie with Jiang. After Liu and Zhang retire, he will stand alone as the leader who had brought them about.

In addition to weaving a network of supporters, Jiang has gone out of his way to cater to the interests of the PLA—in itself an indication of the military's extraordinary stature. He shows great respect to military leaders and to the PLA as a whole. He has cultivated the retired leaders—the "elders"—of the PLA who continue to exercise influence over their former subordinates, some of whom have risen to top commands. He has been unusually generous with promotions—in one fell swoop in May 1994 Jiang promoted to the rank of full general eighteen lieutenant and major generals who held key posts. He makes frequent and well-publicized visits to military units and reportedly shows personal concern for their well-being. He pays great respect to the traditions of the PLA. He makes all the right statements about the need for military modernization and the PLA's requirements. He chairs meetings of the Central Military Commission. And he consults with PLA leaders on critical issues such as Taiwan.[16]

Nothing that Jiang has done has been more important to mobilizing PLA support than the increase of its budget, which has risen every year since 1988. In part, this should be attributed to the pressing needs of the PLA, not only for purchasing new weapons, but also for improving the plummeting living standards of the troops. In part, it stems from the economic upsurge of the 1990s and the obligation of the leaders to make good on their promise that the PLA would also benefit from it.

But an important reason has doubtless been Jiang's desire to show his goodwill towards the PLA in the most concrete way. Even then, although official spending has

---

[15]Shambaugh, pp. 211–232.
[16]*Ibid.*

almost doubled since 1986, the increase, when adjusted for inflation, has only been about 4 percent. However, the real budget of the PLA, according to the best estimates, is 4 to 5 times the official figure,[17] given the hidden allocations and the PLA's earnings from commercial enterprises, research institutes, and weapons sales. Thus, under Jiang the money at the disposal of the PLA has increased significantly.

All these moves have both reflected and complemented the growth of Jiang's personal stature. Despite his anemic image, he has demonstrated substantial political skills—first of all, by the mere fact of survival at the top in the treacherous alleys of Chinese politics, initially with the backing of (an increasingly feeble) Deng, and then without it. For someone who had started out from an inordinately weak base, this is a remarkable feat, which could hardly have been achieved by a political weakling. That could be seen, for example, in the anti-corruption campaign of 1995, which Jiang Zemin used to purge the Beijing party chief and his supporters for criticizing Jiang's appointment of close colleagues to top posts.[18]

While Jiang will never be a model of a charismatic leader, he obviously did not turn out to be—despite universal predictions to the contrary—the exact opposite either. His leadership style may be low-keyed and colorless, but perhaps this is what the Chinese system needs after the turbulent leadership of Mao and the dynamic one of Deng, and is not entirely a disadvantage. While he has not lighted fires, he has managed to create a stable coalition behind his moderate policies, which has not been seriously threatened by opposing alignments. As he continues to consolidate his position, Jiang is not likely to be challenged by a rival leader in a bid for power that is not ignited by major policy failures. Thus, Jiang's political future will depend primarily on the success of his policies, not his personality.

Aside from the specific steps that Jiang has taken, he has also benefited from sources of strength that are inherent in the nature of the PLA. At bottom, the PLA is a party-army with professional characteristics, and both components work in Jiang's favor. As a party-army, the PLA has a long and strong tradition of subordination to the party leadership. It has never deviated from this tradition as an institution. Although the first reason for its subordination was the personal authority of the paramount leader, it also stemmed from the traditional supremacy of the party over the military. As leader of the party, Jiang commands the obedience of the army at least to some extent by virtue of his position, regardless of his individual traits.

Strengthening obedience is the growing professionalism of Chinese officers as the armed forces continue their long march to modernization. In this context, the significance of military professionalism is threefold. It instills the ethos of compliance with orders emanating from higher levels without which no modern army can function effectively. It turns officers away from political pursuits that interfere with their specific tasks, as these become more and more complex. It makes them particularly opposed to intervention in political struggles, which can only end

---

[17]David Shambaugh, "China's Military in Transition: Politics, Professionalism, Procurement, and Power Projection," *The China Quarterly*, No. 146, June 1996, pp. 287–288.

[18]Tai Ming Cheung, *China Under Jiang Zemin*, p. 8.

up by splitting the commanders of the PLA and jeopardizing its existence as a national army. As every senior Chinese officer knows from personal or recounted experience, the PLA's intervention in the Cultural Revolution—the antithesis of professional behavior—had almost destroyed it.

On balance, the new potential of the military for political influence does not imply that Jiang is under constant threat that the military will withdraw their support in some form—somewhere along the entire spectrum from backing a rival to intervening with force. Far from it. The combination of his growing strength and their long-standing style guarantees military support for Jiang as established leader under ordinary circumstances. For his leadership, ordinary circumstances mean, in a nutshell, that China's economic performance is successful enough to ensure social stability, and that foreign policies stay within the nationalistic consensus on such key issues as Taiwan and relations with the United States. What this combination does not guarantee for Jiang is what it did for Mao and Deng—that military support will be forthcoming under any circumstances. Since it is not likely that Jiang—or his successor—will get such a guarantee, they will be dependent on the military in a way that, for all its limitations, is new in Chinese politics.

THE MILITARY AND POLICYMAKING

Below the paramount leader, the influence of the military has increased in the nation's highest policymaking bodies—the Politburo and its Standing Committee. Also a by-product of China's new politics, this increase has brought the military into the center of an arena that until now had been dominated by party leaders. However, as on other levels, its influence has been contained by countervailing trends.

The reason for the increase is twofold: one part has to do with the ruling style of the paramount leader, the other with policymaking procedures. Under Mao, policymaking was a one-man show, less on internal issues and more on foreign policy and security. In the first few years, he was more receptive to the views of other leaders, but from the late 1950s Mao became despotic. What went on in deliberations of the Politburo is not known, but it is known that in larger party forums Mao's colleagues avoided direct confrontations. After the collapse of the Great Leap Forward had convinced them not to respond to Mao's revolutionary exhortations, they chose subtle and roundabout ways to resist implementation—a tactic that fuelled Mao's anger and culminated in the Cultural Revolution. From the early 1960s, Chinese leadership politics were dominated by a series of conflicts. In this environment, there were few set procedures for open deliberations on policies.

Under Deng, policymaking was a much more open process. For political expediency and by personal inclination, he tended to bring his senior colleagues into the process. In formal and informal meetings, as well as in written communications, Deng solicited the opinions of senior colleagues and presumably took those opinions into account in formulating the Politburo's decisions. Still, Deng had the stature to be the ultimate source of such decisions, especially on sensitive foreign and security issues.[19]

Things are different under Jiang. The policymaking process under his chairmanship is undoubtedly more diffused among members of the Politburo and its Standing Committee, and final decisions are much less the sole prerogative of the top leader. This is because Jiang is essentially an arbiter and consensus builder rather than an initiator and leader.[20] Since his standing does not endow him with the privilege of claiming final wisdom on all affairs of state, he apparently does not freely intrude into areas that are the responsibility of his colleagues. And since he does not tower above them, the process of making decisions is more collective and open to influence by other leaders than ever before.

The military are in the strongest position. Not only do they have leverage over a paramount leader who is critically dependent on their support, they are also in an advantageous position in relation to other members of the Politburo. In contrast to Mao's time, and to a lesser extent Deng's, these members do not have their own constituencies in the PLA going back to revolutionary days but need the PLA's support for their own interests. In these circumstances, they will surely try to placate the PLA's leaders.

Does this mean that the PLA has gained a dominant voice in national policymaking? The answer encompasses three separate areas. The first pertains to national affairs that lack a particular military dimension, such as the economy, regarding which it is not likely that army chiefs carry more weight than other Politburo members. The second relates to the internal development and activities of the PLA, regarding which military chiefs virtually have a free hand in setting policy. The third concerns foreign affairs, which are of direct or indirect interest to the military, and regarding which they have gained greater influence.

In the first area, as professional military men, PLA leaders probably do not claim a singular prerogative to make decisions on matters in which they have no particular expertise. In any case, it is not likely that other Politburo members would passively accept such intrusion. Furthermore, in the history of political-military relations at the highest level in China, PLA leaders are not known to have had special influence on decisions outside their professional sphere. To the extent that they did, it was in their capacity as national leaders who saw themselves as standing above strictly institutional concerns. Since these concerns grow increasingly complex and constantly demand higher levels of expertise and attention, and since the generation of military leaders who could intervene in other areas as national figures has passed from the scene, the likelihood of inordinate military intervention in matters that do not directly bear upon the armed forces is remote.

---

[19]Michael D. Swaine, *The Role of the Chinese Military in National Security Policymaking*, RAND, MR-782-1-OSD, 1998, pp. 11–12.

[20]Shambaugh, "Jiang Zemin and the PLA," p. 215.

With one major exception. If the leadership's economic policies falter badly, causing widespread and prolonged social instability, there is every possibility that the military will not remain on the sidelines. The most likely form of intervention will be transfer of support from Jiang to a rival who holds out the promise of rectifying the situation. Here lies another difference that divides Mao and Deng from Jiang, or whoever succeeds him.

Mao, because of his stature, could withstand a horrific catastrophe like the Great Leap Forward. No rival emerged to challenge him (Peng Dehuai tried to effect a change of policy only, and was careful not to criticize Mao personally), and military intervention was never in the cards. The only major crisis faced by Deng was the Tiananmen Affair, but the military intervened on his side, and the rise of a rival was inconceivable. In the case of Jiang, such a scenario cannot be ruled out, since his stature will not protect him from the effects of an economic crash, and the same will most likely apply to his successor. This vulnerability is another feature of the new politics.

In the second area, the military chiefs have more autonomy to run the PLA than they did under Mao or Deng.[21] Under Mao, after the PLA embarked on its first period of modernization in the early 1950s, a de facto division of labor inevitably developed to some extent at the highest level, and the military chiefs took charge of the PLA's daily work. However, Mao remained the real supreme commander, and when he chose to intervene, he did so with impunity. The PLA's fateful shift from professionalism to Maoist military doctrine under his aegis in the late 1950s is the most striking example. Deng was also an active commander-in-chief, and although he was much more considerate of professional military sentiments, he still made the final decisions. The setting of priorities between economic and military development and the allocation of resources to the PLA accordingly throughout the 1980s attest to his supremacy.

In contrast, Jiang has neither the qualifications nor the authority to be an active commander-in-chief who is deeply involved in guiding the PLA. Professional decisions are made by PLA leaders in the Central Military Commission or in other organs, depending on the decision, while Jiang follows their lead and gives his approval.[22] Since Jiang has no pretense to effectively command the PLA, and since he surely does not want a confrontation with its leaders, this arrangement is satisfactory to both sides and seems to be working smoothly.

In foreign affairs, the military have not only acquired a new capacity to exert influence,[23] but are also driven to do so as never before. This drive derives from two factors—one visceral and one strategic—which did not carry the same force in earlier periods.

---

[21] Swaine, *The Role of the Chinese Military in National Security Policymaking,* Chapter 4.

[22] Swaine, *The Role of the Chinese Military in National Security Policymaking,* p. 15.

[23] Swaine, *The Role of the Chinese Military in National Security Policymaking,* Chapter 3.

The first factor is nationalism. Always the most powerful factor in Chinese foreign policy, nationalism acquired new force after the collapse of the Soviet Union and the disintegration of communist ideology. These developments changed the preoccupation of Chinese leaders in two ways: they confronted the Chinese as never before with the question of their role in the world as a rising superpower, and not as a secondary player in the political game between two superior powers; and they shifted the emphasis entirely away from China's defunct global revolutionary mission to an exclusive concern with its role on the world scene.

The new nationalism is a direct result of this single-minded preoccupation. It is driven by the view, whether genuine or affected, that China is again subjected to encroachments on its independence, which are reminiscent of past imperialist humiliations, and which obstruct the pursuit of its rightful place in the world. It is shared by all Chinese leaders but has particular relevance for the military.

The reason is that the military see themselves as chief protectors of China's territorial interests and national honor. This self-image is inseparable from the pride and patriotism that are universal hallmarks of the military profession. In the PLA, nationalism has a particularly sharp quality because Chinese officers function in an intense patriotic milieu, which continuously inculcates them with nationalistic values and imbues them with a sense of mission as protectors of these values.

This sense of mission is strengthened by Jiang Zemin's as yet inadequately proven revolutionary credentials. Under Mao or Deng, the military would hardly venture to claim a unique role in defense of nationalistic objectives. Jiang, however, still has to demonstrate that he is a worthy standard-bearer of Chinese nationalist aspirations. Until he does, the military will tend to view him with some suspicion and presumably consider it necessary to keep a close watch on foreign policymaking so as to ensure that Jiang does not compromise China's core principles. On territorial questions, of which Taiwan is by far the most important, nationalistic aspirations and the means for achieving them coalesce—a combination that invests the military with a central role.

The second factor propelling the military into foreign affairs is the international situation that has emerged after the end of the Cold War and the collapse of the Soviet Union. This has caused changes in China's strategy and has placed new responsibilities on its military leaders.

During the Cold War, the overriding concern of Chinese leaders was the perceived threat from the superpowers—the United States in the 1950s, both the United States and the Soviet Union in the 1960s, and the Soviet Union until the late 1970s. During this period, the sole mission of the PLA was to defend China against an invasion. This was to be accomplished by relying on the Maoist doctrine of "people's war": drawing the invading forces into China's vast interior, grinding them down by a protracted guerrilla war of attrition, and then driving them out with a full-scale counterattack. The essence of this strategy was that the war would be fought on China's territory, and its conventional armed forces were developed (or undeveloped) accordingly—as a massive, technologically backward but highly motivated and well-led force. Given the limited nature of the PLA's primary mission and its stark deficiencies in power projection capabilities, it did not expect to fight far from China's borders, and then only in dire contingencies for China's defense.

The PLA's mission began to change in the mid-1980s, when the Chinese officially acknowledged what had already been their operational belief for several years: that China no longer faced the threat of a major war. If war did break out in the future, they said, it would be limited and local. This momentous change signaled the virtual abandonment of Maoist doctrine—except in the unthinkable eventuality of a full-scale ground invasion of China—and a recognition that the PLA had to prepare for war outside its borders. For this, a new force-building policy was formulated: the development of rapid reaction units that were more mobile, better trained, and better equipped than the rest of the army. This policy was accelerated in the 1990s, with emphasis on the navy, air force, and elite ground force units.[24]

These concepts acquired even broader and more direct relevance after the collapse of the Soviet Union. The elimination of a threat that had constrained them for years, together with a reduced U.S. presence in the Pacific, presented the Chinese with fresh opportunities for pursuing a vigorous foreign policy in their neighborhood. Combined with the new nationalism and the PLA's improved capabilities, this pursuit has raised the possibility that it might eventually draw in China's armed forces. In these circumstances, the military have a direct interest in influencing decisions that might force them to fight. Although such a possibility remains remote, there is one striking exception.

That is Taiwan. On this issue, core questions of China's territorial integrity, national honor, and international stature come together in an explosive mix which catapults the military straight to the center of the policymaking arena. On this issue the military see their full participation as essential to ensuring China's firm nationalistic stand towards Taiwan and the United States, as well as to deciding on actions that might lead to armed conflict. Their role in the crisis of 1995/1996, which was reportedly decisive and encompassed both diplomatic and military moves, demonstrates the centrality of the military on the Taiwan issue.

Does this put them in the forefront of hard-line leadership elements advocating confrontation over Taiwan, even at the risk—however far-fetched—of a military face-off with the United States? This is not the case, because of two counter-trends: the past record and present calculus of PLA leaders.

Chinese commanders in the past sent troops into battle only when they calculated that success was assured and that there was minimal risk of escalation—two critical elements that do not exist in the Taiwan situation.[25] For one thing, the PLA does not have the capability to overwhelm Taiwan's defenses in a quick operation. If it brought all its assets to bear, China could probably conquer Taiwan in the end, but

---

[24]Paul H.B. Godwin, "From Continent to Periphery: PLA Doctrine, Strategy and Capacities Towards 2000," *The China Quarterly*, No. 146, June 1996, pp. 464–487; and Allen S. Whiting, "The PLA and China's Threat Perceptions," *ibid.*, pp. 596–615.

[25]Ellis Joffe, "How Much Does the PLA Make Foreign Policy?" in David S. G. Goodman and Gerald Segal (eds.), *China Rising: Nationalism and Interdependence*, London: Routledge, 1997, pp. 66–69.

only after a long war that would destroy the island. In this scenario, the damage to China's global stature, regional relations, and economic development would be incalculable. The PLA would bear the brunt of the blame for the invasion and would be treated as an international pariah. Its modernization would be severely retarded. Until the PLA acquires a capability to take Taiwan quickly—which will not be for at least a decade—China's professional military leaders would have to take collective leave of their senses to sanction a massive invasion.

Short of such an invasion, there is a range of military options the Chinese could choose—from low-level military moves designed to intimidate, as in the crisis of 1995/1996, to various forms of blockading Taiwan. However, given the intensity of emotions surrounding the Taiwan issue, and the high stakes in terms of national prestige, China's military chiefs are surely aware that once they initiate substantial military operations, the risk of escalation is high. This is because they might be compelled to escalate until the leadership can claim to have achieved its objectives, and no one will be able to foresee the consequences. This must be a forceful argument against any significant military action.

If Taiwan declares independence, PLA leaders are likely to support military action, despite the dangers. In such situations, particularly over Taiwan, the political and military leaderships will come together in a nationalistic drive to uphold what they see as China's sovereignty. If this leads to military action, it will not be the result of aggressive PLA influence, but of broad leadership consensus.

While Taiwan and relations with the United States have been of primary concern to PLA leaders, those leaders have also tried to influence other foreign issues—Hong Kong and the South China Sea, nuclear testing and nonproliferation, arms sales, and multilateral security. They have reportedly pressured the Foreign Ministry to resist Japanese calls for increased transparency, have voiced reservations, if not opposition, to multilateral security initiatives in Asia and transparency in China's defense posture, and have taken a firmer stand in favor of arms sales.[26]

PARTY AND ARMY

Although the military can influence the political leadership only at the apex of China's power structure, its potential for doing this has been bolstered by changes in the institutional relationship between the party and the army. The crux of these changes is that the close integration that had existed between them in the Maoist period and, though not as close, under Deng, has given way to an increasing separation. Consequently, the dichotomy between them as two entities with different, and sometimes conflicting, interests is growing.

This does not mean that the army's ultimate subordination to the party and its leader is in doubt, although it is subject to the reservations arising from the new politics.

---

[26]Swaine, *The Role of the Chinese Military in National Security Policymaking*, p. 34; and Eric Hayer, "China's Arms Merchants: Profits in Command," *The China Quarterly*, No. 132, December 1992, pp. 1101–1118.

What it does mean is that political controls in the armed forces have been weakened. This has been manifested most clearly by the declining importance of the hierarchy that enforces these controls, headed by the traditionally all-powerful General Political Department. As a result, the party's presence in army units is less intrusive and army commanders have more leeway to pursue their specialized missions—within the broad framework set by the party, but without undue interference on the ground.

The organizational framework for the exercise of political controls has in theory not changed since the early 1950s. It consists of party committees, political commissars, and political departments which run parallel to the military chain of command, and whose tasks are supervision and education. Ideally, leadership in the armed forces is based on the principle that, except in emergencies, decisions in military units are made by party committees and implemented by commanders and commissars according to their respective functions.

This ideal was never fully realized. Once Chinese officers developed professional attitudes, they resented and resisted this system on the grounds that it was incompatible with combat effectiveness in modern warfare. Their resistance was strongest in the second half of the 1950s, but declined after the reassertion of Maoist doctrine in the armed forces. In subsequent years, officers tended to keep a low profile in the face of Maoist hostility to military professionalism, and voiced their opposition to excessive political controls only when these blatantly interfered with their military responsibilities—as during Lin Biao's politicization of the PLA in the early 1960s—or when circumstances were favorable to redressing the balance between politics and professionalism—as after the downfall of Lin Biao.

When the drive to modernize the PLA was launched by Deng, the intrusive aspects of political control were watered down radically. They were revived briefly after the Tiananmen Affair, when the wavering of some officers prompted the leadership to launch a forceful campaign to ensure the PLA's political loyalty. Led by Yang Baibing, this campaign highlighted the importance of politics in the armed forces and cast suspicion on professional officers. Its shrillness undoubtedly alienated these officers and contributed to Yang's downfall, following which attacks on the professional military ceased and political control reverted to the Deng formula: the leadership views it as important but does not let it interfere with the prerogatives of military commanders.[27]

Since then, this formula has been preserved. On the one hand, party and military leaders have put enormous emphasis on political and ideological education, especially of high-ranking officers, in order to ensure that the army at all levels will follow commands without question. The main themes of education are the importance of upholding party supremacy over the army, supporting its policies, and maintaining PLA traditions, especially under the threat of corrosive forces stemming

---

[27]Ellis Joffe, "The Chinese Army: Coping with the Consequences of Tiananmen," in William Joseph (ed.), *China Briefing: 1991*, Boulder, CO: Westview Press, 1992, pp. 37–55.

from the market economy.[28] Providing they do not infringe on military duties, these efforts can hardly be opposed by professional officers, especially since one purpose of political education is to improve the combat quality of the troops.

On the other hand, however, little has been heard of the role of party committees as supreme decisionmaking organs in military units, and the division of responsibilities between military commanders and political commissars in carrying out its decisions. Since it was primarily through these two functions of political work that the party had intruded most markedly into the PLA and kept a tight grip on it, their downgrading is a clear indication of the PLA's growing organizational separation from the party and its continued transformation into a professionally oriented institution.

This is hardly surprising, since separation is driven by three powerful factors. One is the drastic decline in the importance of the communist party as the traditional epicenter of the Chinese political system. Although theoretically still occupying this position, under the new politics two developments have greatly eroded its power and legitimacy: the system has become much looser, due to decentralization and the loosening of central authority over bureaucratic and regional power centers; and the party has lost much of its effectiveness, due to the de facto end of ideology, erosion of power, and corruption. The party's decline could not but have undermined the authority of party members in the armed forces, especially the formerly omnipresent political commissars and other full-time functionaries. Since the hierarchy of these functionaries runs parallel to the military chain of command and is supposed to supervise it, its weakening has surely strengthened the distinctive professional identity and insular character of the military organization.

Nothing has strengthened this identity more than the second factor—the modernization of the PLA, which is geared to preparing it "to fight a modern war under hi-tech conditions." After laying the groundwork in the 1980s, the Chinese embarked on an accelerated program of upgrading in the early 1990s and have pursued it in a sustained and comprehensive fashion. They have closed the long-standing gap between doctrine and operations by abandoning Maoist concepts of "people's war" and preparing for realistic limited engagements that will employ modern forces and conventional strategies. They have improved their rapid deployment forces, air force and navy, by buying small quantities of new weapons from Russia, mainly modern aircraft, submarines, and air defense systems. They have advanced their logistics, force structure, training procedures, and joint service operations. And they are increasing and refining their missile delivery systems.[29]

All these efforts have not yet brought the PLA much closer to the level of the most modern armies. The PLA continues to suffer from major deficiencies which will take

---

[28]See, for example, *Jiefangjun bao*, March 22, 1996, in *FBIS*, April 17, 1996, pp. 22–24; and *Jiefangjun bao*, June 8, 1996, in *FBIS*, June 20, 1996, p. 35–37.

[29]See, for example, Godwin, "From Continent to Periphery"; Ron Montaperto, "China as a Military Power," *Strategic Forum 56*, Washington, DC: National Defense University, Institute for Strategic Studies, December 1995; *Aviation Week and Space Technology*, May 12, 1997; and "Selected Military Capabilities of the People's Republic of China," Washington, DC: U.S. Department of Defense Report to Congress, April 2, 1997.

years to rectify.[30] But the Chinese leadership is determined to modernize in order to build armed forces commensurate with its long-term aim of gaining preeminent status in East Asia and a pivotal global role. To this end, it will have to mount a steady long-term effort that will transform the PLA into an ever more complex and effective fighting force. The condition for success is the continuous cultivation of an officer corps that is competent professionally and preoccupied completely with its specialized missions. The military leadership is well aware of this condition and has put its fulfillment at the top of the PLA's priorities.[31] The result is that the professional officer corps, which under Deng began to emerge from two decades of Maoist stagnation, has entered a period of flowering under Jiang. At the same time, the party is wilting and its representatives in the armed forces have lost much of their vigor. The party still "commands the gun," but the gulf between them has never been as large.

The third factor fostering this gulf is the PLA's massive involvement in economic pursuits. Growing rapidly since the late 1980s, this involvement has given rise to a huge empire that embraces every major profit-making activity. Its potential effects on furthering the erosion of party control are wide-ranging. These are some of the possibilities: an attachment of military units to economic projects at the expense of political activities; the participation of party functionaries in financial undertakings to the neglect of their duties; the disregard of party directives by officers who put profits first; the disintegration of revolutionary *elan* in the money-making climate of a modernizing China; the erosion by corruption of the ethic of duty and devotion; the formation of alignments among officers based on common economic interests; the gaining by the military of income independent of government allocations.[32]

Some of these effects are equally detrimental to military professionalism. The leadership is clearly aware of this, and has taken steps to curb the excesses of economic involvement and to insulate elite units from potentially damaging economic activities. However, it is questionable whether they have been able to insulate the party from such effects. Given its uncertain state in the era of reforms, it is likely that the PLA's economic ventures have further loosened the ties that had bound the party to the army.

The loosening of these ties has not cast doubt on the subordination of army units to the party leadership. Orders are transmitted by the leadership to the military high command and then passed down to the PLA. The compliance of the PLA depends on two conditions: that PLA chiefs themselves follow orders, and that the military

---

[30]*New York Times,* December 3, 1996; *International Herald Tribune,* January 29, 1997; Michael D. Swaine, "Don't Demonize China," *The Washington Post,* May 18, 1997; and Eric McVadon, "Strait Shooter," *Far Eastern Economic Review,* January 16, 1997.

[31]James Mulvenon, "Professionalization of the Senior Chinese Officer Corps: Trends and Implications," RAND, MR-901-OSD, 1997.

[32]James Mulvenon, "Soldiers of Fortune: The Rise of the Military-Business Complex in the Chinese People's Liberation Army, 1978–1998," Dissertation, University of California, Los Angeles, 1998; Tai Ming Cheung, "China's Entrepreneurial Army: The Structure, Activities, and Economic Returns of the Military Business Complex," in Dennison Lane et al. (eds.), *Chinese Military Modernization,* pp. 168-197; and Ellis Joffe, "The Chinese Army and the Economy: Effects of Involvement," *Survival,* Summer 1995, pp. 24–43.

command and control system is effective. As long as these conditions are met, the obedience of the army to the party leadership is assured.

Nonetheless, the loosening of these ties has reduced the capacity of party controls to perform what ideally had been its most crucial function in the past. This was to guarantee that military units not only obeyed orders from above, but that these orders were implemented by their commanding organs in line with party policy and spirit. It was, furthermore, to guarantee that if military imperatives did not conform to party preferences, priority would be given to these preferences. Finally, it was to instill in officers and men the belief that such choices were proper and unquestionable.

The inability of the party to perform these functions has enhanced the freedom of the PLA to conduct its affairs according to military imperatives alone. This, in turn, has two implications for its political position. First, by downgrading the latent conflict between professionalism and politics, it has strengthened the internal cohesion of the armed forces. Second, by highlighting the role of the military chiefs as leaders of an institution with distinct and specialized interests, it has strengthened their hand in dealings with the party leadership.

IN CONCLUSION

The political role of the military in the transition to the post-Deng period is shaped by two factors: developments outside the PLA, which have given rise to Chinese politics with new characteristics; and developments inside the PLA, which are gradually changing its old characteristics. This combination has increased the capacity of the PLA to exert political influence, but has also set limits on it. The balance between these trends will not be finally worked out until these two factors settle into fixed patterns, and this is a process that will take years to unfold. Until then, the PLA as a political force will be pulled in different directions in line with the circumstances of changing situations.

EPILOGUE

This paper was completed in the summer of 1997. Since then, the trends and counter-trends discussed here have been brought into sharper focus by developments in both the PLA and the political arena. Highlighted by two major events—the 15th Congress of the Chinese Communist Party in September 1997 and the Ninth National People's Congress in March 1998—these developments indicate that the various trends noted in the paper have coalesced into a new mode of party-army relations. This mode is defined by several features that may seem contradictory but that, in fact, complement each other.

The first is that the Chinese military have gained an unprecedented potential for wielding political power. However, this potential is tempered by a growing military professionalism and by countervailing political factors. It is manifested mainly in the military's new capacity to influence high politics and policies, not in routine involvement in political or administrative affairs. The military's imprint on policymaking is unprecedented but selective.

This new mode stems from the vast changes that have occurred on the Chinese political scene and in the PLA, beginning in the twilight of the Deng period and accelerated under Jiang. Before discussing the implications of these changes, it will be useful to review them briefly.

The first change stems from Jiang's relationship with the military. Lacking the ingredients that had given Mao and Deng enormous personal authority—charisma, achievements, and connections—Jiang cannot count on the unconditional support of the military in possible crisis situations as Mao and Deng could during the Cultural Revolution and the Tiananmen demonstrations.

Normally, Jiang can get the military's support by virtue of his institutional position, concessions to the military, and newly woven political ties. This probability has been strengthened after the long-delayed retirement at the 15th Party Congress of the two senior military commanders, Liu Huaqing and Zhang Zhen. As revolutionary veterans, Liu and Zhang had towered over Jiang in military stature and, having been put in their positions by Deng Xiaoping, owed Jiang no political debts. The new PLA chiefs, Zhang Wannian and Chi Haotian, belong to Jiang's generation and owe their appointments to him, as do numerous other new commanders.

Nonetheless, if Jiang falters badly, the military might back a rival. Since such a switch was inconceivable in earlier years, the military now has a greater capacity than ever before to influence the fate of the paramount leader, which makes him more amenable to military views.

This capacity also extends to national policymaking. Under Mao, policymaking was a one-man show; under Deng it was much more collegial, but Deng still had the stature to make major decisions. Jiang, in contrast, does not tower over his colleagues, and his standing does not endow him with supreme authority in all affairs of state. As a result, the policymaking process is spread among members of the ruling group.

The military are well placed to influence this process. They have leverage over a paramount leader who needs their support. They also have leverage over other top leaders. In contrast to their revolutionary predecessors, these leaders do not have their own constituencies in the army going back to the old days, but they too need the army's backing for their special bureaucratic interests, which makes them receptive to its influence.

This influence is increased by the separation between the party and army organizations, which is driven by three powerful forces. One is the decline in the importance of the communist party as the epicenter of the Chinese political system and its concurrent weakening in the armed forces. Another is the modernization of the armed forces and the flowering of military professionalism at the expense of political intrusion. The third is the involvement of the military in economic pursuits and the consequent loosening of the party's organizational grip over military units.

This separation has not cast doubt on the subordination of the army to party leadership, but it has fostered the freedom of PLA leaders to conduct their affairs in accordance with military imperatives, and has all but terminated the long-standing conflict between politics and professionalism. This, in turn, has solidified the internal cohesion of the army and has strengthened the position of its chiefs in relation to the party leadership.

These changes account for the extraordinary potential that the military now has for intervention in the political arena. However, this is only one side of the new party-army relationship. The other is that the PLA's actual participation in politics is limited and declining. Its commanders have backed away from political entanglements and have focused more than ever on professional concerns. There are several reasons for this.

One is Jiang's dependence on military support, which ensures that he will not stray too far from their preferences, and does not require their close involvement in policy areas that do not impinge directly on their interests. The removal of a military representative from the Standing Committee of the Politburo at the 15th Party Congress reflects both the trend toward military detachment from nonmilitary affairs and their confident acquiescence in this detachment. Undergirding this acquiescence is the presumed belief of the military that Jiang as chairman of the Politburo's Standing Committee will not go against PLA views on matters that had previously been discussed at the Central Military Commission, of which he is also chairman. On the contrary, Jiang can be expected to advocate the adoption of these views.

In any case, the military can do this themselves through their representatives on the Politburo—the two senior military leaders Chi Haotian and Zhang Wannian. At the 15th Congress they replaced Liu Huaqing, who retired, and Yang Baibing, who had been kept on the Politburo for reasons of face even after he was dismissed as director of the General Political Department and secretary-general of the Central Military Commission in 1992. Since Yang had had no influence since then, his replacement by a top PLA leader has, in fact, increased the weight of the military on the Politburo. With these representatives watching out for its interests, the PLA can focus on its professional pursuits.

Also facilitating this is a generational change which has been going on for several years and was formalized by the Party and National Congresses. In the PLA, this change has elevated professional commanders whose primary concern is to oversee the long-term transformation of the Chinese army into a modern force. Most have risen through the ranks of the ground forces, have (limited) combat experience, have stayed out of politics, are familiar with the imperatives of modern warfare, and have displayed professional competence in their careers. The party and government have brought into top positions leaders whose sole concern is the development of the economy. Most are technocrats who are university-educated and have worked their way up in the economic and technical bureaucracies.

These leaders are oriented towards their bureaucratic specializations. In contrast to the founding fathers who, as national figures, did not respect institutional boundaries and blurred the distinction between military and political spheres, they will tend to stick to their particular bailiwicks and to sharpen this distinction.

This distinction is augmented by the growing importance of economic leaders in the party's top policymaking organs, which will increase in coming years under the premiership of Zhu Rongji and the weight of China's economic problems. This does not necessarily point to a contest for influence between military and economic bureaucracies. Their relationship is based more on a convergence of interests than on rivalry. Both want the economic reforms to succeed, and PLA leaders are neither qualified nor inclined to interfere in their management, provided the reforms produce results and the armed forces get appropriate allocations.

For these reasons, the military has used its new power selectively. In the political arena, it has provided crucial support to Jiang Zemin during the final years of the Deng period and since then, but has stayed out of politics. In economic affairs, its imprint has been minimal. In foreign affairs, the military has apparently limited its involvement to issues that relate directly to its concerns, most notably Taiwan. In running the PLA, the military has gained unprecedented autonomy.

The new party-army relationship has largely obviated the relevance of approaches used in the past to explain aspects of this relationship. The symbiosis approach was based on the notion that the political leadership had co-opted the military into the policy process in order to neutralize them, but now political and military leaders are going their separate ways. The political control approach was based on the notion that the party had exercised complete control over the military, but now such control is limited by the new freedom of the military. The professionalism approach was based on the notion that there had been an enduring conflict between professional and political priorities, but now this conflict has been resolved by the supremacy of professionalism.

The PLA after Mao and Deng is different and more complex. It is still a party-army relationship, but is increasingly professional and autonomous; it is still loyal to the party as an institution, but is not unconditionally subservient to a particular paramount leader; it is still committed to non-intervention in the political arena, but has vast potential power to do so. Will it stay in or out of politics?

The answer will be determined by specific circumstances. Unlike Mao and Deng, Jiang cannot count on the unqualified support of the military in all circumstances. Mao retained the support of the PLA despite the colossal disaster of the Great Leap Forward and the dismissal of defense minister Peng Dehuai. Deng retained the PLA's allegiance—exemplified in his waning days by PLA support for his chosen successor, Jiang Zemin—despite the deleterious effects of the unwanted Tiananmen intervention that Deng had forced upon it.

Jiang does not have this invaluable advantage. He will retain the PLA's support as long as he delivers the goods: economic advance and social stability, which in the military translates into funds and conditions for modernization. However, if there is a sharp and prolonged economic downturn followed by widespread unrest, the PLA might withdraw its support from Jiang.

This could be manifested anywhere along the spectrum from backing a rival to intervening with force. However, since resort to force is a remote possibility in top-level Chinese leadership conflicts, military intervention could conceivably commence with pressure for drastic policy changes and culminate in the transfer of support to a rival. Since no political leader after Deng can survive without such support, this scenario adds a novel dimension to the potential power of the military in Chinese elite politics.

Also novel is that the PLA is less likely than ever before to be used in political struggles. Mao and Deng had used it despite the reluctance, and even resistance, of PLA leaders. For Jiang or his successor, this will be immensely difficult, if not impossible. The PLA never moved in force into the political arena on its own initiative, but only because Mao or Deng had ordered it: Mao during the turmoil of the Cultural revolution, and Deng during the Tiananmen demonstrations. Mao and Deng could do so because of their personal stature on the Chinese political scene and in the PLA, even though senior commanders opposed the interventions.

The new commanders are less political than their predecessors and will be even more reluctant to intervene, while the successors to Mao and Deng do not have the stature to force intervention upon them. Consequently, China's leaders will be circumspect in calling for PLA intervention and PLA commanders will not respond to such calls automatically. Only an extreme emergency can change this—for example, if the regime is threatened by mass demonstrations that are provoked not by policy failures but by demands for rapid political change, which the People's Armed Police is unable to handle. In that case, the army will most probably protect the party.

4. PLA RAPID REACTION FORCES: CONCEPT, TRAINING, AND PRELIMINARY ASSESSMENT

Andrew N. D. Yang and Col. Milton Wen-Chung Liao (ret.)[1]

BACKGROUND

In June 1985, the Central Military Commission (CMC) of the Chinese Communist Party (CCP) held an extended meeting to develop consensus views on future warfare and set the direction of formulating new military strategies. At that meeting, paramount leader Deng Xiaoping, also Chairman of the CMC, made the following concluding remarks:

> there will not be large-scale warfare in the foreseeable future. Factors preventing imminent and large-scale warfare are increasing; such factors are: (a) neither the Soviet Union nor the United States has achieved advantages in terms of military deployment and therefore prevent either side from making the first move; (b) the third world countries are not cooperating with these two powers in terms of military deployment, in fact, people in the third world countries support peace more than war; (c) the importance of technological advantage and sustaining economic competition have replaced the importance of sustaining war-winning military capability and becoming new contested terrain in world competition.[2]

Deng's remark not only impelled Chinese military thinkers to discredit a large-scale warfare scenario, but also replaced conventional war fighting thinking that called for "fighting an early war (*zhao da*), a large-scale war (*da da*), and a nuclear war (*da hezizhan*)" with a call for "military construction under peaceful era."[3]

In the eyes of PLA military thinkers, the world is still unsafe and unstable even though a large-scale nuclear war is highly unlikely, thanks to the collapse the of Soviet empire and the East European communist bloc. Regional instability and potential conflict are characterized by territorial dispute, ethnic and religious conflict, and arms proliferation. Such unstable and war-inducing regional factors

---

[1]Andrew N. D. Yang is Secretary General of the Chinese Council of Advanced Policy Studies in Taiwan, Republic of China. He is also a research associate at Sun Yat-sen Center for Policy Studies, National Sun Yat-sen University, Kaohsiung. Mr. Wen-chung Liao is currently a project leader at the Chinese Council of Advanced Policy Studies; his main focus is on Chinese military modernization and PLA weapon systems. He is the Asian Associate Councillor of the Atlantic Council of the United States.

[2]Peng Guangqian, "Deng Xiaoping's Strategic Thought," in Michael Pillsbury (ed.), *Chinese Views of Future Warfare*, pp. 3–10.

[3]Hong Baoxiu, "Deng Xiaoping's Theory of War and Peace," in Michael Pillsbury (ed.), *ibid.*, pp. 19–23.

also exist in China's peripheral areas, such as Xinjiang, Tibet, Taiwan, and the South China Sea. In order to sustain as long a peaceful period as possible and effectively prevent or win a regional conflict, military operations in the peaceful era should be designed to enhance two military capabilities: first, the capability of modern strategic weapon systems to exert effective deterrence; and second, to develop highly competitive, high-technology-based rapid reaction forces (RRF) (*kuaisu fanyin budui*) to cope with future small-scale, highly intensive regional combat and military operations. Under these two major principles of military construction, the CMC has given orders to learn the lessons of Western rapid deployment forces as the basis of developing PLA's RRF. The PLA began its RRF development in the early 1980s. Since then, RRFs have been set up in PLA Army, Air Force, and Navy units, as well as Army special forces, Army aviation, Marine Corps, and airborne units.

However, lessons of Allied Forces operations in the Gulf War in 1991 forced the CMC to consolidate the development of RRF. In 1992, a special force named "Resolving Emergency Mobile Combat Forces" (REMCF) was created and placed directly under the CMC's control. This special force was given the tasks of border defense, dealing with internal armed conflict, maintaining public order, and conducting disaster relief missions. REMCF has been developed in two phases. Phase one was initiated at the beginning of 1992. Each Group Army corps of every Military Region (MR) selected an infantry division to be the designated REMCF for dealing with emergency situations in every Combat Region (CR).[4] Phase two was implemented in 1994, continuing the development of a second batch of REMCF and enhancing the capability for "quick fighting, quick resolution" under the conditions of high-tech regional warfare. The two phases of the REMCF development program will be completed by the end of 1998, with an estimated 300,000-man REMCF force to be established and directly controlled by CMC.

It is the authors' purpose to analyze the origins, force structure, doctrine, weapons system capabilities, and training for these special units, as well as to assess their general capability in terms of meeting designated task requirements in so-called "future regional warfare."

CONCEPTION FOR RRFs

The impetus for the PLA's desire to develop RRFs was the 1991 Gulf War. In analyzing the lessons of that conflict, the PLA came to the following conclusions: (a) modern war is high-tech war, and technology can not only fulfill tactical and combat missions but can also fulfill strategic objectives; (b) regional warfare can serve as a viable means for political resolution and render large-scale warfare unnecessary; (c) the existence of high-tech weapon systems holds out the possibility of "quick resolution" by conducting long-distance, high-power, and precision attacks; and (d) high-tech weapon systems have changed the needs of force composition and resulted in new types of combined operation.

---

[4] *Chinese Communist Annual Report 1997*, Military Section, Chapter 9, p. 68, Chinese Communist Press Publications, 1998.

In addition to emphasizing notions of applying high-tech weapon systems, the PLA's conceptualization of "rapid reaction" capability emphasizes adaptation to warfare scenarios and developing quick responses to varying battlefield contingencies. In other words, adaptation and responsiveness are two key elements of developing a rapid reaction force, and these elements require sufficient and comprehensive preparation for unexpected contingencies and future war scenarios.

Key elements in PLA's rapid reaction concept are

a.   Training

b.   Speed

c.   Strength

d.   Effectiveness.

The key elements in rapid reaction effectiveness are

a.   Emphasizing an "active defense strategy"

b.   Emphasizing "inferiority vs. superiority" and "weak vs. strong"

c.   Emphasizing mobility and the strategy of attacking.

Policies for Enhancing the RRF's Operational Readiness

To increase RRF combat capability and cope with the requirements of future high-tech regional warfare, the PLA has operationalized policies to enhance RRF's operational readiness. These measures include:

a.   Identifying possible targets and intensively assessing conditions of the war zone (*zhanqu*);

b.   Applying imported technology to store, assess, and revise every operational plan and implementing these plans in training and exercises so as to adjust combat capability in different war scenarios;

c.   Deploying motorized vehicles, electronic warfare equipment, and modernized logistic support to RRF units to enhance force mobility, long-range deployment, electronic countermeasure and electronic counter-countermeasure capabilities; and

d.   Reducing the timeframe for transforming levels of operational readiness, such as transforming operational readiness from level 4 to level 1 in a time period not exceeding three days.

In addition to promulgating policies to enhance an RRF's operational readiness, CMC also issued a document entitled "Regulations for Constructing REMCFs" as a blueprint for REMCF training. It includes the following points:

a.   Require every soldier in RRF units to undertake special training, such as swimming, skiing, and mountain climbing. Apart from basic training, infantry soldiers with service above two years should acquire proficiency with every weapon system assigned to the company units;

b. Consolidate "three attacks and five defenses" (*san da wu fang*) and conduct the "three real trainings"[5] to enhance capabilities of anti-guided and precision weapon systems, and C^{4}I etc.; and

c. Enhance command and control efficiency.

To implement these training directives, the PLA has conducted various exercises since 1995, concentrating particularly on long-range and intraregional rapid mobile deployment. For example, RRF units in different MRs have been selected to conduct long-range and mobile combined exercises. To this end, RRF combined exercises were carried out in 1995 and 1996 in the Gobi desert, the Tibetan and Xinjiang highlands, and in the southwestern tropical forests to enhance the RRF's adaptative survival capabilities.

The RRF's training and exercises in 1996 emphasized the following objectives:

a. Ground forces focused on broader retaliatory capabilities for attacking in mountain regions with combined forces and amphibious landings;

b. Naval forces focused on Ro/Ro [roll on-roll off] amphibious landings, electronic warfare, air-sea combined operation, anti-submarine warfare, anti-air operations, and ship-to-ship guided missile training;

c. Air forces focused on long-range interregional air attack, long-range mobile transit, air-to-sea attack, airborne training; and

d. People's Armed Police (PAP) and reserve units focused on responding to mobilization calls and providing logistic support.

Current Status and Operational Capabilities of RRFs

Army Special Forces. The PLA has established a regiment-level Army Special Force (ASF) in every MR as an RRF unit, directly under the MR headquarters command. The principal officers of an ASF, including the commander, political commissar, and chief of staff, are full colonels. Officers above the platoon level are required to be university graduates and receive further education in the Army Command Academy. In every Group Army, a battalion-level special reconnaissance task force has been set up under the Group Army HQ's command. Officers and men of this ASF are selected from reconnaissance and technical units of every Group Army. The wash-out rate is about 50 percent after receiving further tests and training. In addition, every MR has established special training facilities for their RRF units. Special skills, such as martial arts, are also included in the training.[6]

---

[5]"Three attacks" refers to attacks on helicopters, tanks, and airborne troops. "Five defenses" means to defend against nuclear/biological/chemical attacks, electronic countermeasures, and precision-guided weapon systems. "Three real trainings" refers to deploying real troops, conducting real operations, and using live ammunition in training.

[6]Zhou Mon-wu, "Elevation of PLA Special Force," in *Military Digest,* October 1997, pp. 2–3.

Each RRF unit is equipped with the most advanced weapon systems, equivalent to special forces in Western countries. They also possess remotely piloted vehicles (RPVs), night-vision goggles, and GPS satellite communication systems. The total strength of ASF may be as high as seven regiments and twenty-four battalions, or approximately 25,000 personnel.

Army Aviation Unit. The Army Aviation Unit (AAU) was established in April 1986. Its main task is to deploy helicopters and light aircraft to support ground operations. The AAU is directly under General Staff Department (GSD) command, and has been seen in many combined exercises in Northern China (Huabei) performing anti-tank, special forces insertion, and electronic countermeasure operations. The AAU possesses small numbers of S-70C Sikorsky helicopters, which have been converted into command, control, and communication platforms. The AAU's main helicopter fleet includes 200+ Z-8 transports, 100+ Z-9A, and 30+ Mi-8/Mi-17. Harbin Aircraft Co. has also developed a gunship variant of the Z-9A (WZ-9) for the AAU. Twelve of the Gazelle helicopter gunships were procured in 1988 and deployed to the 38th Group Army in the Beijing MR. In 1993, China Aviation Technology Import and Export Co. entered into a joint venture with Singapore Aviation Industry to co-produce EC-120 light helicopters, and China has ordered 150 EC-120 for the AAU. The PLA is planning to import helmet-mounted night-vision devices to enhance the AAU's night-fighting capability.

As shown in recent PLA exercises, the AAU has taken part in various combined exercises such as anti-tank, personnel transport, command post relocation, reconnaissance, and electronic countermeasures missions. This demonstrates that PLA is trying to improve its fighting capabilities in a high-tech environment.

Marine Corps. The PLA Marine Corps was established in December 1954 and was consolidated into a full-fledged amphibious landing force in 1979. The Marine Corps is attached to the South Sea Fleet, headquartered in Zhanjiang. Equipment and weapons systems for the Marine Corps include: Type 63 amphibious landing tanks, Type 7711/7712 amphibious armored personnel carriers, Type 54 artillery, and HJ-73 anti-tank missiles. Between 1995 and 1996, the PLA has conducted several amphibious landing exercises and deployed helicopters, hovercrafts, and other amphibious landing equipment, demonstrating that the Marine Corps seeks to enhance its amphibious landing capability through combined exercises.

Airborne Troops. The 15th Airborne Corps of the PLAAF is composed of three airborne brigades. The 43rd brigade, stationed in Kaifeng, Henan Province, is attached to the Jinan MR. The 44th brigade, stationed in Yinshan, Hubei Province, is attached to the Lanzhou MR. The 45th brigade, stationed in Huangpi, Hubei Province, is also attached to the Lanzhou MR. The airborne troops are accompanied by the 13th transport division of the PLAAF. The airborne troops are directly under CMC control. Strategically, the airborne troops are considered to be a reserve force, yet in tactical terms the airborne troops are deployed as an advance force. It could be reconstituted as an air mobile rapid attack force. The airlift capability of the PLAAF is composed of 10 IL-76 heavy lift, Yun-8, and Yun-7 transports, as well as Mi-17, Mi-8, S-70c, Z-8, and Z-9 helicopters. In terms of weapon systems, the airborne troops are equipped with BMPs [Russian armored personnel carriers], Hongjian-8 anti-tank missiles, Hong Yin 5A anti-air missiles, and Russian-made flame-throwers. In recent years, the airborne troops have developed several technical combat units, including reconnaissance, communication, artillery, and anti-chemical units.

The airborne troops have conducted exercises in different types of terrain, as well as all-weather, daytime, and night conditions. The exercises are normally conducted at the company level. The timeframe of each exercise is three days and troops are given a two-day food ration. The exercise missions include occupying and defending strategic key points, sabotaging airfields, anti-air attack, anti-reconnaissance, and survival course training. In the 1996 Taiwan Strait exercises, an airborne battalion was parachuted to Dongshan Island, supporting a Marine amphibious landing exercise.

RRF Exercises Since 1996

RRF exercises were conducted in each of the following MRs:

Nanjing MR

a.   Nanjing MR conducted three-phased exercises from March 8 to 21, 1996 in the Taiwan Strait:

Phase I: 2nd Artillery fired Dongfeng-15 (M-9) missiles off Keelung and Kaohsiung harbors.

Phase II: Air Force and Navy conducted air attack and missile firing exercises, electronic warfare, low-level penetration air attacks, blockade, and air-sea combined exercises.

Phase III: Air-sea-land forces conducted amphibious landing exercises on Pingtan Island. AAU helicopters, Su-27 fighters, and airborne troops were also deployed.

b.   In September 1996, the East Sea Fleet conducted combined exercises off the Zhejiang coast, including air-sea combined exercises and anti-submarine warfare.

c.   In October 1996, another air-sea combined exercise was conducted with emphasis on air superiority and sea control capability.

d.   In November 1996, an air-land combined exercise was conducted in Anhui. A simulated "Blue force" was attacked by a "Red" tank regiment. The Air Force deployed Su-27 and J-8II jet fighters to attack the "Blue" force. All land forces were rapidly transported by railways and vehicles to the exercise region.

Jinan MR

a.   From late October to early November 1996, an air-land combined exercise was conducted in the Anyang and Tongbao mountain regions of Henan Province. Artillery, communication, anti-air, Army Aviation, and Air Force units all took part. "Red" troops attacked a simulated Republic of China (ROC) army company that was occupying strategic points.

b.   From 19 to 31 October 1996, an air-sea combined exercise was held in the Yellow Sea. Submarines, destroyers, and Navy fighters and bombers simulated "Red" vs. "Blue" war games.

Guangzhou MR

a.   In September 1996, an electronic warfare exercise was held in the Yangquan and Yangjiang area of Guangdong Province. Electronic countermeasure (ECM) and counter-countermeasure (ECCM) courses were practiced.

b.   In mid-October 1996, an amphibious landing exercise was held in Zhanjiang, Yangjiang (Guangdong Province) and the Wutong region of Guangxi Province. Forces included an infantry division, communication, ECM, and logistic support units, two landing ships (LSTs) and an aviation unit of the South Sea Fleet, including 10 H-6 bombers, F-7s, and Su-27s of the Southern Air Force regiment. Troops were deployed to the exercise region by motorized transport. The exercise also included "Red" vs. "Blue" war games.

c.   March 12–20, 1996, a missile and artillery live-fire exercise was held in Nanao (Guangdong Province). Nuclear submarines, destroyers, and conventional submarines from North Sea Fleet, East Sea Fleet, and South Sea Fleet were deployed to the exercise zone. Naval aviation collaborated with naval ships in conducting missile attacks, ECM, and anti-submarine warfare drills.

Beijing MR

A large combined exercise was held in Hebei Province and Inner Mongolia in late August 1996. Forces included a tactical missile unit, an infantry division, a logistics unit, and Army aviation helicopters, as well as A-5, J-7, H-5, H-6, and ECM aircraft of Northern Air Force regiment. The main drills of this exercise sought to enhance long-distance rapid-deployment capability, air-land counter-attack capability, and logistics support capability.

Shenyang MR

a.   In mid-September 1996, an amphibious landing combined exercise was held on Changshan Island of Bohai Bay. Forces included a mechanized infantry division, LSTs, landing craft (LCMs), and Air Force aircraft.

b.   In late October 1996, an air-land combined exercise was held near Harbin. Forces included a mechanized division, artillery and tank brigades, and communication units. The exercise emphasized long-distance rapid-deployment capability and force coordination.

c.   Between late October and early November of 1996, an exercise was held in Liaoning and Jilin. Forces included an artillery brigade, an anti-aircraft brigade, a tank division, a helicopter unit, and a logistics support unit. Exercise drills included long-distance rapid deployment, counter-attack, live firing, and logistics efficiency.

Lanzhou MR

a. A 1000-mile railway transport rapid-deployment exercise was held in Lanzhou MR in August 1996. The purpose was to enhance mobile deployment capability.

b. In late August 1996, a Northern Air Force bomber and fighter regiment (including Su-27s) from various MRs were assembled in Gansu Province and conducted an air attack exercise, targeting a simulated ROC Ching Chuan Kan airbase that was built in Dingxin. The exercise included air combat, air-to-ground attack, and the firing of a new type of surface-to-air missile

c. From late September to early October of 1996, a "Red" vs. "Blue" combined exercise was held in Shanxi Province. The exercise simulated a counter attack against an invading Russian army.

The number of exercises conducted from 1995 to 1996 far exceeded the number of exercises conducted between 1992 and 1994. The focus of the exercises shifted from anti-airborne, anti-amphibious landing to amphibious landing and airborne offensive operations. In particular, the 1997 exercises emphasized amphibious landing capabilities and urban combat tactics. New types of equipment, such as Ro/Ro cargo ships, imported Russian BMPs, and flame throwers were deployed in supporting these operations. However, the PLA is still incapable of conducting a cross-strait amphibious landing. Specifically, the current two-brigade Marine Corps attached to the South Sea Fleet is not capable of a successful amphibious landing attack against Taiwan. Recently, the CMC relocated several Army divisions, along with airborne troops, to assist amphibious landing exercises. This is a clear indication that PLA is trying to enhance its amphibious landing capability. These intentions were further verified by a seminar in October 1996 held at PLA-NDU (National Defense University) to discuss the effectiveness of combined forces in amphibious landings. Aside from building new LSTs, LCMs, and Ro/Ro amphibious landing ships, the PLA has developed a "wing in the ground" vehicle with the assistance of Russian engineers and specialists. In 1997, PLA successfully produced the DXF100 "wing in the ground" vehicle. New heavy-duty tires for this vehicle are currently being tested.[7]

---

[7]China successfully developed a first-generation "wing in the ground" (WIG) vehicle in January 1997, code-named DXF100. The DXF100 vehicles were tested in Taihu in Jiangxi Province. Its maximum lifting capacity is 4.5 tons, and it is capable of flying 1–5 meters above water with a maximum speed of 120 km per hour. It was designed and developed at the China Institute of Technology and Development, the Beijing Institute of Aerodynamics, and the Hubei Institute of Amphibious Aviation. These design and development centers are developing and producing new civilian-use and military-use types of WIG vehicles. Types H and I are for military cross-sea amphibious landing operations. The H model weighs 140 tons and can travel at speeds up to 400 km per hour. It has a lift capacity of 30 tons of cargo, or 250 fully armed soldiers. The I model weighs 400 tons and can travel at speeds of 500 km per hour. It can carry a full battalion (500 men and equipment). Its heavy lift capacity, low radar detection profile, and high speed make WIGs very attractive and lethal for future cross-strait military operations. (Liao Wen Chung, unpublished report on Asia Aviation Exhibition in March 1998. During his visit to the air show in Singapore, Mr. Liao interviewed Russian technicians specializing in WIGs and learned that there are a number of Russian scientists and engineers working closely with Chinese specialists in China to develop various types of WIGs.)

CONCLUSION AND COMMENT

1. The creation of RRFs and REMCFs is a new development in PLA force structure. A 100,000-man RRF was established in 1994, bolstered by a 300,000-strong REMCF in 1997, made up of the Army's 91st division and 121st division, the Navy's 5th amphibious landing detachment, and 15th Airborne Corps. The establishment of RRF and REMCF is in line with PLA's force reduction policies. Clearly, RRF and REMCF will be the backbone force of the PLA in the near future. At the same time, these new units not only increase the complexity of the MR system but also create challenges in terms of force coordination, logistic support, and C3I. However, it could be argued that the establishment of RRFs and REMCFs is an inevitable step in the rationalization of the PLA since 1985, when the slogans of professionalization and fighting and winning a high-tech regional warfare were first introduced.

The impetus for these changes was manyfold. Lessons learned from the 1989 Tiananmen operation demanded that the PLA improve its operational efficiency, should it be called upon to deal with similar crises in the future. PLA commanders are fully aware that their forces are not capable of dealing with much more complex crises, such as military confrontation in the Taiwan Strait or South China Sea. The 1991 Gulf War provided the PLA with an opportunity to learn the ways and means of implementing RRF concepts, theories, and force operations. China's reform and open-door policies along with the ending of the Cold War also provided ample opportunities for PLA commanders and senior officers to conduct intensive and comprehensive interactions with external advanced military forces, which allowed PLA officers to learn modern high-tech warfare operations.

At the same time, the establishment of RRFs and REMCFs has clearly affected the PLA's existing force structure and operational doctrines. Frequent and aggressive combined exercises have encountered little criticism within the PLA, an indication that there is widespread consensus among top commanders and officers in supporting these new force units. The adding of the General Equipment Department and streamlining of the National Defense Commission for Science and Technology are other indications of gradual structure change.

2. RRF and REMCF are mission-oriented task forces designed to meet PLA's revised strategic perceptions for the post–Cold War era and to deal with domestic and peripheral potential threats if necessary. It is still too early to say the PLA can effectively deal with new threat contingencies should RRF or REMCF be confronted by equally strong or stronger adverse forces. The establishment and development of the PLA rapid reaction forces can also be linked to the PRC's changing threat perceptions, especially vis-à-vis the Taiwan Strait, South China Sea, Tibet, and Xinjiang. In the Taiwan Strait scenario, Beijing's military preparation takes into account the possibility of confronting U.S. and Japanese military forces. To this end, these forces have conducted various exercises in the East China Sea since 1995. Judging from these exercises, the PLA is vigorously practicing combined force operations with emphasis on long-range mobile rapid-deployment and amphibious landing capabilities. The PLA even constructed a simulated "Taiwan Special Region" near the mountainous areas of Anhui Province, where it conducted "Red" vs. "Blue"

war games, including amphibious landings, as well as airborne, air-to-sea, sea-to-air, and ECM/ECCM operations. Deployment of rapid reaction forces is subject to adequate air and sea lift capabilities. Evidence suggests that the PLA is increasingly upgrading these lift capabilities by building new Ro/Ro ships and importing heavy lift transport, such as the IL-76MF. The development of hydrofoils is another example of enhancing lift capability for rapid reaction operations.

5. "COSTIND IS DEAD, LONG LIVE COSTIND! RESTRUCTURING CHINA'S DEFENSE SCIENTIFIC, TECHNICAL, AND INDUSTRIAL SECTOR"

Harlan W. Jencks[1]

INTRODUCTION

> Remember this: the most likely behavioral consequence of any new directive from Beijing is that which will be necessary to subvert the directive.
>
> —*Michael Oksenberg*[2]

This is a work in progress about a work in progress. I attempt to describe the current status of restructuring in the Chinese government, military, and economy, with special reference to the Defense Scientific, Technical, and Industrial (DSTI) sector. I do not expect the PRC government to complete its restructuring for several years. There is a very real possibility that the restructuring will never be "completed" at all. Like countless other national campaigns since 1949, it may just fade away. At this writing, in August 1998, I can only hope to lay out some broad outlines, identify a few trends, and pose some of the many interesting questions.

Few official PRC sources offer much detail, and nobody knows what ultimately will happen. On the other hand, the most remarkable single fact about the current organizational upheaval is that Chinese officials are telling the truth—not lying, not stone-walling, not spouting official "boiler-plate." Further, they seem to be unusually willing to share their personal views, speculations, and fears with foreign interlocutors. Perhaps this is partly an (officially directed?) effort to reassure current and potential foreign investors. This paper is based largely on

---

[1]Harlan W. Jencks is a Research Associate at the Center for Chinese Studies at the University of California, Berkeley, and an analyst at Lawrence Livermore National Laboratory. In 1993, he retired from the U.S. Army Reserve (Special Forces) with the rank of colonel. He has published extensively on international security issues related to China and the Far East. He is the author of *From Muskets to Missiles: Politics and Professionalism in the Chinese Army, 1945-81*, Boulder, CO: Westview Press, 1982; and co-editor with William Potter of *The International Missile Bazaar: The New Suppliers' Network*, Boulder, CO: Westview Press, 1993. This work was funded, in part, by the Nuclear Transfer and Supplier Policy Division of the U.S. Department of Energy. The views expressed are those of the author. They should not be construed as reflecting the views of the Department of Energy, Lawrence Livermore National Laboratory, or any of the Laboratory's other contractors.

[2]Michael Oksenberg, electronic mail message to the author, July 23, 1998.

hearsay, rumors, and speculation gleaned from the press, private discussions, and e-mail exchanges. I am grateful to many colleagues and friends for sharing the information they have picked up from their Chinese travels and contacts.[3]

Currently, the most accurate way to judge the state of the reorganization seems to be to observe which Chinese officials are doing what kinds of business, what offices they occupy, and which organizations they claim to represent. It is still unclear whether slight changes in organizational names, or the translations thereof, mean anything. It is telling that Chinese interlocutors have told several of my informants that they themselves do not know what organizations they represent or will represent in the near future, nor what higher-level organizations they are currently answerable to. There is plenty of evidence of stress among Chinese officials and of widespread efforts to resist and/or short-circuit the restructuring. Indeed, discontent and resistance seem to be the best documented aspects of the restructuring effort.

THE 15TH PARTY CONGRESS AND PRE-NPC RUMORS

At the 15th Communist Party Congress in September 1997, President Jiang Zemin announced that the main agenda for the next five years was "reform" of state-owned enterprises (SOEs). There are some 305,000 SOEs, employing 70 percent of China's urban work force (about 109 million workers) and generating 30 percent of total industrial output.[4] Under the slogan "Grasp the Large, Release the Small," SOEs were to be reorganized, rationalized, down-sized, and "marketized." Reformed enterprises would have to sink or swim in the free market (and go bankrupt when they sank). Bloated staffs would have to be cut. It was recognized that this would be a painful process, which might well provoke social, and even political, unrest. To take charge of the reform, Vice Premier Zhu Rongji was (correctly) expected to be elected Premier at the National People's Congress (NPC) in March 1998. Jiang made it clear that the reform was especially aimed at the large, inefficient, loss-making industrial SOEs. Shortly after the Party Congress, the Commission of Science Technology, and Industry for National Defense (COSTIND) reportedly hosted an important meeting at Qingdao (October 21–24, 1997). There also was an "Armed Forces Equipment Working Meeting" in Beijing on December 1, 1997.[5]

During January and February 1998, rumors about the reorganization abounded. The most startling rumor was that COSTIND would be abolished. Since its creation in 1982, COSTIND had occupied a unique position in the PRC government, with one organizational foot in the civilian DSTI complex directly under the State Council, and the other in the People's Liberation Army (PLA)

---

[3]I am indebted to Dennis Blasko, Tai Ming Cheung, John Frankenstein, Bates Gill, Richard Latham, John Lewis, Ellis Melvin, and Xue Litai, as well as others who preferred not to be identified.

[4]Pamela Yatsko, "Owner Power," *Far Eastern Economic Review*, May 21, 1998, p. 14.

[5]*China Defence Science and Technology Information* (published by the China Defence Science and Technology Information Center of COSTIND, Beijing), January-February 1998. I am grateful to Ellis Melvin for drawing this report to my attention.

directly under the Party Central Military Commission (CMC). COSTIND was charged with coordinating the often contradictory requirements of the PLA and the defense industry. It supervised the research, development, testing, and evaluation (RDT&E) of all new PLA weapon systems and equipment. In theory, COSTIND set priorities, allocated resources, and mediated the conflicting interests of the operational PLA and the DSTI system. By all accounts, it never fulfilled these responsibilities very successfully. COSTIND and its pre-1982 organizational predecessors (the Communist Party's National Defense Science and Technology Commission and the State Council's National Defense Industrial Office) were subject to constant criticism for their chronic failure to expeditiously complete RDT&E programs and to get modern systems into mass production for delivery to operational troop units.[6]

In recent years, expert observers have disagreed as to whether COSTIND was (or indeed, ever was) really strong or important. Notably, Jonathan Pollack of RAND has long held that COSTIND was a loose aggregation of organizational parts. While some parts were important, many were simply sinecures for semi-retired bureaucrats, or fronts used by well-connected individuals to do private business. If one accepts this argument, it could further be argued that COSTIND's abolition was not particularly important, depending upon what happened to its few really important components.[7]

The reorganization was intended to fix other problems in the DSTI system as well. These include the massive debts run up by defense SOEs, including the problem of "triangular debt." Many enterprises have huge debts but also have huge accounts payable that they can't collect from other enterprises, which have the same problem. Another problem was the long development times, and chronic failures to deliver, associated with weapons programs. Design of the FB-7 attack aircraft, for example, began in the 1970s. The FB-7 finally flew for the first time in 1987, but is still not in operational service. Another example is the seemingly endless variations of the F-8 fighter.[8] Still another reason for reform was the evident failure of "defense conversion," which was supposed to create all kinds of technological and managerial "spin-ons," for the defense industry.

The implications of the restructuring seemed to reach endlessly in all directions. Clearly, if the industrial sector were to be truly reformed, the financial and banking sectors would have to be reformed in parallel. One reason the SOEs were drowning in red ink was that for decades state banks advanced loans to loss-

---

[6]Richard Latham, "China's Defense Industrial System: Order or Disorder?" paper presented at the Asian Studies on the Pacific Coast Conference, University of Hawaii, June 1975; Harlan W. Jencks, *From Muskets to Missiles*, chapter 6; and Benjamin C. Ostrov, *Conquering Resources: The Growth and Decline of the PLA's Science and Technology Commission for National Defense*, Armonk, NY: M. E. Sharpe, 1991. Also see the voluminous more recent work of Tai Ming Cheung, Bates Gill, Richard Bitzinger, and John Frankenstein.

[7]A similar argument was advanced by David Bachman, "Defense Industrialization and State Making in Mao's China," paper presented at the Center for Chinese Studies (CCS), China Symposium, University of California, Berkeley, March 3–4,1995.

[8]Kenneth W. Allen, Glenn Krumel, and Jonathan D. Pollack, *China's Air Force Enters the 21st Century*, RAND, MR-580-AF, 1995.

making enterprises. That was closely related to the problem of "triangular debt." President Jiang's 15th Party Congress speech clearly implied that the process would be painful and disruptive.

As comprehensive as it appears, the reorganization will not address some chronic problems. One is poor coordination among China's various DSTI organizations. Another is the bureaucratic, risk-averse corporate culture throughout the military-industrial complex.

Besides COSTIND, foremost among the DSTI organizations to be affected were the "Big Five" military-industrial SOEs under the State Council: China National Nuclear Corporation (CNNC), Aviation Industries Of China (AVIC), China Ordnance Industry Corporation (OIC, better known as the Northern Industrial Corporation [Group], or NORINCO[G]), China State Shipbuilding Corporation (CSSC), and China Aerospace Industry Corporation (CASC) (see Table 1). The reorganization would also presumably involve the defense electronics industry, largely housed within the Ministries of Electronics Industry (MEI) and of Posts and Telecommunications (MPT). All the research academies and institutes under all of the above seemed likely to be implicated, as well as such important civilian institutions as the Chinese Academy of Sciences (CAS) and the State Science and Technology Commission (SSTC). If the latter were implicated, then it seemed reasonable to expect the research institutes of China's universities (and therefore the Education Ministry) also would be involved.

Table 1

Major Former Defense Scientific, Technical, and Industrial Organizations Involved in the Reorganization

Chinese Name (Alternate)	Usual translation (Alternate)	Abbreviation (Alternate)
Zhongguo he gongye zonggongsi	China National Nuclear Corporation	CNNC[a]
Zhongguo hangkong gongye zonggongsi	Aviation Industries of China (Chinese General Company of Aeronautics Industry)	AVIC[a]
Zhongguo Beifang gongye zonggongsi (Zhongguo bingqi gongye zonggongsi)	China North Industries Group (Ordnance Industry of China) (China Ordnance Corporation)	NORINCO[G][a] (OIC)[a]
Zhongguo chuanbo gongye zonggongsi	China State Shipbuilding Corporation	CSSC[a]
Zhongguo hangtian gongye zonggongsi	China Aerospace Industry Corporation (Chinese General Company of Astronautics Industry)	CASC[a]
Dianzi gongyebu	Ministry of Electronics Industry	MEI
Youdianbu	Ministry of Posts and Telecommunications	MPT
Guofang kexue jishu gongye weiyuanhui (Ke Gong Wei)	Commission of Science, Technology, and Industry for National Defense (National Defense Science, Technology and Industry Commission)	COSTIND (NDSTIC)
Guojia kexue jishu weiyuanhui	State Science and Technology Commission	SSTC
Zhongguo gongcheng wuli yanjiuyuan	Chinese Academy of Engineering Physics	CAEP

Source: *China Directory, 1998.* (Kawasaki, Kanagawa, Japan: Radiopress, November 1997) and *China's Defense-Industrial Trading Organizations*, Defense Intelligence Reference Document PC-1921-57-95, U.S. Defense Intelligence Agency, 1995.

[a]"The Big Five" Defense Industrial Corporations.

If COSTIND was to be abolished, all the RDT&E facilities and institutions under its direct control would be swept up in the restructuring. Foremost among these would be the Chinese Academy of Engineering Physics (CAEP), responsible for China's nuclear weapons program. Other COSTIND facilities include the space-launch complex at Xichang, Sichuan, and the nuclear test site at Lop Nur, as well as lesser test ranges and facilities for everything from strategic missiles to small arms.

In early 1998, the Beijing rumor mill was remarkably consistent on many details.[9] By the time the NPC actually convened at the beginning of March 1998, the rumor mill said that President Jiang and soon-to-be Premier Zhu were about to embark on a traumatic top-to-bottom makeover of China. They would (1) accelerate SOE reform, potentially laying off tens of millions of workers; (2) accelerate banking reform, cutting off loans to marginal or unprofitable enterprises; (3) cut the PLA by hundreds of thousands; (4) eliminate COSTIND, create a fourth PLA General Department, and create a new ministry/commission for defense industries; (5) reform PLA enterprises; and (6) dramatically expand infrastructure expenditures to absorb excess labor.

The general expectation on the eve of the NPC was that COSTIND would be replaced by two new institutions. A new "General Equipment Department" of the PLA would be formed, taking over the Bureau of Military Equipment and Technology Cooperation (BOMETEC) and the Equipment Bureau, both from the General Staff Department (GSD); plus some functions of the General Logistics Department. A new "Ministry of National Defense Industry" (MNDI) was expected to take control of the "purely military" parts of the defense industrial SOEs, plus the RDT&E-related functions of COSTIND. It was generally expected that "Big Five" factories that were mostly producing civilian goods would be cut loose to become local collectives or private enterprises. But how could factories and enterprises be divided up if they were producing both civilian and military goods? There was lots of speculation about "painting stripes across factory floors" to separate civilian and military production lines. Then there was the even bigger problem of trying to divide up traditional SOE work-units (*danwei*), which include workers and retirees, spouses and children, day-care centers and schools, hospitals and housing.

ANNOUNCEMENTS AT THE NINTH NATIONAL PEOPLE'S CONGRESS

Opening the NPC on March 2, outgoing Premier Li Peng announced that 22 ministries and commissions would be left intact, while 15 would be eliminated. In a declaration issued by Li Peng just before the NPC convened, COSTIND was first listed under the 15 *bumen* to disappear, but a few pages later a new COSTIND, listed under its old name, appeared. The new COSTIND would assume the former

---

[9]The following is based largely on articles in *The South China Morning Post* in February 1998 by Willy Wo-Lap Lam. Also see *Kuang chiao ching (Wide Angle)*, February 16, 1998. Personal contacts from U.S. National Laboratories reported rumors among their colleagues in the nuclear sector.

COSTIND's functions of defense industrial management and take over the State Planning Commission's defense department (*guofangsi*). It also would take direct control of the ("Big Five") military industry corporations (*jungong zonggongsi*). Premier Li left the really bad news to State Council Secretary General Luo Gan. On March 6, Luo told deputies that, "The streamlining reforms amount to a revolution. . . . Reform is bound to face resistance and risk, but we must press ahead. There is no other way out."[10] Luo roughly outlined a plan to reduce the central government's 40 ministries and commissions to 29. Four ministries and commissions were to be substantially reorganized or newly established, and three would be renamed (see Tables 2–4). Fifty percent of government cadres (about 4 million people) would supposedly lose their jobs, along with 3.5 million industrial workers.

Table 2

Ministry-Level Bodies Abolished by the 9th NPC

1. Ministry of Power Industry
2. Ministry of Coal Industry
3. Ministry of Metallurgical Industry
4. Ministry of Machine Building Industry
5. Ministry of Electronics Industry
6. Ministry of Chemical Industry
7. Ministry of Internal Trade
8. Ministry of Posts and Telecommunications
9. Ministry of Labor
10. Ministry of Radio Film and Television
11. Ministry of Geology and Mineral Resources
12. Ministry of Forestry
13. State Physical Culture and Sports Commission
14. Commission of Science, Technology and Industry for National Defense
15. State Commission for Restructuring Economy

SOURCE: REUTERS, 100541Z MAR 98.

Table 3

Ministries That Changed Their Names

1. State Planning Commission (renamed State Development and Planning Commission)
2. State Science and Technology Commission (renamed Ministry of Science and Technology).
3. State Education Commission (renamed Ministry of Education)

SOURCE: REUTERS, 100541Z MAR 98.

---

[10]Quoted by Daniel Kwan, "Cutback tackles red-tape malaise," *South China Morning Post*, March 7, 1998. Luo's speech is authoritatively translated and summarized in *China Daily*, March 12, 1998.

Table 4

New Ministry-Level Bodies

1. Ministry of Information Industry
2. Ministry of Labor and Social Security
3. Ministry of Land Resources
4. State Commission of Science, Technology and Industry for National Defense

SOURCE: REUTERS, 100541Z MAR 98.

Announced measures were neither as consistent nor as radical as what had been rumored before the NPC, which probably indicated last-minute compromise and horse-trading. SOEs would be restructured, but the defense industry, at least, would not be "privatized." In mid-March a former CNNC official said that just before the NPC, a "last-minute" meeting, involving powerful COSTIND supporters, reached a compromise which resulted in the announced reorganization.[11]

COSTIND IS DEAD. LONG LIVE COSTIND!

On March 6, 1998, Luo Gan said that: (1) the major military industry corporations (the "Big Five") will eventually be organized into enterprise groups (*qiye jituan*); (2) the State Aerospace [Industry] Bureau (*guojia hangtian ju*) and the China Atomic Energy Authority (*guojia yuanzineng jigou*) will remain as constituted. These are both governmental regulatory agencies which have been, in reality, subordinate to China Aerospace Corporation and China National Nuclear Corporation, respectively. According to Luo, they will continue to represent the country abroad but "be under [New] COSTIND for internal activities"; (3) new COSTIND "will coordinate with relevant units related to the Central Military Commission" (i.e., the PLA) for production and supply, research, and long-term planning; (4) new COSTIND will be in charge of development plans, regulations, and management for all military industry enterprises; and (5) new COSTIND will coordinate with the State Economic and Trade Commission (SETC) for military-to-civilian conversion plans. *China Daily* reported that the SETC will assume responsibilities for the now-dismantled ministries of Machine Building, Chemical Industry, and Metallurgical Industry.

The new State Commission of Science, Technology, and Industry for National Defense has come to be referred to as "The New COSTIND," though I personally prefer "SCOSTIND" (State COSTIND). There is also confusion over the re-use of the name inside China. Some people reportedly refer to SCOSTIND as the *"Kegong Bu"* to distinguish it from old COSTIND, which remains *"Kegong Wei."* In light of the pre-NPC rumors, it is ironic that, to avoid confusion, common usage is making SCOSTIND a ministry (*bumen*), even if the NPC did make it a commission (*weiyuanhui*).

---

[11]Bates Gill, electronic mail message to the author, March 18, 1998.

Soon after the NPC, it became apparent that, in crucial ways, SCOSTIND will be less potent than its predecessor. It is completely separate from the CMC and PLA, answerable only to the State Council, and is expected to be in place by the end of 1998. Initially, at least, SCOSTIND will control defense industry, including the research academies and institutes (RAs and RIs, respectively) of the "Big Five." Whether it will actually conduct RDT&E (as opposed to administrative housekeeping for the RAs and RIs) remains unclear. SCOSTIND's concerns appear to overlap those of the new Ministry of Science and Technology and the Chinese Academy of Sciences.

Liu Jibin, the newly appointed director of SCOSTIND, has considerable experience in defense industry and in financial management. A 1962 graduate of the Beijing Aeronautical Institute's Department of Engineering Economy, he worked his way up in the Songling Machinery Plant to become corporation deputy manager in 1982. He then became a deputy chief of engineering at the former Ministry of Aeronautics Industry in 1985, and a vice minister from 1985–1988, when he was concurrently Director of the State Administration of State-Owned Property. For the past decade (1988–1998), he has served as Vice Minister of Finance. No military service is mentioned in his official biography.[12] The PLA-SCOSTIND disconnect was starkly demonstrated in April 1998, with the publication of at least three different versions of an interview with Minister Liu.[13] In explaining SCOSTIND's role and his plans for it, he spoke in general terms of "the needs of national defense," but never once mentioned meeting PLA needs or requirements. Indeed, he never so much as mentioned the PLA!

According to Liu, "the newly formed Commission for National Defense Science, Technology, and Industry has three functions. The first function is administration over national defense industry formerly under the administration of the old commission. The second is administration over national defense construction formerly under the administration of the National Defense Department of the State Planning Commission. The third is taking up all the functions of the former five big corporations, namely, the Nuclear Industry Corporation [CNNC], the Aeronautics Industry Corporation [CASC], the Astronautics Industry Corporation [AVIC], the Ordnance Corporation [NORINCO], and the Shipbuilding Corporation [CSSC]." Although he mentioned "scientific research" in passing, Liu's description of SCOSTIND's functions heavily emphasized "organizing, conducting, and coordinating" industry, rather than RDT&E.

Liu said, "This commission is composed of seven people from seven departments including the five big corporations for the military industry, the State Planning Commission, and the Ministry of Finance."[14] On April 17, Xinhua News Agency

---

[12] *South China Morning Post,* March 19, 1998.

[13] The longest version was "Interview With Minister of National Defense Science," Beijing Central Television Program One Network in Mandarin, April 21, 1998, in FBIS-CHI-98-114, April 24, 1998. Also see Gao Jiquan, "Shoulder Heavy Responsibilities, Accept New Challenges—Interviewing Liu Jibin, Newly Appointed State Commission of Science, Technology, and Industry for National Defense Minister," *Jiefangjun bao*, April 9, 1998, p. 5, in FBIS-CHI-98-119, April 29, 1998.

[14] "Interview With Minister of National Defense Science."

identified six people named vice ministers of COSTIND,[15] whose former affiliations with the Big Five and the State Planning Commission were all known. There will be a small bureaucracy (totaling less than 100 people) under Minister Liu and his six vice ministers, but at least we now know who the top seven are, and can guess pretty confidently at their responsibilities (See Table 5).

Table 5

COSTIND Minister and Vice-Ministers

Name	New Position	Former Position(s)
Liu Jibin	Minister	Vice-Minister of Finance
Zhang Junjiu	Vice-Minister	President, COIC & NORINCO
Xu Penghang	Vice-Minister	President, CSSC
Luan Enjie	Vice-Minister and Director, State Astronautics Bureau	Vice-President, CASC
Zhang Huazhu	Vice-Minister and Chair, China Atomic Energy Authority	Vice-President, CNNC
Zhang Hongbiao	Vice-Minister	Vice-President, AVIC
Yu Zonglin	Vice-Minister	Director, National Defense Department of the State Planning Commission

After an embarrassing delay, SCOSTIND finally occupied offices in the Xuanwu District of Beijing in mid-May.[16] It appears that SCOSTIND is meant to regulate, rather than manage or control. There have been rumors in Beijing that the Commission actually has a limited life-span—three years, or possibly five years until the next NPC in 2003—during which it will oversee the reorganization of the "Big Five" and then disappear. SCOSTIND is a strictly civilian organization, which must "coordinate" with the PLA—presumably with the new General Armaments Department (GAD, see below). An important indication of SCOSTIND's limited authority is that many of old COSTIND's most important people and organizations have gone to the GAD. All of old COSTIND's military personnel will be, or have been, transferred to the GAD as well. SCOSTIND appears to have no authority over the electronics industry at all. SCOSTIND-subordinate defense industries will continue to use old COSTIND's former RDT&E ranges, including Lop Nur and the missile and space-launch bases, but these facilities too will be controlled by the GAD.[17]

A well-informed U.S. government official takes the apparently contradictory view that SCOSTIND will, in effect, re-centralize control over military industries which have become decentralized and out of control. Possibly, however, both views are correct: Perhaps SCOSTIND is supposed to reassert central control and reorganize the "Big Five" into Industrial Enterprise Groups (IEGs, see the next section), then release the IEGs into the market and revert to being a small regulatory agency and/or disappear altogether. The new Ministry of Information Industry (MII) is supposed to execute just such a "centralize-reorganize-release-regulate" sequence

---

[15]"State Council Appoints, Removes Personnel," *Xinhua Domestic Service*, April 17, 1998, in FBIS-CHI-98-108, April 18, 1998.

[16]SCOSTIND's Office is at Huaneng Dajie, Guang'anmennei Nan Jie, Xuanwu Qu, Beijing 100053.

[17]Dennis Blasko, electronic mail message to the author, April 22, 1998.

in the telecommunications industry.[18] If that is the intent, we may be forgiven some skepticism about the likelihood of MII or SCOSTIND overcoming the universal tendency of bureaucracies to perpetuate themselves and expand their powers and purview.

CONSOLIDATED MILITARY-INDUSTRIAL ENTERPRISE GROUPS

Luo Gan's statements that SCOSTIND would be "in charge of development plans, regulations and management for all military industry enterprises" was a surprise, as was the provision that SCOSTIND would "coordinate with the State Economic and Trade Commission for military-to-civilian conversion plans." In other words, SOEs producing both military and civilian products would not, after all, be split up immediately. Perhaps that was the intent right up to the eve of the NPC, but the actual dividing up of the *guofang gongye* enterprises apparently was postponed. That suggested, as did other reports, that the current shakeup is just the first of several that Zhu Rongji has in mind. Others will take place over the next decade to accomplish the "final" restructuring. It is worth recalling that the last major shakeup of the DSTI sector, by the 5th NPC in May 1982, was followed by sixteen years of still more reorganization.

Reportedly, Premier Zhu has been enamored with the South Korean *chaebol* as a model for industrial organization and has not abandoned the model entirely, despite the *chaebol's* prominent role in Korea's 1997–98 economic melt-down. According to the *Far Eastern Economic Review*, "The [Chinese] government is nurturing 1,000 of the larger enterprises—of which 120 will be turned into big business groups. . . . The remaining 304,000 small and medium-sized state enterprises are being 'released' from state ownership and left to sink or swim."[19] According to government economic advisor Lu Yansun, "China will push ahead with its plan to build South Korean-style conglomerates while carefully avoiding the pitfalls highlighted by the Asian economic crisis." He explained that, "Some South Korean firms fell because of heavy debts and improper deals with the government." He did not say how China, with its rampant corruption and massive SOE debt, could or would be any different, though he added that Beijing was also learning from the success of "Taiwan's smaller-scale companies, many of which tended to be technology-based with low liability."[20] John Frankenstein observes that while the *chaebol* does sound more like an SOE than a small, specialized Taiwan Chinese family firm, "there is one crucial difference: the *chaebol*, like the smaller Taiwanese firms, are very good at handling market information. With a few exceptions, the SOEs are still in the dark ages here."[21]

Chinese views of the South Korean model are evolving, however. Most PRC leaders now emphasize that mergers should be voluntary, not forced from above.

---

[18]John Lewis, discussions at Stanford University, August 7, 1998.

[19]Pamela Yatsko, "Owner Power," *Far Eastern Economic Review*, May 21, 1998, p. 14.

[20] Justin Jin, Reuters (Beijing), March 12, 1998.

[21] John Frankenstein, electronic mail message to the author, March 13, 1998.

Also, unlike the *chaebol*, even conglomerates should have a core competence, not trying to manage everything from ski resorts to ship-building. Perhaps most important, Zhu Rongji, although a proponent of conglomeration, is adamant that state-owned firms, even those earmarked for mergers, must operate on a profitable basis.[22]

While it is not at all clear what these consolidated Industrial Enterprise Groups (IEGs, *Gongye qiye jituan*) will look like, they clearly will not be "privatized." They will continue to be state-owned, though supposedly with considerably more economic and operational autonomy. The IEGs will supposedly integrate research and development, marketing, and production. If that is so, the research institutes (RIs) of the old Big Five corporations will remain under SCOSTIND and the Enterprise Groups.[23] What will be the difference between the five bureaus of SCOSTIND and the former Big Five corporations? The Big Five seem to meet all the announced criteria for the new IEGs, so why are they being reorganized at all? A reasonable guess is the political factor: Perhaps Beijing simply wants to regain control of industrial enterprises which have become far too independent. But reining in the Big Five and regulating the new IEGs will require that SCOSTIND exercise considerable authority, which, as observed above, is questionable. For example, at some point, the IEGs probably will be required to rid themselves of nondefense enterprises, but it remains to be seen whether they will give up profitable noncore production. If an enterprise in the aviation industry is making money exporting cosmetics, it is difficult to imagine an aviation IEG giving that enterprise up without a fight.

If the reform goes through, there will be a significant number of layoffs and plant closures. Supposedly, the aviation industry will be laying off around 200,000 people, the ship-building industry 90,000, the ordnance industry 90,000 or more, and the nuclear industry some 30,000. So far, some of these layoffs have been "phony." Some workers are being paid to stay home and do nothing, but they are going into the statistics as layoffs. If a factory is transferred to another enterprise, for example, its jobs are counted as layoffs by the losing enterprise, even though the workers are still working at the same jobs, in the same place, for a different organization. There has been a lot of talk about retraining and re-educating workers and cadres prior to laying them off, but very little has been done.

All Chinese, especially managers, are extremely averse to closing factories and laying people off. There have been few cases of allowing a factory to go bankrupt and close down, and those few cases have all been small enterprises. Even after the announced reforms, there is likely to be massive overstaffing. Right now the DSTI complex has perhaps triple the work force it actually needs to be productive, so the planned one-third cut would still leave twice what is needed.

---

[22]James Kynge and James Harding, "Beijing firm in the belief that bigger is better," *Financial Times*, March 11, 1998.

[23]This paragraph and the following discussion on IEGs owe much to discussions with a U.S. government official who preferred anonymity.

Below, I suggest four scenarios for SCOSTIND's future. Case I represents the most drastic reform; Case IV would be a worse situation than before the restructuring was announced.

Case I: SCOSTIND would form "administrative councils" to preside over the division of assets and dissolution of existing Big Five enterprises, and reorganize them into IEGs. The councils, and possibly SCOSTIND itself, would then disappear, leaving the IEGs to operate in the market.

Case II: After organizing the IEGs, as in Case I, SCOSTIND would continue to exist, and its bureaus would function as regulatory agencies, somewhat analogous to those of a Western government. For example, the Atomic Energy Bureau/CAEA of SCOSTIND would have regulatory functions which, in the United States, are fulfilled by the Department of Energy and the Nuclear Regulatory Commission. The Aviation Bureau of SCOSTIND would function something like the American FAA. The Space Bureau of SCOSTIND would function somewhat like NASA; and so forth. They would regulate the Enterprise Groups. The latter, each under its own headquarters, would compete in the market.

Case III: The bureaus of SCOSTIND would be essentially the same as the "Big Five" corporation headquarters were before, with nothing really changed but the names. For example, the Atomic Bureau of SCOSTIND would be essentially the same as CNNC was before, the Aviation Bureau essentially would be the same as AVIC, etc. "Enterprise Groups" would be the same subordinate enterprises that were formerly subordinate to the Big Five, and they would have the same relationships with their respective bureaus that they formerly had with the Big Five corporate headquarters. "Enterprise Groups" would just be a collective designation for the enterprises, plants, and research institutes subordinate to a given bureau.

Case IV: The SCOSTIND bureaus would exert control, as in Cases I and II, but there also would be five or more new IEG headquarters as in Case II. This would be the worst possible outcome because it would add a new layer of bureaucratic control but not allow for any more market behavior on the part of the IEGs or the individual enterprises, plants, and research institutes.

Organization of the Enterprise Groups

The subsidiary corporations, factories, and research institutes of the Big Five could be reconfigured in a variety of ways. Four representative configurations, across the spectrum of possibilities follow:

1. Enterprise Groups could be set up as comprehensive, *chaebol*-style conglomerates. These would combine industrial research, development, production, and marketing, like large conglomerates in the West (e.g., Lockheed-Martin or UTI). They might have integral financial organizations, like Mitsubishi or Hyundai. Minister Liu seemed to indicate this sort of model in the *Jiefangjun Bao* version of his interview: "To separate government

functions from enterprise functions [we must] reorganize some former national defense industrial corporations into a number of large-scale enterprise groups transcending regions, departments, professions, or trades and capable of integrating production with scientific research and intensive operation and combining military industrial production with civilian industrial production."[24] This configuration would require either Case I or Case II regarding the role of SCOSTIND and its bureaus.

2. The Enterprise Groups could simply be set up geographically. For example, all of the former AVIC assets north of the Yellow River under one IEG, those between the Yellow River and the Yangzi in another, those south of the Yangzi in a third, and those in the Sichuan basin in a fourth. This is approximately what is, in fact, happening in the petrochemical industry.[25]

3. The present Big Five subsidiary corporations, factories, and research institutes could be subdivided into IEGs that were specialized in terms of their end products. In the nuclear industry, for example, one IEG might specialize in mining, smelting, and uranium conversion; another in enrichment and fuel fabrication; another in reactor construction; and so forth. Aviation industry IEGs might specialize in engines, airframes, avionics, and so forth.

4. All the subsidiary corporations, factories, and research institutes formerly under the Big Five could be assembled under five IEGs (i.e., Cases III or IV, above).

An important determinant of the future of technical innovation in China will be what happens to the research academies and institutes (RAs and RIs) which, under the old system, were under the Big Five. These could be placed directly under the jurisdiction of the bureaus of SCOSTIND in Cases II, III, or IV. They could be placed under the Enterprise Groups in Cases I, II, III, or IV; or they could be placed under some other organization entirely, possibly the Chinese Academy of Sciences, the PLA General Armaments Department, or the new Ministry of Science and Technology. Or research academies and institutes could be divided up among some or all of these. In any of these cases, individual research organizations might be abolished, consolidated, or reorganized.

THE PLA AND THE REFORMS

The General Armaments Department

Unquestionably, the biggest organizational "winners" so far are the PLA and its new general department. Although the latter's Chinese name, *Zong zhuangbei bu*, best translates as "General Equipment Department," Xinhua News Agency calls it the "General Armaments Department," and that name seems to be gaining general acceptance. Being internal to the PLA, creation of the GAD was not announced at

---

[24]Gao Jiquan, "Interviewing Liu Jibin."

[25]Jean-Francous Tremblay, "Chinese Restructuring Outline Emerges," *Chemical Engineering News*, April 27, 1998. p. 21. Also see Bruce Gilley, "Slick Maneuvers," *Far Eastern Economic Review*, July 2, 1998, pp. 60–61.

the NPC, but by the CMC on April 5.[26] On April 13, at the National University of Defense Technology in Changsha, General Cao Gangchuan was identified as GAD Director, and Lieutenant General Li Jinai as its Political Commissar.[27] Formerly, Cao had been First Deputy Chief of the General Staff prior to taking over as the last Director of old COSTIND in December 1996. He probably presided over the demise of old COSTIND with a good deal of satisfaction, for he was one of those PLA leaders who was frustrated and angered by COSTIND's chronic failures. Cao is well-connected, having served successfully under Zhang Wannian, Chi Haotian, and Fu Quanyou during his time in the GSD. The prevailing organizational turbulence was demonstrated at the official ceremony on March 27, when Cao and nine others were promoted to the rank of General; he was identified as Director of the officially nonexistent COSTIND.[28]

As expected, the GAD took over the Bureau of Military Equipment and Technology Cooperation (BOMETEC, which oversees foreign military aid and sales from PLA stocks) and the Equipment Bureau from the General Staff Department, plus some functions of the General Logistics Department (see Table 6). It appears increasingly likely that the GAD will have important RDT&E responsibilities, but it remains to be seen which laboratories, RAs, and RIs will be directly subordinate to the GAD. It will be interesting to see whether the GAD takes over the various PLA service branch research institutes. Officials of the Chinese Academy of Engineering Physics (CAEP), which runs the nuclear weapons program, are

Table 6

**Organizations and Responsibilities Taken Over by
the PLA General Armaments Department (GAD)**

From the General Logistics Department (GLD)
    Xinxing Corporation
    Motor Vehicle & Boat Transport Units

From the General Staff Department (GSD)
    Equipment Bureau
    BOMETEC
    "703 Group" (Chinese Arms Control Leading Group)

From (old) COSTIND
    Former Office Building
    All military personnel
    All test sites and ranges (including Xichang Space Launch Facility)
    China Defense S&T Information Center (CDSTIC)
    Science and Technology Committee
    Export vetting responsibilities
    "Purely military" R&D (probably including CAEP)
    Some production (?)

---

[26] *South China Morning Post,* April 6, 1998.

[27] *Xinhua,* April 13, 1998.

[28] The promotion order had been signed on March 4, just before the NPC convened. See "CMC Holds Ceremony for Promotion to Army and Police General Rank," *Xiandai junshi (CONMILIT),* No. 256, May 1998, p. 2.

confident that CAEP will come under GAD control, though the decision officially still is pending at this writing. The GAD seems certain to retain or take over control of all the test ranges formerly controlled by old COSTIND, including the Lop Nur Nuclear Test Site, the Xichang Space Launch Center, and various missile and ordnance test sites and ranges. It may take over some of the RAs and RIs of the Big Five corporations, but will probably not take over nearly all of them.[29]

At a minimum, the GAD is supposed to generate ideas and initiatives for new weapons and equipment, but it remains to be seen whether those ideas get turned into actual innovations by in-house GAD research and development, or whether that will be the sole responsibility of SCOSTIND. Given the poor track record of old COSTIND, and the completely civilian personnel and outlook of SCOSTIND, it is difficult to imagine that the PLA will give up the R&D capabilities it had under the old system—they existed, after all, as a hedge against the failures and delays of COSTIND. Put another way, if all the PLA's current R&D organizations are formally transferred to SCOSTIND, we can expect to see PLA service branches reconstituting in-house RDT&E capabilities, whether formally authorized to do so or not.

It remains to be seen how much leeway the GAD will have to purchase foreign technologies or end-items to satisfy PLA requirements. There will still be limits— both economic and political—on foreign acquisitions. Historically, however, the PLA has been more interested in quick results and much less concerned with "self-sufficiency" than either old COSTIND or the industrial establishment.

PLA Enterprises

Altogether, PLA organizations operate some 10,000 business enterprises. Some of the best known are Polytechnologies Corp. (run by the General Staff Dept.), Carrie Corp. (General Political Dept.), Xinxing Corp. and 999 Corp. (General Logistics Dept.), and Lantian Corp. (Air Force). Over the past decade, the CMC has repeatedly tried to get the PLA out of nonmilitary businesses, and to get most of the PLA out of business altogether.[30] An obstacle to PLA divestiture has been that business profits support a significant, but extremely ill-defined, portion of its operating expenses. Larger state defense budget allocations will be necessary if the PLA is to forgo all its business activities, although the magnitudes of such budget increases are unknown, and are subject to exaggeration (in pursuit of many differing agendas) by Chinese and foreign observers. An important first step was accomplished by shifting the GSD's former R&D, acquisition, and export vetting responsibilities to the GAD. That freed the GSD (whether all GSD officers liked it or not) to concentrate on plans, operations, and training.

---

[29]Dennis Blasko, John Lewis, Xue Litai, et al.

[30]See, especially, the call for the PLA to get out of business by Generals Liu Huaqing and Zhang Zhen in *Jiefangjun bao*, August 1, 1993.

It has long been recognized that PLA entrepreneurial activity promotes corruption and is corrosive to discipline and professionalism.[31] Getting the Army out of the marketplace is even more vital during the current SOE reforms. Left to compete unchecked with newly "freed" former SOEs, heavily subsidized and legally privileged PLA businesses are bound to drive many of the former into bankruptcy.

Accordingly, at a July 22, 1998 meeting of PLA and People's Armed Police commanders, nominally convened to crack down on smuggling and corruption, President Jiang Zemin initiated the most serious attempt yet to get the uniformed services out of the marketplace. Xinhua quoted Jiang as saying, "The Army and armed police forces must earnestly carry out checks on all kinds of commercial companies set up by subsidiary units, and without exception from today must not engage in their operation."[32] The next day, Chief of the General Staff General Fu Quanyou ordered "every unit and every cadre" to implement Jiang's decree "without conditions." As Associated Press observed, "The way that the order to shut military businesses was linked to the crackdown seemed to confirm suspicions that the PLA is mixed up in smuggling—a huge illegal business involving everything from oil to luxury cars that is estimated to cost the government and firms as much as $12 billion a year in lost revenues."[33]

The wording of Jiang's decree was sufficiently ambiguous that the precise extent to which the PLA is to "get out of business" remains unclear. For example, demobilized soldiers may be allowed to continue operating businesses; active PLA units may be able to retain ownership, provided they do not actually operate enterprises. Another possibility is that officers involved in army companies will be transferred to reserve status but will continue in their jobs. Meanwhile, speculation is rife regarding the "buyout" prices and budgetary compensation PLA units will demand.[34]

Jiang Zemin's effort to get the PLA out of business is clearly related to the SOE reform, but there are many other reasons, some no doubt more important. Among these are improved tax collection, the fight against corruption and smuggling, the effort to professionalize and modernize the PLA, and Jiang's need to establish his personal authority over the armed forces.[35]

---

[31]Allow me modestly to cite my prediction of all this in "Organization and Administration in the PLA in the Year 2000," in Richard Yang (ed.), *SCPS PLA Yearbook, 1988/89*, Kaohsiung, Taiwan: Sun Yat-sen Center for Policy Studies, National Sun Yat-sen University (published in the United States by Lynne Rienner Publishers), 1989, pp. 52–54.

[32]Quoted by Reuters, July 23, 1998.

[33]Associated Press, July 24, 1998.

[34]Tai Ming Cheung, "The Future of the Chinese Military Business Complex," *South China Morning Post,* July 26, 1998; Susan V. Lawrence, "Out of Business," *Far Eastern Economic Review,* August 6, 1998, pp. 68–69; and Willy Wo-Lap Lam, "PLA to get $28b for businesses," *South China Morning Post,* August 3, 1998.

[35]For a superb analysis, see Ellis Joffe, "Getting China's Military Out of Business Is a Tall Order," *International Herald Tribune,* August 10, 1998.

HOW IS THE NEW SYSTEM SUPPOSED TO WORK?[36]

Beginning about 1990, RAs and RIs no longer automatically received central (State Planning Commission and COSTIND) funding. They have become responsible for seeking out and contracting their own research project funding. By 1998, because the successful RIs had independent funding sources, neither the industrial corporations nor even the Research Academies controlled them, let alone supervised their work. The central authorities relied upon their ability to allocate research funding to control the RAs and RIs indirectly. Old COSTIND allocated funds to RIs to develop the various components of a given weapon system or piece of equipment, while COSTIND itself was primarily responsible for systems integration. Not surprisingly, systems integration has been chronically weak. By 1998, the system actually had become more decentralized than it is in the United States—even major projects lacked comprehensive oversight and management. Instead, COSTIND was constantly convening coordination meetings. This RDT&E system produced poor weapons systems and equipment, which the Equipment Bureau of the PLA refused to buy—withholding the large production contracts that COSTIND insisted were necessary to fund an effective RDT&E complex. This provoked constant feuding between COSTIND and the PLA.

Under the new (post-March 1998) system, the GAD has taken over essentially all of old COSTIND's RDT&E responsibilities, plus the former GSD Equipment Bureau's acquisition authority. GAD, SCOSTIND, MST, and/or CAS may "house," and loosely regulate, the research academies and institutes. The GAD will allocate research and development funds to RAs and RIs, coordinate R&D, and supervise (or directly conduct) systems integration, testing, and evaluation. Once it finally is satisfied with a prototype, the GAD will then call for production bids from the defense-industrial IEGs under SCOSTIND. The GAD thus will act as initiator, allocator of funds, coordinator, and finally as customer. The new system at last addresses the disconnect between the operational PLA and the RDT&E system.

It remains to be seen whether the new system can develop prototypes that the industrial enterprise groups are willing and able to produce. A magnificent prototype weapon might meet the PLA's operational needs, but not be producible. It might, for example, require so much in the way of expensive materials and exotic manufacturing processes that the industrial enterprises would refuse to bid for them.

OBSERVATIONS, IMPLICATIONS, AND QUESTIONS

All sources agree that the restructuring is to proceed from the top down. Five months after the NPC, however, we are barely beginning to discern the organization and responsibilities of SCOSTIND and the GAD, and their relationships to the CMC, State Council, and other S&T-related ministries and

---

[36]The following is drawn mostly from discussions with Xue Litai at Stanford University, August 7, 1998. Its general outlines were anticipated by Jonathan Pollack at the CAPS-RAND San Diego Conference.

commissions. Nobody seems to know whether SCOSTIND will be permanent, or what the "administrative councils" are supposed to do.

Will there be more or less centralized control over defense research and industry? Will there be more or less military-civilian integration in science, technology, and industry? Although the former defense conversion program failed, Liu Jibin indicated in his interview that there will be no drastic break between military and civilian science and technology. The Chinese are still searching for ways to facilitate spinoff of military technologies into the civilian sector and vice versa. Will there be more or less assured funding for military-industrial programs?

Structural reforms do not address such fundamental problems as gaps in the technological base (e.g., some kinds of electronics, metallurgy, and material science). There is an almost total absence of quality-control consciousness in China's industrial work force. (As John Frankenstein writes, *"Chabuduo Xiansheng* lives!") Structural reforms alone do nothing about a corporate culture that is risk-averse, plodding, and not only non-innovative but actually anti-innovative. Corporate managers don't like innovation because it causes change, which they find upsetting. Chinese management style, particularly in heavy industry, is rigid, authoritarian, and "feudal."

During the current transition, old COSTIND still seems to be functioning, as do all of the "Big Five" and the MEI. Defense industrial enterprises are still involved in hundreds of domestic and foreign contracts, and somebody has to be watching the store. Thus, some transactions and organizations have an old supervisor plus one or more new supervisors, while other transactions and organizations may have no supervision at all. This confusing situation must be causing delays in some projects, including military ones. Consider the obvious problem of funding: If AVIC is still trying to supervise the aviation industry during the transition, does it still have the authority to disburse funds to, say, Xi'an Aircraft Corporation (XAC) for work on the FB-7? If not, does XAC have access to other funds, or does the project stall while managers search for funds from COSTIND or SCOSTIND or somewhere else?

Banking reform is critically interrelated with the industrial reforms. It probably will be necessary to write off billions of RMB in bad debts, and then crack down on loan criteria to make sure that new bad loans aren't made.[37] Similarly, much depends on the continuing internal reform of the PLA, especially the latest effort to get rid of "PLA Inc."

The social and political implications of the reform are staggering, in terms of economic cost, social disorder, bureaucratic resistance, and political instability. Perhaps the most fundamental question of all is how much unrest is going to be provoked, and what the authorities will have to do to maintain social order. If they are forced to choose between social chaos and abandoning the reforms, they are likely to choose the latter. In any case, it is certain to be a messy, prolonged

---

[37]Bruce Gilley, "Breaking the Bank," *Far Eastern Economic Review*, July 16, 1998, pp. 66–68.

exercise in "feeling the stones to cross the river." The only certainty is that it will not work out entirely as planned.

6. PLA AIR FORCE LOGISTICS AND MAINTENANCE: WHAT HAS CHANGED?

Kenneth W. Allen[1]

"In the PLA Air Force headquarters' Equipment and Technical Department, one can press a key on a computer keyboard and immediately know any Air Force plane's maintenance conditions. If one stands at the meteorological center's computer terminal, one sees changes in global meteorological conditions. All airfields in the country have been equipped with a multi-functional computerized and telecommunication command guarantee system that shows the Air Force flight guarantee system has reached a new standard. In order to meet the needs of modern air combat, the Air Force has made significant headway in developing a flight training guarantee system and relevant facilities and has maintained a quality flight training rate at and above 99.8 percent for several years running."[2]

"The Chinese Air Force has reached world standards for its aircraft lifespan ascertainment technology, completing research on 17 aircraft models and their airborne equipment. This has saved almost 3 billion yuan and added almost three million hours in flight duration. The application of research results for assessing aircraft service life has better solved major, tough problems concerning structural fatigue and breakup of some 1,000 aircraft, which were endangering flight safety."[3]

INTRODUCTION

During the 1990s, China's People's Liberation Army Air Force (PLAAF) has embarked on a long, uneven, but positive, journey to modernize its obsolescent force of 1950s and 1960s vintage combat aircraft. They have developed, produced, and purchased new aircraft, formed a rapid reaction force, trained with new combat tactics, computerized some of their command centers, and revised logistics and maintenance support capabilities, as noted in the two quotes above.

---

[1]Kenneth Allen is a Senior Associate at the Henry L. Stimson Center in Washington, D.C. Previously, he was Executive VP of the US-ROC Business Council and served 21 years in the United States Air Force, including assignments in China, Taiwan, Japan, Berlin, and Headquarters Pacific Air Forces. He has written extensively on China's Air Force.

[2]Li Dehua and Rong Qingxiang, "Air Force Training Guarantee Systems Being Modernized With Each Passing Day," *Jiefangjun bao*, October 17, 1995.

[3]Sun Maoqing, "Air Force Technological Advances," *Xinhua*, June 5, 1996. Sun Maoqing and Xu Zhuangzhi, "Aircraft Lifespan Technology Reaches World Standards," *Xinhua*, May 12, 1998.

In order to modernize rapidly, the PLAAF and China's aviation industry have had to seek foreign assistance for entire weapon systems, subsystems, or technical support. For example, the Air Force has purchased two regiments of Russian-made Su-27 fighters and several IL-76 transports. They have also developed an airborne refueling capability and have been negotiating with Russia and Israel for airborne early warning aircraft. In addition, China's aviation ministry will soon begin co-producing around 200 Su-27s and is developing the indigenous F-10 fighter with Russian and Israeli assistance, as well as producing an upgraded F-8II with Russian equipment.

The Air Force has established a "Blue Army" aggressor unit, which has exercised with numerous units, and has also established a training base to develop new combat tactics, in order to fight in a future high-tech war. Besides new training techniques, one of the biggest challenges the PLAAF faces is providing logistics and maintenance support to keep its current fleet of aircraft flyable and integrating these new aircraft into the inventory. The PLAAF has set up computerized logistics and maintenance command centers at unit and higher levels, and has begun to develop logistics and maintenance support systems for a rapid deployment force. The PLAAF has also established a Davis-Monthan–like facility to store over 1,000 aircraft removed from the inventory. Most important, they are in the process of moving from concept to implementation in these areas.

It should be noted here that there are increasingly divergent views among PLA-watchers about the scope and pace of China's military modernization. For example, a 1997 publication entitled *Chinese Views of Future Warfare* includes the translation of about 40 authoritative articles written in China about the revolution in military affairs (RMA). This book, edited by Michael Pillsbury, provides a good look at where the Chinese would like to be in 10–20 years. In addition, Major Mark Stokes, a former U.S. Air Force assistant attaché in China, has produced an as-yet unpublished paper entitled "China's Strategic Modernization: Implications for U.S. National Security." This document is one of the most detailed studies of China's specific R&D programs designed to rapidly modernize the PLA in terms of information dominance, long-range precision-strike capabilities, and aerospace defense modernization.[4] Both of these books point to major changes in the PLA's future weapon systems and war fighting capabilities.

Although there are divergent views about the PLAAF's future capabilities, the purpose of this paper is to review information available on current changes in the PLAAF's logistics and maintenance systems in response to general guidance from the General Logistics Department (GLD) and to changes in the Air Force's tactical training concepts and techniques. The paper begins by defining exactly what is meant by the terms "logistics and maintenance" in the PLAAF. Next, there is a discussion about a set of general guidelines, which were laid out by the General Logistics Department in 1995, for modernizing the overall People's Liberation Army's (PLA) logistics system. In order to understand how and why the PLAAF is trying to

---

[4]Michael Pillsbury (ed.), *Chinese Views of Future Warfare*. See also Mark A. Stokes, "China's Strategic Modernization: Implications for U.S. National Security," Maxwell AFB, AL: USAF Institute for National Security Studies, October 1997.

change its logistics and maintenance practices, there is a discussion of some of the changes in PLAAF aviation training. Having laid out these changes, there is a discussion of how the various Military Region Air Force (MRAF) logistics departments have responded to the new challenges.

WHAT IS PLAAF LOGISTICS AND MAINTENANCE?

The best way to understand what is meant by the terms logistics and maintenance in the PLAAF is to examine the logistics and maintenance organizational structure.

The PLAAF's Logistics Department's basic mission is to provide supplies for construction, operations, training, and daily life. The Logistics Department's command staff includes a director, political commissar, two deputy directors, a chief of staff, and two deputy chiefs of staff.

The Logistics Department has 18 subordinate departments, bureaus, divisions, and offices that are responsible for individual aspects of the overall logistics system. These include the Headquarters, Political, Finance, Quartermaster, Health, Armament, Transportation, Fuels, Materials, Airfield Construction, Airfield and Barracks Management, Air Materiel, and directly subordinate Supply Departments; the Audit, Engineering Design, and Research Bureaus; the Administrative Division; and the Production Management Office. Each of these 18 offices is represented throughout the chain of command from the General Logistics Department (GLD) through the Headquarters Air Force all the way down to the lowest unit level.

The PLAAF's Equipment and Technical Department, which combined the Aeronautical Engineering Department and the Equipment Department in 1994, is responsible for determining how much and what types of equipment should be procured; for general management of equipment; for aircraft and engine maintenance, repair, and procurement; for aviation maintenance/repair research; and for aircraft ground-support equipment. The Logistics Department, not the Equipment and Technical Department, is responsible for maintenance of equipment belonging to anti-aircraft artillery (AAA), surface-to-air missiles (SAMs), radar, communications, or airborne troops/units. Although the Equipment and Technical Department is not responsible for SAMs or AAA maintenance, it is responsible for air-to-air missiles. Of note, within the PLAAF, the term technical merely refers to maintenance.

The Equipment and Technical Department has a director, at least two deputy directors, and at least seven second-level departments/offices/divisions: General Office, Political Department, Equipment Department, Field Maintenance Department, Procurement Department, Factory Management Department, and Administrative Division. Each of these offices has a mirror-image element down to the lowest unit level.

Whereas the PLAAF's Headquarters, Political, and Logistics Departments have always been subordinate to a higher-level PLA department (i.e., the General Staff Department/GSD, General Political Department/GPD, and General Logistics Department/GLD, respectively), the Equipment and Technical Department, and its predecessor Aeronautical Engineering Department, did not have an equivalent higher-level department until the establishment of a fourth general office—the General Equipment Department/GED—in April 1998.

The PLAAF has 21 repair factories which employ 40,000 workers and carry out major aircraft and engine overhaul, as well as major and intermediate repairs. The aviation units also have repair factories, which are responsible for intermediate and minor repairs.

Maintenance personnel are trained in several ways, depending upon their rank (officer or enlisted). Officers are trained at the Aeronautical Engineering College in Xi'an, or at one of two Maintenance Technical Training Schools. There are eight Aviation Maintenance Training Regiments to train new enlisted maintenance troops.

Finally, there are PLAAF academies, schools, and training units to train logistics and maintenance personnel. In addition, the Logistics Department and the Technical and Maintenance Department also have several subordinate research institutes, such as the Maintenance, Aviation Medicine, Fuels, Clothing, Aviation Munitions, Four Stations Equipment, Capital Construction, and Aviation Repair Research Institutes.

General Logistics Department Guidance

The PLA's GLD conducted extensive analysis of logistics operations during the 1991 Gulf War and has tried to implement those portions that meet Chinese capabilities and requirements. While most open-source PLA articles on logistics are short on substance, a May 1995 article in the Liberation Army Daily by Major General Yang Chengyu, Director of the GLD's Headquarters Department, did a good job of laying out the challenges and concepts of a modernized logistics system for the PLA.[5] In the article, General Yang stated:

> The widespread application to the military arena of modern high-technology has brought many new features to local wars. These can be seen mainly in areas such as the rapid mobility of combat operations, the tight coordination of participating service arms, and the intensity of the logistics defense struggle. These new features of high-tech local wars are putting new and greater demands on logistical support. Technology determines tactics, with combat determining logistics. In order to adapt to the logistical demands of high-tech local wars, the PLA must adapt to the needs of rapid mobility combat by raising our contingency mobility support capacity. In rapid mobility operations by support units, communications and transportation are the keys.[6]

General Yang specified that:

> laterally, the PLA needs to set up a joint logistics support system for all service arms, combining centralized supply with specialized supply. Vertically, the PLA needs to set

---

[5]Yang Chengyu, "On Adapting to the Demands of High-Tech Local Wars by Striving to Improve Our Military Logistics Support Capability," *Zhongguo junshi kexue* [*Chinese Military Science*], No. 2, May 20, 1995, pp. 95–98.

[6]*Ibid.*

up a three-grade reserve supply system combining strategic, campaign, and tactical logistics. In order for the PLA to set up joint logistics, we will need to act in line with our realities, having our own features and distinctions. In recent years, we have tested a network-type support division for our three service arms, achieving good results and accumulating the necessary experience. Grounded in that, we still need to create terms in all areas for gradually expanding our centralized supply status and factors. While our general need and model for all three service arms is centralized supply of interchangeable materiel, centralized repair of interchangeable materiel, centralized repair of interchangeable equipment, centralized reception and treatment of all wounded, and centralized organization of communications and transportation, all service arms need to preserve their individuality premised on adhering to generality.

General Yang continued, "While China is a great nation, our economy remains undeveloped. So with such glaring conflicts between military spending supply and demand, how are we to intensify our logistical combat readiness to raise our sustained support capability during wartime?

- First, we need to solidly establish an overall concept. The only way that the defense establishment will have a solid material base and the PLA's logistical support will have sustained follow-up support is if the national economy continues to grow.

- Second, our logistics personnel need to establish a firmer sense of management efficiency.

- Third, we need to give precedence to supporting priority force building.

- Fourth, we must have the capacity to effectively defend rear areas in order to protect our logistics survivability.

- Fifth, the matter of a centralized but flexible and discretionary logistics organization and command merits conscientious study."

Finally, in a special *Liberation Army Daily* report in February 1996, an unidentified author stated that joint operations under high-tech conditions is a brand-new issue for the PLA, which lacks practical experience in this area. The PLA is also exploring theory, and there are many problems it needs to study and solve. Specifically, the PLA must emphasize research on basic theory of joint operations, then sum up, put in order, and disseminate the results. Based on the operational ideas of "overall operations and key strikes," the PLA must establish, under high-tech conditions, the basic idea of "three service arms united for joint-force logistics support," and mutually adapt it to guiding principles, command strategy, coordination and combat, and other theories, and gradually make it systematic.[7]

The essence of these articles is that the PLA intends to integrate the individual services' logistics systems into a joint, coordinated logistics system as much as possible. However, before it can do this, it has to first understand the theory, decide how it fits into the PLA, disseminate the information, then test it out in small areas before it implements the concepts PLA-wide.

---

[7]Xu Genchu, "On Logistics Support in Joint Operations," *Jiefangjun bao*, February 6, 1996.

CHANGES IN PILOT TRAINING LEAD TO LOGISTICS CHANGES

The PLAAF has definitely made progress in many of the areas General Yang described, but lags behind in others. While there is little information available about the PLAAF's logistics and maintenance support, most open-source Chinese articles discuss the PLAAF's history, operational concepts, aircraft, and training.

On the training and operational side, the PLAAF has established a "Blue Army" aggressor unit to simulate hostile forces against the "Red Army" both offensively and defensively. Furthermore, PLAAF pilots have intensified their training under different weather conditions, at lower altitudes, and, most significantly, over water. They have also practiced rapid deployment to fixed and auxiliary airfields. As a result, the Air Force has had to adjust its logistics and maintenance training and operations to meet these new challenges. These include computerizing individual logistics and maintenance operations, and then networking the computers within the unit and among different units at the same and higher levels. It also has meant establishing small logistics and maintenance teams capable of deploying by rail or air at a moment's notice to accompany the unit's aircraft deployment. As a result, the Air Force has had to adjust its philosophy concerning acquisition, storage, and distribution of spare parts. Therefore, before discussing the logistics and maintenance changes, it is appropriate here to first discuss the major trends in pilot training and where these trends are taking the PLAAF.

According to PLAAF commander Liu Shunyao, Air Force aviation units during 1996 exceeded their annual training plan requirements by 1.8 percent and flight safety has remained up to the world's advanced level for 16 consecutive years.[8] The A-class regiments, which have higher combat capability, now account for approximately 90–95 percent of the flight units' combat regiments, and 75 percent of pilots have now been trained to fly in all types of weather.[9] In addition, all air division and regiment leaders are special-grade or first-grade pilots, and one-half of the pilots in the flight units are college-educated. Flight units generally carry out training in difficult subjects, including night-time flight in complicated weather conditions, guided-missile targeting practice, shooting missiles, operating marker lights, training over the ocean, low-altitude and super low-altitude flight, and emergency mobility maneuvering. The overwhelming majority of the flight units' combat regiments

---

[8] One has to take comments about the PLAAF's annual training plan and safety record with a grain of salt. For example, the PLAAF was involved in the huge joint exercises opposite Taiwan in early 1996, which obviously increased the planned flight training. In addition, according to a 1996 *Xinhua* report, a series of arresting cables was installed at various units, which safely arrested more than 140 aircraft that either aborted takeoff or overshot the runway during landing. This report indicates that there were numerous accidents that took place before the arresting cables were installed. In addition, General Cao Shuangming, the PLAAF's commander from 1992–1994, was relieved of duty because of an excess number of aircraft accidents during this time.

[9] Training at the "transition training bases" lasts for one year (100 to 120 flying hours). The pilots begin flying the F-5 for basic airmanship, then transition to the F-6 or F-7. Upon graduation, the pilots are expected to be capable of flying in "three weather conditions" (i.e., day and night visual flight rules [VFR], and day instrument flight rules [IFR]). "Four weather conditions" adds night IFR flights. Thereafter, annual flying hours vary according to the type of aircraft: bombers (80 hours), fighters (100–110 hours), and the A-5 ground-attack aircraft (150 hours). See *Dangdai Zhongguo kongjun* [*Contemporary China's Air Force*], Beijing: Zhongguo shehui kexue chubanshe, 1989, pp. 503–504.

conducted live-ammunition targeting practice in a combat environment. This type of training accounted for 45 percent of the planned annual training time.[10]

General Liu stated that the Air Force recently built and put into operation a modern, comprehensive air tactical training base marked by an actual combat environment and advanced facilities. This is of great significance to rapidly improving the air force's tactical training quality and fighting capacity and signals a new stage in China's Air Force tactical training. A salient characteristic of this tactical training base is that it enables training for Air Force units under actual combat conditions and on a real battlefield. The training base has air and ground tactical training ranges, simulated runways built to scale, a surface-to-air missile base, anti-aircraft gun positions, radar and radar support vehicles, and simulated "enemy" command posts, ammunition depots, and oil depots that look like real ones. There are also a large number of simulated tanks deployed in groups in combat positions. During an early-1997 live-fire exercise involving various types of planes, command posts of eight air divisions studied, tried, and explored tactics and conducted training on the base's advanced training systems and facilities.

General Liu described a command and control center at the training center, where training directors and air division commanders received air combat reports, directed air battles, communicated with combat planes in the air, monitored units in the exercise, and directed units' deployment and movement in a timely fashion through use of the base's advanced facilities. A monitoring, control, and appraisal system installed in the base's central command hall received timely information about planes, including flight path, course, speed, altitude, and other parameters through monitoring, automatic video recording, radar, and flight orientation systems, so as to provide training directors with accurate information for training results appraisal.

According to an April 1997 Liberation Army Daily article, since its founding in the late 1980s, a PLAAF unified "Blue Army" aggressor unit, composed of special-grade and first-grade pilots and equipped with advanced equipment, has "fought" numerous air battles with each and every Air Force fighter plane and attack plane unit on "battlefields" in the blue sky, in order to help improve the PLAAF's high-tech combat effectiveness and improve their knowledge of tactics. The specialized "Blue Army" has emerged as a strong "enemy unit" in simulated air battles and brought about a lot of changes to the Air Force's tactical training. In light of China's training realities, the specialized "Blue Army" unit has launched both fierce offensives and sudden raids, created true-to-life modern air fighting situations, provided no fighting

---

[10]Sun Maoqing, "Training Improves Air Force Combat Effectiveness," *Xinhua*, March 26, 1996. Sun Maoqing and Man Dongyan, "Air Force Flight Units Fulfill Training Mission," *Xinhua*, December 23, 1996. Neither of these articles attributed the excess training to the exercise conducted opposite Taiwan during March 1996 and the large exercise in the Gobi Desert in September, but this is the most likely cause. Sun Maoqing, "Make Efforts To Build Modernized People's Air Force," *Liaowang*, No. 15, April 14, 1997, pp. 20–21. Zhang Nongke and Zhang Jinyu, "Air Force Builds Modern, Comprehensive Tactical Training Base," *Jiefangjun bao*, October 28, 1996. There are often conflicting stories that relate the amount of training that PLAAF flying units receive. For example, an article in Volume 5 of the 1995 *Zhongguo Kongjun* [Air Force] magazine, discusses the number of sorties flown (54,506) over the eight-year period of 1987 through 1994 by an unidentified air division on the Leizhou Peninsula. However, assuming the division has at least 72 aircraft, this equates to 6,813 sorties per year or 94 sorties per aircraft per year and less than two sorties per week.

procedures or clues beforehand, and conducted simulated true-to-life confrontations.[11]

The concepts developed at the training base and through the "Blue Army" are now being moved to the unit level, where several units have begun to turn these new combat theories and concepts into live-ammunition exercises. According to the Liberation Army Daily article, the PLAAF has obtained some initial results in important combat study areas, such as maneuverable combat, air attack, fighting for air supremacy, and night attack and defense. A new training syllabus has taken shape characterized by adaptability to combat situations based on future high-technology developments. The development of flight simulators as a means of efficient, high-technology training has also been fruitful. The simulation capabilities of the Air Force have evolved from electromechanical simulation to laser, electronic, and computer simulation; from technical simulation to tactical and campaign simulation; and from the simulation of a single armament or aircraft type to integrated simulation of the main battle arms combined with multiple aircraft types and various forms of weaponry. The modern scientific training methods have replaced the traditional ways of military training.

On several occasions, Air Force units have formed offensive and defensive exercise teams with Navy, Army, and Air Force anti-aircraft, radar, and missile troops to conduct exercises modeled on future warfare. According to one article, "Flying low-altitude bombing raids over the sea is usually thought of as deadly, since the sea and the blue sky look almost the same to the pilot. Pilots have flown large bomber groups less than 100 meters above the vast sea, which is indistinguishable from the sky by color, and have achieved good results in hitting all targets that were spotted. Some troops have successively organized experimental learning projects, such as flying close to strategic points at sea, launching surgical air strikes against enemy troops, waging offensive air campaigns, conducting over-the-horizon air combat, and imposing air and sea blockades."[12]

In his interview, General Liu pointed out that a large two-day "offensive-defensive" exercise utilizing the PLAAF's "Blue Army" unit was held in the Gobi Desert during September 1996. It was made clear from the very beginning that this exercise was to be conducted under unknown conditions, almost all the subjects in this exercise were new to its participants, including the "multi-typed aircraft joint offensive-counteroffensive against battlefield targets under electronic confrontation conditions" and the "coordinated attacks against airport targets and air interceptions in an multi-typed aircraft composition formation." According to a prescribed principle, anything related to an exercise would not be leaked to either party in confrontation, including the deployment during different stages of the exercise, combat tasks, battlefield targets, or flight routes. The exercise, which began with the crash of a fighter en route, included mid-air fights to seize control of the air, air raids

---

[11]Zhang Nongke and Zhang Jinyu, "'Grindstone' Confronts 'Iron Wings' in Blue Sky—Air Force Forms Unified 'Blue Army' Unit for First Time To Confront Airmen Units in Rotation," *Jiefangjun bao*, April 28, 1997.

[12]Sun Maoqing, "Training Improves Air Force Combat Effectiveness."

by attack planes and bombers after avoiding enemy intercepting planes, and airborne landing operations and anti-airborne landing operations.[13]

According to a May 1998 Xinhua report, the PLAAF recently completed producing the first full set of aviation lifesaving equipment for pilots flying over five different types of combat environments—sea, desert, tropical forest, cold zones, and plateaus. Because of China's vast territory, its complex climate and geographical environment, and its various aircraft models, accommodating a wide variety of lifesaving supplies in a small combat aircraft cockpit is a difficult problem. For a long time, there were no full sets of specialized aviation lifesaving equipment for Chinese military pilots flying over various geographical regions. As a result, pilots who were forced to parachute lacked a reliable guarantee for their safety. This new full set of aviation lifesaving equipment consists of specially prepared survival rations; a device for sterilization, purifying water, and producing heat; a multipurpose survival knife; other daily necessities; and communications apparatus such as a survival radio set, a signal flare, a target indicator, and various monitoring devices. In addition, according to the needs of different regions, the set is provided with protective emergency supplies such as a cold-proof sleeping bag, an inflatable rubber boat, an oxygen cylinder, snake and shark repellents, emergency energy-preserving pharmaceuticals, and a helicopter hoisting rescue device. Besides meeting the needs of future high-tech battlefields, this functionally steady and reliable survival equipment has widened the safety margin for the survival of Air Force pilots who bail out. It also has extensive applications for civil aviation, rescue and relief operations, and open-country surveys.[14]

The Air Force tested this survival equipment in its first aerial rescue test on the Qinghai-Tibetan Plateau in early 1998. The rescue operation involved parachuting onto the plateau, then conducting self-rescue and survival until the pilots were rescued. The exercise showed that the new equipment could help pilots survive three days after parachuting, making it possible for them to wait for the arrival of rescuers. According to the report, the PLAAF has also conducted rescue tests in jungles, deserts, frigid zones, and on the sea.[15]

THE PLAAF'S LOGISTICS AND MAINTENANCE CHALLENGE

As the Air Force moves toward a leaner force with rapid deployment capabilities, it is in the process of trying to diversify its logistics patterns in several areas, including emergency resupply, prepositioning of supplies at key airfields, cooperation among front-line and rear area airfields, and cooperation among the different arms and

---

[13]Zhang Nongke and Zhang Jinyu, "Summary of the Air Force's Pioneering Offensive-Defensive Confrontation Exercise With Participation of Varied Types of Aircraft (Arms of the Service) Under Unknown Conditions," *Jiefangjun bao*, September 25, 1996.

[14]Xu Zhuangzhi and Zhu Liang, "China: Military Develops Survival Equipment for Pilots," *Xinhua*, May 22, 1998.

[15]"PRC Air Force Performs Plateau Aerial Rescue Test," *Xinhua*, June 10, 1998.

services. The Air Force calls these concepts logistics and maintenance "guarantee" systems, and has divided them into six categories as follows:[16]

1. **Providing emergency guarantees.** In order to fulfill combat tasks, the Air Force has established an emergency mobile supply system. As a result, airfields can receive emergency logistics support once their own logistics guarantee systems are knocked out. Moreover, emergency guarantees can be extended to such areas as setting up temporary airfields, repairing damaged key airfields as well as damaged aircraft takeoff and landing facilities within a short time, and guaranteeing field oil supply and emergency air transport.

2. **Providing partial guarantees in advance.** To fight a high-tech air battle, the Air Force has to supply key combat goods and materials to the front-line and backbone airfields in advance. Since it takes a long time for air units stationed in front-line airfields to prepare for the second round of operations and to receive emergency goods and materials, and since transport lines are often vulnerable to enemy attack, the Air Force has to supply key combat goods and materials, all types of aircraft maintenance equipment, high-tech weaponry repair instruments, and so on in advance, in order to gain the logistics initiative and save time.

3. **Providing guarantee to key airfields.** All types of aircraft are involved in a modern air battle, and various types of aircraft are to be assigned to, or temporarily landed at, key airfields. The Air Force's logistics departments should supply necessary personnel, technology, goods and materials, instruments, and equipment to key airfields that undertake to maintain various types of combat aircraft to ensure maintenance and combat effectiveness.

4. **Providing independent guarantees to different areas.** The Air Force should divide a combat zone into independent guarantee areas in light of its jurisdictional and topographical characteristics and supply routes; clearly define responsibilities, tasks, and requirements for independent guarantee areas; properly strengthen logistics force of independent guarantee areas; and organize guarantee operations on the basis of independent guarantee areas under normal circumstances.

5. **Providing guarantees among departments concerned.** To ensure effective guarantees to the front-line and second-line airfields, airfields located in the hinterland should cooperate with the front-line (and second-line) airfields. The ·front-line and second-line airfields and airfields located in the hinterland should "help each other" by establishing either permanent or temporary relations of mutual guarantee and support. In peacetime, the two types of airfields should cooperate with each other in training and clearly define tasks of mutual guarantee and support. In wartime, they should provide timely and mobile guarantee and support to each other in light of actual needs. Moreover, they should provide either specific or comprehensive guarantee and support to each other with regard to personnel, goods and materials, instruments, and equipment.

---

[16]Liu Youfeng and Wang Bin, "Diversified Logistics Pattern," *Jiefangjun bao*, February 27, 1996. The term guarantee (*baozheng*) literally means all of the necessary people, support, and supplies from any part of the logistics or maintenance organization.

6. **Providing guarantees based on overall cooperation.** To provide logistics guarantees to a high-tech air battle, logistics departments of all arms and services should closely cooperate with one another and with the localities concerned in providing combined logistics guarantees. To this end, the Air Force's logistics departments should seek and rely on the unreserved support of the Army's and the Navy's logistics departments with regard to common goods, materials, and services. Once they land at the Army's or the Navy's airfields, the Air Force units should rely on logistics support provided by those airfields. The Air Force units should also seek the cooperation of the Navy's logistics departments in searching for or rescuing aircraft crew at sea.

TESTING THE CONCEPTS

One of the PLAAF's biggest challenges has always been moving from concept to implementation. In this regard, it has probably accomplished more in the 1990s than it has at any other time in its history, especially in terms of computerizing its operations. The following paragraphs discuss how the Nanjing Military Region and the Guangzhou, Lanzhou, and Chengdu Military Region air forces have dealt with various logistics issues over the past couple of years.

According to a January 1996 Liberation Army Daily interview with Wang Chuanwu, Director of Nanjing Military Region Logistics Department, the logistics departments of the Nanjing Military Region, the Navy, and the Air Force, as well as the units of the 2d Artillery and the State Commission of Science, Technology, and Industry for National Defense (COSTIND) have adopted a prearranged task method and instituted a theater unified logistic command system for the first time in the history of the PLA. In two major exercises held in the Nanjing theater in 1995, the logistics departments and affiliated organizations of the three services were placed under the unified logistics command for logistics support to all exercise units.[17]

General Wang stated that since the beginning of 1995, the logistics departments and affiliated organizations of the three services in the Nanjing theater conducted extensive studies of the situation, based on guidance from the Central Military Commission. Because the present tri-service situation is characterized by barriers between departments and regions, duplicate functions, scattered forces, and reverse flow of materials, Nanjing Military Region logistics officials spent a month jointly inspecting the warehouses, hospitals, and material supply depots of the three services in the theater. The officials also looked at local support elements in terms of the principle of tri-service joint support, army-civilian integration, and close proximity and convenience to facilitate logistics support. As a result, they set up a number of "joint support areas" so that the "lines" of the navy, the "points" of the air force, and the "surfaces" of the ground forces could be integrated and complement one another.

---

[17]Ding Jianwei and Yan Jinjiu, "Nanjing Theater Exercises Unified Command Over Logistic Service of Three Armed Services To Effect Joint Support," *Jiefangjun bao*, January 8, 1996.

During 1995–1996, the Guangzhou Military Region Air Force (MRAF) Logistics Department party committee focused on the features of coming wars, investing more human, material, and financial resources to build a modern battlefield support system faster. Between 1993–1996, the Logistics Department performed the support mission for six crucial exercises, with its flight support success rate at a reported 99.9 percent. By early 1996, it was using over 20 scientific and technological achievements, such as an automated campaign logistics command system, an automated computerized rapid receive-and-send flight data system, a computerized rapid fuel supply and refueling system, and a night-flight power-failure emergency power-generation system.[18]

The Guangzhou MRAF Logistics Department also allocated a million yuan (USD120,000) in 1995 to build a modern administration and command system. Its automated command system consists mostly of a battlefield logistics command room networked within the headquarters. The unit also has a visual band-wave network for a long-range point-to-point and point-to-multiple-point visual conferencing function. The command room's major control workstation and the three-tier (high, medium, and low) long-distance communications network is used for jobs such as data transmission and graphics faxing, as well as giving the battle logistics command micro-networking and unit visual dialogue for better command support. At the end of 1995, the Logistics Department carried out a large-scale logistics support graphic projection exercise. This exercise generally takes two days, but was completed in only three hours, being more efficient than manual operations.

The region's logistics warehouses have also instituted a microspeed on-line network for providing flight data and material. In 1993, the Guangzhou Air Force Logistics Department upgraded the low-capacity computer system used by its grass-roots flight data warehouses, renewing and replacing all of its flight data unit microcomputers. As of 1996, all battlefield station flight data units in the region, from flight data [navigation material] planning, preparation, and supply to maintenance and management, were using micro-automated multipurpose management, achieving five times better efficiency. Support units such as ordnance, fuel, and transport were also all using micro-multipurpose management, having a counterpart network with logistics units, which forms an integrated support network for command and operations.

The Air Force has made efforts to transform airfield and support stations that provide ground service only into those that provide services to both Army and Air Force units. To meet the needs of air fleets on long-distance, mobile combat missions, all major airfield and support stations have several times organized air transport support personnel and railroad transport support personnel to hold three-dimensional, joint air-ground support exercises on the basis of existing equipment.[19]

---

[18]Fan Qilin and Xia Hongqing, "On Demanding Support from Science and Technology—the Guangzhou Air Force Logistics Department Is Having Marked Success in Building a Modern Battlefield Support System," *Jiefangjun bao*, April 1, 1996.

[19]Zhao Xianfeng and Zhang Jinyu, "Lanzhou MR Air Force Improves Logistics Support for High-Tech Air Battles," *Jiefangjun bao*, December 6, 1995.

As a result, during 1995, the Air Force for the first time ordered that large transport planes carry support personnel and equipment to accompany air fleets in emergency mobile combat support exercises. Previously, virtually all of the Air Force personnel and equipment were moved by road or rail. Of the two Air Force transport divisions, one supports operations out of Beijing, and the other, which has acquired several new Russian-built IL-76s, supports the Air Force's 15th Airborne Army. To provide better mobile accompanying support, the Air Force has set up 11 emergency support teams specializing in nine types of support, including rapid airfield construction, emergency airfield facilities repair, battery charging, oxygen production, refrigeration, and so on. These teams can either individually set out to provide support by whatever transport means are available (air, ground, or rail) or set out to provide mobile support as a large support team.

Aviation units have made efforts to turn airfield and support stations from those that provide steady and conventional support into those that provide rapid and highly efficient support. They have also made great efforts to tackle a number of long-standing problems that undermine support efficiency, including backward plane refueling technology, backward bomb loading technology, and so on, with the result that some airfields have built computerized automatic refueling systems and all major airfields have stopped using refueling trucks and commissioned pipeline refueling. For example, in 1997, an Air Force emergency mobile field pipeline team conducted a pipeline laying exercise in the Nanjing Military Region. The team used close to 5,000 steel pipes weighing over 100 kilograms each to connect an oil depot with an air station 30 kilometers away, crossing a hill of over 400 meters, a town, 27 railroads and highways, 12 caves and bridges, and over 800 meters of rivers and marshes. The exercise was held against the background of a forward airfield, whose oil depot had been bombed.[20]

The Lanzhou MRAF Logistics Department and subordinate unit personnel tested some of these new concepts during two 1995 "joint high-tech ground and air attack exercises." During one exercise, Air Force logistics personnel supported more than 30 combat planes of four categories and seven types, which took off from an unidentified airfield within three minutes of each other. The other exercise involved three categories and six types of combat aircraft, including attack planes, large transport planes, armed helicopters, and transport helicopters.

During the exercises, Lanzhou MRAF aviation units made efforts to turn airfield and support stations from those that provided logistics support for only one category of combat planes in the past into those that provide support for all categories and all types of combat planes. Since different categories and different types of combat planes are to participate in future air battles in one air fleet, units have worked out different types of support plans, renovated and transformed existing combat planes' service equipment and facilities, and imported advanced foreign logistics support equipment and facilities, with the result that airfield and support stations can now

---

[20]*Jiefangjun bao*, May 15, 1997.

provide logistics support for different categories and different types of combat planes.

During the mid-1990s, the Chengdu MRAF increased investment to speed up the modernization of the logistics support system of Air Force stations in Tibet: POL and ammunition reserve bases were built and their supportive warehouses and logistics support systems were built or improved; aviation control centers and modern logistics command systems were connected with the operational logistics command offices by system networks; construction of logistics support facilities for rear-area airports was stepped up; the conditions for logistics support for airports were improved; and aircraft parking areas were enlarged.[21]

In the late 1980s, they succeeded in developing air transport, forming a three-dimensional (air, land, and rail) multi-directional transport system equipped with various types of aircraft that increased the transport capacity more than sevenfold. To increase their capacity to provide logistics support for air warfare using advanced technology, they also built military medicine reserve bases for Army units stationed in Tibet and increased much-needed and conventional equipment, to form a complete system of combat-readiness logistics support.

As a result of these changes, during June 1996, a Tibet-based Chengdu MRAF station took only two hours to accomplish preparations for redeployment of a certain air unit, instead of the 20 days it took to accomplish the same task in the past. Li Maifu, director of the Chengdu MRAF Logistics Department, stated that was the result of strengthening the Tibet-based Air Force's comprehensive logistics support system. In the past, various Tibet-based Air Force stations lacked reserve bases, so when air units entered and were stationed or trained in Tibet, all supplies had to be shipped there from the hinterland by road, which took a long time and was subject to seasonal changes.

JOINT SERVICE AND CIVIL-MILITARY LOGISTICS REFORMS

According to a 1997 *Xinhua* report, the establishment of the socialist market economy has constituted an unprecedented challenge to PLA logistics, but it has also provided an opportunity for new developments. This has especially been the case for the fuel and transportation sectors. For example, in the Jinan theater of operations, the supply of fuel oil, which accounts for 60 percent of the total consumption of materials in a modern war, was less than 35 percent of the planned amount because of market constraints. Therefore, between 1987–1995, the Jinan MR reformed the supply of fuel oil by "joint supply of the three armed services" and the "army-civilian joint supply." As a result, the reformed system now provides a fuel oil support mechanism among the three services, the local military logistics departments, and the local economy.

---

[21]Deng Guilin, "Air Force in Tibet Sets Up a Comprehensive Logistics Support System," *Jiefangjun bao*, July 17, 1996.

The *Xinhua* article also described how the Nanjing theater of operations took the lead in establishing a new military-civilian defense transportation network and joint supply structure for vehicles and equipment between 1987 and 1995. This system consolidated national defense, increased the capability of ensuring adequate military transportation support, facilitated local civil transportation, and promoted local economic construction. According to statistics, in the 8th Five-Year Plan period alone, the state and local governments invested a total of more than 2 billion yuan (USD240 million) in building and renovating more than 50 national defense highways and dozens of bridges and tunnels. In the construction of the Lanzhou-Xinjiang double-railway line and the Baoji-Zhongwei, Nanning-Kunming, Beijing-Kowloon, Guangzhou-Meixian-Shantou, and the Northern Xinjiang trunk railways, more than 200 national defense requirements were met. In civil aviation, national defense requirements were also met in the building and expanding of the Lhasa, Kunming, Guizhou, and other airports.[22]

PLAAF STORAGE AND MAINTENANCE FOR OLD AIRCRAFT

According to a March 1997 Hong Kong report, U.S. reconnaissance satellites discovered in June 1993 that China had gathered over 1,000 combat aircraft at an airfield in central China, which turned out to be an exceptionally large aircraft depot to accommodate retired planes. The storage center has three purposes: taking over and storing retired aircraft from all Air Force units nationwide; routine maintenance for those planes still functioning well; and renovating old and broken aircraft.[23]

It is not easy to store aircraft for a long time, since special conditions are usually required to achieve this. China can store aircraft only in a cave or in the open as it is doing now, making the job extremely difficult. When aircraft are stored in the open for a long time, the sheeting that seals the cabins will gradually age and break, leading to pools of water in the cabins on rainy days. Excessive humidity in a cave can erode aircraft components.

To overcome the above difficulties, the storage center has developed a set of new maintenance and management measures. For those aircraft stored in the open, new materials are used to mothball the whole aircraft, thus maintaining the critical temperature and humidity to keep the aircraft from rusting and harboring harmful bacteria. Personnel subject precision instruments to partial de-oxygenation treatment. The new techniques can protect aircraft stored in the open from damage

---

[22]"Striding Toward Modern and Great Logistics—All Armed Services Implement Chairman Jiang Zemin's Five General Guidelines: Guideline on Providing Effective Logistics Support," *Xinhua*, August 2, 1997.

[23]"Storage Center for Fighters, Transport Planes, Bombers of Air Force," *Ta kung pao*, March 17, 1997. Of note, the U.S. Air Force hosted a small PLAAF delegation in the United States in 1988 to discuss maintenance and logistics. The delegation visited Hill Air Force Base, Davis-Monthan Air Base, and the Smithsonian's aircraft refurbishment center in Washington, D.C. One of the leaders of the delegation returned to Beijing and became the head of the Air Force's new aviation museum at Shahezhen airbase located just north of Beijing. During the visit, the delegation made known that they were planning to build a huge aircraft storage facility similar to Davis-Monthan, but were having problems finding a place with the proper climate and necessary space. The most likely place was going to be near the historic capital of Luoyang, in Henan Province.

for as long as six years. When the aircraft are unsealed, they will have preserved their original performance and function.

For aircraft stored in caves, the center has adopted another, totally different mothballing technique. Through experimentation, the center has attained the ability to control airframe humidity, temperature, and anti-rust oil molecule density, adjusting them at any time according to changes in the micro-climate of the caves. This method is called "meteorological mothballing."

Apart from maintaining aircraft that function satisfactorily, another job of the center is to dismantle and reprocess old and useless aircraft. In a simple and crude aluminum metallurgical workshop in the center, workers have dismantled more than 100 aircraft and smelted more than 100 tons of aluminum ingots from abandoned parts. The useful parts and components are then delivered to the air-materiel maintenance plant for renovation. The plant currently can repair more than 300 items and annually recovers hundreds of aircraft engines, as well as 10,000-odd pieces of other equipment.

PLAAF'S SU-27 FIGHTERS: IMPLICATIONS FOR LOGISTICS AND MAINTENANCE

One of the most important events in the past 50 years was China's 1990 agreement with Russia to purchase Sukhoi-27 fighters and then to produce up to 200 of the aircraft in China. The Su-27s produced in China will be known as the J-11. Although reports have differed about the exact number of Su-27s, the first batch of 26 aircraft arrived at the 3rd Air Division in Wuhu in 1992, followed by a second group of 24 more at Suixi airfield in 1996. Initial reporting indicated that China would acquire a total of 72 aircraft in three batches of 24 each, and would then begin to produce approximately 200 more at the Shenyang Aircraft Factory over the next 10–15 years at a cost of $2 billion U.S.

According to ITAR-TASS reports from November 1997, Russia was completing transfer to China of technical documents for the licensed production of 200 Su-27CK fighters.[24] Following the handover of the documents, China will scrutinize them and coordinate all technical issues with the Sukhoi design bureau, which was to send a group of Russian specialists to China before the end of 1997. There is a special provision in the "Chinese contract" stipulating that any changes in the Su-27CK fighter produced in China can be made only with Russian consent. The provision is valid not only for the period of batch production of Su-27CK jets in China but also for the entire time of their use by China's Air Force. The contract prohibits any exports of Su-27CK that have not been authorized by Russia. China has no opportunity to modernize and export Chinese-made Su-27CK without Russia's participation. Russia supplies to China the AL-31F engines and the complete set of the on-board radio-electronic equipment.

---

[24]Nikolay Novichkov, "China To Begin Licensed Production of Su-27CK Fighters," ITAR-TASS in English, November 13, 1997.

The ITAR-TASS report specified that the first licensed Su-27CK jets should take off before 2000. In line with world practice in licensed production transfer, China will assemble its first Su-27CK from knockdown kits of units, assemblies, and systems beginning in 1998. The kits will be supplied by the Komsomolsk-on-Amur aircraft production association (KnAAPO), the head Russian manufacturer of SU-27CKs. The KnAAPO should hand over to China the necessary technological equipment and assembly fixtures worth $150 million U.S. The company is also to deliver some 30 percent of all assembly units for China's 200 Su-27CKs. Under the contract, Russia is obliged to deliver AL-31F engines and the whole set of the on-board radio-electronic equipment for all the jets to be assembled. The maintenance service for AL-31F engines will be set up in China. The "Chinese contract" will require certain restructuring to expand the production of some parts in Russia. This is explained by the fact that a number of companies producing aircraft instruments were in former Soviet republics and many of them are unable to ensure supplies to Russia.

The latest reporting from *Jane's Defense Weekly* indicates that the first 50 Su-27s, worth $450 million U.S., will be assembled from Russian components, with the first two projected to come off the assembly line by the end of 1998. The share of Chinese components will keep growing, finally reaching 70 percent. The AL-31F turbofan engine is not among the items that will be license-produced by China. According to sources who toured the Shenyang factory, manufacturing capabilities and procedures are generally poor and could take several years before successfully assimilating Russian manufacturing technologies. A clause in the Su-27 contract stipulates that if Shenyang fails to meet the annual production target of 10–15 aircraft, then Russia's facility at Komsomolsk will provide the substitute aircraft.[25]

What does all of this mean for the PLAAF's capability to provide logistics and maintenance support for the Russian-built Su-27s and IL-76s? First of all, the PLAAF typically stores at least one year's worth of spare parts for each aircraft at the base housing the aircraft. This should not be that big of a problem for the IL-76s, which are concentrated at a single base in south central China. However, the Su-27s are a different story.

For now, the Su-27s are located at only two bases (Wuhu and Suixi), and reportedly have not been physically deployed to any other bases for exercise purposes. Therefore, the Air Force has been able to concentrate its spares at only two locations, with others possibly stored at the bases' supporting regional depots. However, if the Air Force follows through on its concept of predeploying spares at deployment bases, then it will have to purchase even more spares. In addition, the majority of these spare parts will have to come from Russia for several years to come, until China can begin producing some of the parts. Even then, some of the most critical parts, such as the engines, will still be made in Russia.

---

[25] "Beijing Builds Su-27 fighters from Russian Kits," *Jane's Defense Weekly*, June 10, 1998, p. 12

Based on conversations with U.S. aviation company representatives, the PLAAF's philosophy for purchasing spare parts has not changed for over 20 years.[26] Basically, they always have at least one, and usually more, spares on hand for any given item. Ordering parts is initiated from the unit's purchasing department through a third party, such as the Aviation Ministry's import and export arm. The spares requested are compiled by the different departments (avionics, electrical, mechanical, etc.). The departments in general take the full recommended spares and add to them based on the number of airplanes in the fleet. The total amount of money needed to purchase the spares is then requested by the purchasing department. The entire process can take six months or more before the parts are received. Ordering of special items is difficult and at best might take 2–3 weeks to process the initial order. In general, foreign-acquired aircraft are operated seldom and the Air Force as a whole puts all of its emphasis on allowing no defects for these airplanes. From the outside, their operation is extremely safety-oriented with well-trained crews. However, the negative side is the lack of documentation, procedures, records for cost of operations, and the personal desire to achieve and improve these items.

Another U.S. aviation-related company representative discussed possible limitations of supplying petroleum, oil and lubricants (POL) for foreign aircraft. Whereas the PLAAF can purchase and store sufficient nonperishable hardware, the situation is different for various perishable POL supplies. For example, many of the specialized lubricants have a shelf life of only 90 days from the time they are produced. By the time the supplies are actually delivered in China, the shelf life has been reduced to about 60 days. Therefore, the PLAAF will have to maintain a good supply line for its POL products. This problem will be compounded if the PLAAF intends to preposition logistics and maintenance supplies at auxiliary and forward bases.[27]

Finally, the PLAAF has 21 factories, which produce spare parts for older aircraft such as the F-6, A-5, and F-7, and overhaul and repair all of the Air Force's engines and aircraft, except the F-8, which has to be sent back to the Shenyang Aircraft Factory for any modifications or overhaul. Therefore, until the Shenyang Aircraft Factory or the PLAAF is capable of overhauling the Su-27s, all of the PLAAF's Su-27, as well as the IL-76s, will have to be sent back to Russia to be overhauled. Although there is no information available on the time-between-overhauls for China's Su-27s, the PLAAF completely overhauls its B-6 bombers after 800 hours of flight operations. Overhaul of the PLAAF's current fighters and bombers can take anywhere from 6 to 12 months in the repair facility.[28]

---

[26]The U.S. representatives asked not to be identified by name or company, but their responses to specific questions were virtually the same.

[27]The representative asked not to be identified by name or company.

[28]This information is based on the author's visit to several PLAAF bases and repair/overhaul facilities between 1987 and 1989.

CONCLUSIONS

Taken at face value, the above description of the PLAAF's improvements, which were taken primarily from a limited number of PLA publications, is quite impressive. It is obvious that the Air Force has not only identified its weaknesses to fight in a high-tech, or any other, war, but that it is moving in positive directions to modernize its combat tactics and logistics and maintenance support to meet those challenges.

On the positive side, the PLAAF has slowly begun to automate its command posts and to network the various functional organizations from command and control to logistics and maintenance. It has established a tactical training base to develop new tactics and train pilots, and has established and expanded exercises using a "Blue Army" to simulate hostile forces against the "Red Army" both offensively and defensively. Furthermore, PLAAF pilots have intensified their training under different weather conditions, at lower altitudes, and, especially, over water. They have also practiced rapid deployment to fixed and auxiliary airfields, using transports to support the deployment by carrying equipment and support personnel. The pilots have also received flight safety gear for all combat environments. Finally, the Air Force is beginning to study the concepts and work more with the Army and Navy on joint operations and logistics and maintenance support. There are also efforts to enhance civil-military logistics support capabilities.

On the negative side, the Air Force still has many airfields and support stations that provide service to only one type of aircraft and are not equipped to support deployed aircraft for any length of time. In addition, many aviation units still have a number of long-standing problems that undermine support efficiency, including backward plane refueling technology, backward bomb loading technology, and so on. The Air Force, and PLA as a whole, is still a long way from being able to conduct joint logistics and maintenance support. The Su-27s, IL-76s, and any Chinese-made aircraft with extensive foreign parts will continue to provide a logistics and maintenance challenge for the Air Force.

This paper raises many more questions than it answers. For example, more information is needed about the "Blue Army" before its impact can be assessed. Are the pilots under strict ground-controlled intercept (GCI), as is the rest of the PLAAF? When will Su-27s be introduced into the unit? What types of electronic warfare will the aircraft use? Are the exercises set piece, or does the aggressor unit show up without warning?

Other questions raised include how often do PLAAF units deploy and practice from deployed airfields? How many aircraft usually constitute a deployment—division (72 aircraft), regiment (24 aircraft), group (8 aircraft), or a squadron (2–3 aircraft)? What types of logistics and maintenance packages (personnel and equipment) are deployed to support the aircraft? Do most deployments take place within the same military region or between military regions? What type of sortie generation and rapid turnaround capabilities do these aircraft have when they deploy? How proficient are PLAAF pilots in firing their air-to-air missiles? Is most of their night flying done on clear, moonlit nights or when it is pitch black? Do PLAAF units have a direct communications link with ground and naval forces that they are supporting, so that the pilots do not get shot down by friendly fire? What is the current identification friend or foe (IFF) capability on PLAAF aircraft?

In conclusion, it is clear that the PLAAF is slowly moving toward modernizing its fighting force, as well as its support systems. The question is how well it will be able to carry out these reforms, especially in a joint force arena, and how well will it be able to support its weapons systems acquired in full or in part from foreign sources.

7. CHINA'S NATIONAL MILITARY STRATEGY

David M. Finkelstein[1]

I. INTRODUCTION

It is an excellent time to reassess China's national military strategy. The next wave of significant reforms for the Chinese People's Liberation Army (PLA) is already beginning to unfold. The results of the 15th Party Congress (September 1997) and the 9th National People's Congress (NPC) (March 1998) indicate that after many years of study and debate firm decisions have been made to move forward with structural, organizational and other adjustments to China's armed forces. At the Party Congress Jiang Zemin announced a 500,000-man reduction in force size over the next three years. In the wake of the 9th NPC, a fourth General Department was created—the General Armaments Department—and the Commission for Science, Technology, and Industry for National Defense (COSTIND) was elevated to ministry status with a civilian in charge. We should expect to see more changes, although Beijing's timetable is unknown. Some change may be dramatic and public. Most will be quiet and not easily discernable, given the opaque nature of the Chinese defense establishment.

While the recent party congress and NPC serve as significant benchmarks, we must remember that they are points on a continuum of change that the PLA has been undergoing for almost two decades. *What does the PLA hope to achieve and why? How does it plan to achieve its ends?* These very basic questions, which on the surface seem so simple, are probably the most critical questions one can ask in evaluating the Chinese armed forces. They are critical questions because if one does not address these overarching issues it is difficult to make sense of all other developments: command and doctrinal issues, organization and force structure, or hardware development and acquisition, to name a few.

---

[1]Dr. David M. Finkelstein is a specialist in Chinese security affairs at the Center for Naval Analyses Corporation in Alexandria, Virginia. While on active duty, in addition to various tactical and command positions in the field in both the United States and Korea, he served in successive assignments as a China Foreign Area Officer for over sixteen years. Among his many China-related positions, Dr. Finkelstein served on the faculty at West Point, and was billeted as a senior military analyst with the Defense Intelligence Agency and as Assistant Defense Intelligence Officer for East Asia and the Pacific in the Pentagon from 1993 to 1997. He is a graduate of the United States Military Academy, holds a doctorate in Chinese history from Princeton University, and is a graduate of the Army War College. Among his major publications is *Washington's Taiwan Dilemma, 1949–1950: From Abandonment to Salvation* (George Mason University Press, 1993). The views expressed in this paper are strictly his own.

To answer these basic questions we must have a framework and a context. This paper attempts to provide such a context. It will offer a notional national military strategy for China.[2]

There are five major assumptions implicit in this paper. The first is that China does in fact have a national military strategy. That is, there is a rationale behind the ongoing PLA reforms. A second assumption is that outside observers can adduce that rationale, even if imperfectly, from public domain information. The third assumption is that the PLA remains subservient to the party and the state and therefore China's national military strategy is derived from and mutually supportive of Beijing's overarching national security strategy. A fourth assumption is that while there is much that is unique about China's armed forces, there is also a good deal of universality in how defense establishments go about the business of planning at the national level. A fifth assumption is that Western models can sometimes help structure a discussion of Chinese phenomena even to the point of using Chinese terminology comfortably within those constructs.

The U.S. Army War College (AWC) model of military strategy as developed by Colonel Arthur Lykke, USA (Ret.) serves as the superstructure of the following analysis of the PLA.[3] While the PLA would certainly *not* use an American construct to articulate its national military strategy, this model is nevertheless a useful tool for the descriptive and analytic purposes of this paper. I have also borrowed useful frames of reference from the planners on the Joint Staff who produce the Pentagon's national military strategy (which in turn is based upon a derivative of the Army War College model),[4] as well as several universal military concepts such as "center of gravity" and others. Into these "frames" we shall place Chinese "lenses" to articulate a vision of the bigger picture.

II. WHAT IS A NATIONAL MILITARY STRATEGY? (AND OTHER DEFINITIONAL BURDENS)

Before proceeding there are three terms which should be addressed: strategy, national security strategy, and national military strategy.

Strategy. Strategy is an easy word to use but is difficult to define. Most standard dictionaries are more confusing than enlightening on that particular entry because a strategy refers to a holistic system and process. Our interest in the word strategy will focus on its component parts because of their utility in analyzing the whole. The U.S. Army War College utilizes a simple but powerful formula to express what a strategy is and what its critical component parts consist of: *Strategy = Ends + Ways + Means.* In

---

[2]It is notional, obviously, because the PRC has not published a detailed national military strategy.

[3]Colonel Arthur F. Lykke, Jr., USA (ret.) (ed.), *Military Strategy: Theory and Application*, Carlisle Barracks, Penn.: U.S. Army War College, 1993.

[4]See especially the approach taken in Joint Chiefs of Staff, *National Military Strategy of the United States of America: A Strategy of Flexible and Selective Engagement*, Washington, D.C.: U.S. Government Printing Office, 1995; and Joint Chiefs of Staff, *National Military Strategy of the United States of America: Shape, Respond, Prepare Now—A Military Strategy for a New Era*, Washington, D.C.: U.S. Government Printing Office, 1997.

this equation "ends" are our objectives or goals, "ways" are the courses of action we choose to achieve those goals, and "means" are the resources either at hand or which must be developed to enable the courses of action.[5]

There are three important aspects of this model to keep in mind. The first is that the three components of a strategy—ends, ways and means—are interdependent. All of the components must be appropriate to the whole and in proper balance with the others if the strategy is to be successful. The second point to keep in mind is that when we attempt to study someone else's strategy, such as the PLA's, focusing on only one component of the strategy without an understanding of the other two may lead to incorrect or incomplete conclusions. The third point is the utility of this model as an analytic tool. It is almost universal in its applicability and is not limited to military affairs. One can easily use this equation to craft, describe or analyze political or economic strategies. Also, in the realm of military planning it is applicable across the three levels of warfare—the strategic, operational and tactical levels.

National Security Strategy. American analysts often use the term "national security strategy" or "security strategy" in the context of a nation's military concerns or military-related issues. This paper adopts a variant of the much broader definition used by the U.S. Joint Chiefs of Staff. A National Security Strategy (NSS) will refer to the development, application and coordination of all the elements of national power (political/diplomatic, economic, informational, military, sociological/cultural) to achieve a nation's objectives in domestic and international affairs in peace as well as in war.[6] In pursuing the national objectives set forth in a NSS, multiple strategies co-exist: an economic strategy, a political strategy, a diplomatic strategy, a social strategy, and a military strategy, at the very least. There are two points to make about this definition. First, Lykke's equation is still a valid construct. A national security strategy will have to articulate ends (objectives), ways (courses of action), and means (resources). Also, it should be noted that the military element of national power is a subset within the broader national security strategy. This brings us to a description of a national military strategy.

National Military Strategy. A National Military Strategy (NMS) is the military component of a nation's overall National Security Strategy. Its objectives are derived from those within the overarching NSS. It is the role of the national military leadership to ensure that the military element of national power will be available to contribute to the NSS in both peace and war, in the here and now and in the future. The NMS is the vehicle through which the national military leadership articulates,

---

[5]For a fuller discussion of Colonel Lykke's model of strategy and its component parts, see his lead article, "Toward an Understanding of Military Strategy," in *Military Strategy: Theory and Application*, pp. 3–8.

[6]Definitions have changed over the years even within the Joint Staff. For example, in Joint Pub 1-02 (as amended through January 12, 1998), *Department of Defense Dictionary of Military and Associated Terms*, "National Security Strategy" is defined as "the art and science of developing, applying and coordinating the instruments of national power (diplomatic, economic, military, informational) to achieve objectives that contribute to national security." The 1987 Joint Pub 1-02 defined it as "the art and science of developing and using the political, economic, and psychological powers of a nation, together with its armed forces, during peace and war, to secure national objectives."

revalidates, and adjusts the ends, ways, and means of the armed forces to comport with changing NSS objectives, a changing security environment, or changes in the availability of national resources to be applied to the armed forces. A national military strategy is usually influenced by civil political and economic decisions. Consequently, national military strategies are dynamic and require constant review, revision and updating. At the national level of military planning—the level at which a NMS is generated—direction, guidance and policies are articulated in broad terms to steer the armed forces in the correct direction. Specific decisions, programs and detailed planning follow in due course.[7]

Once again, the equation is still valid. A NMS must articulate ends, ways and means. In the case of a NMS, the equation is rewritten to reflect the national level of military strategy: *National Military Strategy = National Military Objectives + National Military Strategic Concepts + National Military Resources*. National Military Objectives (NMOs) will be derived from the NSS. The National Military Strategic Concepts (NMSC) will articulate the courses of action that will be undertaken to achieve the ends. The National Military Resources (NMRs) describe the types of capabilities that will be required to be on hand or be developed to enable the NMSCs.

The starting point for crafting a NMS is the articulation of NMOs. From these are developed NMSCs and NMRs. But before articulating NMSCs strategic planners first must use a critical "strategic filter" that identifies the imperatives of conflict and the possible constraints planners may have to consider. At minimum, the strategic filter: (1) considers political decisions handed down to military planners or brokered between military and civilian leaders; (2) assesses the current and projected security environments (conflict with whom? when?); and (3) performs an analysis of the operational environment (what kinds of conflicts?).[8]

The synergistic relationship between the three elements of a NMS is readily apparent. If the resources are not available to enable the NMSCs, then weighty political-military decisions are in order: either adjust the NMOs and NMSCs or commit the resources (usually funding) to develop the NMRs. But such a zero-sum set of decisions is usually unacceptable for political reasons and impractical for reasons of national security. Consequently, at the national level of military planning, the NMS often encompasses multiple substrategies for different time frames. At a minimum, one strategy must be focused on current capabilities and near-term contingencies. Another should consider the requirements of coping with future potential security problems.[9] Crafting military strategies for the here and now while

---

[7]These concepts are more or less a direct adaptation of the U.S. view of a NMS. For more discussion about the nature of the NSS and the NMS and the relationship between them see Joint Pub 3-0, *Doctrine for Joint Operations*, February 1, 1995, pp. I-4, I-5; and Joint Pub 0-2, *Unified Action Armed Forces*, February 24, 1995, pp. I-2, I-3. See also definitions in Joint Pub 1-02.

[8]The "strategic filter" is not part of the Army War College model, but a modification to it added by the author.

[9]Military strategies can either be capabilities-based or force developmental-based. The former is driven by near-term contingencies and is the basis for operational planning. If one has to go to war today one can only go with what is currently available in terms of resources. Therefore, the "ways" will be driven by the "means." In the case of the latter, an assessment of future threats will dictate the development of

accounting for over-the-horizon problems is a conundrum that is encountered almost exclusively at the national level of military planning and, save success in war, is probably the ultimate test of a nation's generalship.

With the preceding discussion behind us we can now move on to consider the case of China. To do so we will take the iterative and dynamic process that is used to *craft* an NMS, which is a cyclical process, and artificially stretch it out in linear fashion in order to be *descriptive* of its components. (See Figure 1)

III. CHINA'S NATIONAL SECURITY OBJECTIVES

China's national security strategy has been apparent and quite public for over two decades and there is no need to detail it here. The political reports from the 13th,14th and 15th Party Congresses, the work reports associated with meetings of the National People's Congress (NPC), and a good number of key decisions and policies emanating from Central Committee plenums and NPC Standing Committee meetings have all become increasingly public over the years. From the time Deng Xiaoping consolidated his power in the late 1970s, observing and analyzing the ends, ways, and means of that strategy has defined the essence of China-watching. The "Four Modernizations," "Reform and Opening Up," "Economic Construction as the Central Task," and other phrases coined by the Chinese to describe *aspects* of their strategy are all quite familiar even to the casual student of Chinese affairs. But because we are slowly working toward developing a national *military* strategy for China, we must posit and articulate Beijing's national security objectives since they will drive the PLA's national military objectives.

If one were to distill all of the statements of China's national security objectives, both explicit and implicit, that have been publicly declared or adduced over the last few years they could be distilled to three simple words: *sovereignty, modernity,* and *stability.* These words encompass the totality of everything the Chinese nation is determined to achieve. Moreover, these objectives are not only those of the People's Republic of China (PRC), but capture the essence of the Chinese Revolution.

That revolution has been in motion since the last decades of the Ch'ing (Qing) Dynasty. The Chinese proudly proclaim over five thousand years of continuous civilization, and rightly so. But China is new to the business of developing a nation-state—less than one hundred years, less experience than the United States. The history of the Chinese Revolution has been, and continues to be, the story of the difficult transformation of an ancient traditional civilization into a modern nation-state. Every stage of the revolution has more or less sought the same three objectives: sovereignty, modernity, and stability. From K'ang Yu-wei to Dr. Sun Yat-sen to Republican China under Chiang Kai-Shek, to Deng Xiaoping and now Jiang Zemin—all have attempted to achieve these goals. Where there has been divergence has been in the "ways" to achieve those "ends." The great exception was, of course, Mao. While he embraced sovereignty and modernity as legitimate national security

---

resources probably not available at the moment to enable strategic concepts of a very different nature than currently employed. See Lykke, pp. 4–5.

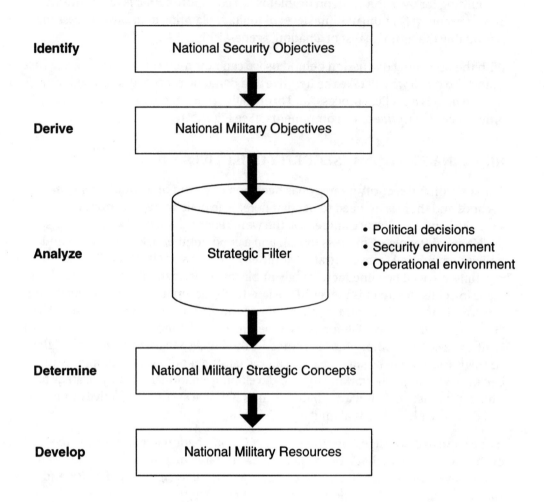

Figure 1—Thinking Through the Elements of a National Military Strategy

objectives, he rejected stability as a goal in his later years. By substituting "perpetual revolution" for stability and by making "class struggle the key link," Mao set back China's progress immeasurably. Consequently, the history of China under Deng Xiaoping is in great measure the story of Deng's efforts to reverse the damage and find a new path to progress.[10] And there is little doubt that China under Jiang Zemin embraces these three national security objectives as well. Let us explore each a bit further.

---

[10]While Deng will rightly be remembered for his pragmatism in economic matters and his quest for China's modernity, his other great achievement often goes unrecognized. Deng brought stability back to elite politics. He dragged both the left and right toward the center and by ceasing to make political infighting a live-or-die, winner-take-all struggle was able to achieve the consensus necessary to move forward with a bold and coherent national strategy for reform.

Sovereignty. One would think that sovereignty, "freedom from external control" to cite *Webster's*, is so fundamental to nationhood that it need not be articulated as a national security objective. For China, this is not the case. Beijing's right to assert and defend its sovereign prerogatives as a nation-state is an enduring preoccupation that even today's leadership brings with them to office. It is not surprising that in the first section of his political report to the 15th Party Congress in September 1997 Jiang Zemin made explicit references to the Opium War (1840), the Eight Power Intervention (during the Boxer uprising, 1900), the war against Japan (1937–1945), and China's "Hundred Years of Humiliation."[11] The prominent position of these references is not mere rhetoric. It underscores that sovereignty is one of the enduring national objectives of the PRC under the Chinese Communist Party (CCP) just as it was for Republican China under the Kuomintang (KMT). This is the result of more than a century of foreign military, political, and economic intervention from without, warlordism and regionalism from within, and the difficulties all Chinese regimes since the fall of the Ch'ing have experienced in just defining and securing the geographical scope of the Chinese nation-state. Consequently, as portrayed by Beijing, Hong Kong's retrocession to China in July 1997 was as much an emotional event for the people of China as it was a political event.

In the context of 1998, how Beijing defines sovereignty is now much broader than merely being the master of its own nation. Today, issues related to sovereignty encompass at least six categories of issues. First, of course, is concern over the return of territories Beijing considers part of the PRC, but over which it exercises no jurisdiction—Taiwan and Macao, for example. Second are issues related to border disputes China still has with some of its neighbors and the problems of demarcation and control. This encompasses a series of bilateral problems. Issue number three concerns areas of China over which Beijing does exert control but whose indigenous non-Han population oppose China's rule, such as in Xinjiang and Tibet. A fourth category is one of multiple competing claims such as in the South China Sea, not just for atolls, reefs and islets but for maritime resources. A fifth issue involves what Beijing views as unwanted and unwarranted foreign concern over, or meddling in, Chinese domestic social and political issues. A sixth category relates to international pressure for China to accede to multilateral instruments and protocols which might constrain Beijing's freedom of action whether or not the instrument in question is or is not problematical.

The constant lectures on China's sovereignty to which foreigners, official and private, friends and adversaries, are subjected are tedious, formulaic and almost archaic. But it would be a mistake to dismiss the seriousness of the message because of the medium.

Modernity. During the last two decades China's quest for modernity has probably been the most studied, analyzed and scrutinized of Beijing's three national security objectives. This is especially the case in economic matters. Modernity, of course,

---

[11] "Jiang Zemin's Political Report" (hereafter Political Report), Beijing Central Television, September 12, 1997, in FBIS-CHI-97-255, September 12, 1997. See Section I: Issues and Prospects at a Time When the New Century Is to Begin.

encompasses much more than economics. It includes social change, political reform, cultural adaptations, intellectual change, and technological and scientific innovation, to name just a few of the aspects of the modernization process.[12] The leadership in Beijing understands this and in fact does have separate objectives and strategies for each of these areas.[13] But for the purposes of this short discussion, we must condense this vast subject to its essential elements.

In the final analysis, the objective of modernity as defined by today's PRC leadership means increasing the economic strength of the nation, enhancing the technological and scientific capabilities of the state, and raising the standard of living of the population. China's leaders have come to believe that its sovereignty and its place among the leading nations of the world in the future will be secured by, and a function of, its economic and technological strength. Hence the centrality of this objective and the emphasis it receives in China's national security strategy.

The 15th Party Congress and the 9th National People's Congress were devoted in the main to articulating the latest adjustments to China's modernization strategy. The Party Congress provided broad statements of "ends" and "means." The NPC, in turn, announced specific plans and policies to enable them. It is worthwhile to quote Jiang Zemin at the NPC in his role as State President for he offers an excellent example of a broad statement of a national security objective in general and a specific "end" or objective for the national security objective of modernity.

> The goals we have set are as follows: When the People's Republic celebrates its centenary, the modernization program will have been basically accomplished and China will have become a prosperous, strong, democratic and culturally advanced socialist country. At that time, our country will rank among the moderately developed countries of the world, the Chinese people will have achieved common prosperity on the basis of modernization and the great rejuvenation of the Chinese nation will have been realized.[14]

Stability. This last of China's three national security objectives is as sacred as the other two. A large dimension of the history of Chinese civilization through the millennia is the history of periods of peace alternating with periods of unfathomable social chaos and violence. Moreover, during the last two centuries stability has been the exception, not the norm. It has been a period of *nei luan wai huan* [domestic disorder and foreign calamities]. The legacies of the White Lotus, Taipings, Nien, Miao Tungan, I-ho ch'uan, warlords, civil war, Red Guards, right through to Tiananmen in 1989 represent what every Chinese regime has feared the most, including the current inhabitants of *Zhongnanhai*. They do not take internal stability for granted.

---

[12]Although now seventeen years since being published, one of the best overviews of modernization as a phenomenon and its course and impact in China since the early modern period remains Gilbert Rozman (ed.), *The Modernization of China*, New York: The Free Press, 1981.

[13]Achieving "socialist democracy" and developing "spiritual civilization" would come under this larger rubric of modernization.

[14]"Text of Jiang Zemin's Speech at NPC Closing Session," *Xinhua*, March 19, 1998, in FBIS-CHI-98-077, March 18, 1998.

Today the greatest challenge for Beijing's leaders is undoubtedly the need to balance their bold plans for modernization with the risk of instability, for it is clear that economic modernization has been accompanied by tremendous social dislocations. Over the past year and a half, one is struck with the frequency of press reports citing labor unrest, worker protests, and acts of civil disobedience. The Chinese leadership is acutely aware of the *maodun* (contradiction) which they have set in motion. Again, we quote Jiang Zemin at the 15th Party Congress:

> . . . it is of the utmost importance to correctly handle the relations between reform and development on one hand and stability on the other so as to maintain a stable political and social environment. *Without stability, nothing can be achieved.*[15]

But the objective "stability" does not just refer to maintaining the internal peace. A second aspect of stability means regime maintenance; that is, maintaining undiluted the authority and monopoly on power held by the Chinese Communist Party (CCP).

The third aspect of "stability" has an outward dimension. It is the firm belief of the Chinese leadership that only a peaceful and stable international environment will permit China to successfully pursue its national objectives. Or, as Jiang Zemin put it, "We need a long-term international environment of peace for carrying out socialist modernization, especially a favorable peripheral environment."[16] And it is precisely because of Deng Xiaoping's great "strategic decision" shortly after the Third Plenum of the 11th Central Committee in 1978 that the world security situation had relaxed and a world war was a remote possibility that he decided the time was right to launch China on its grand experiment in modernization. It is clear that how Beijing chooses to pursue its objective of sovereignty while seeking to preserve a peaceful international environment will also pose a challenge in the future.

These, then, are the three key national security objectives posited as the driving forces behind China's current national security strategy. By striving to achieve these objectives the PRC hopes not only to enhance the state of its domestic conditions but also to be able to strengthen its desired role as the preeminent nation in Asia and as one of five key actors in the future multipolar world order that Chinese theorists argue will revolve around the PRC, the US, Russia, Japan, and Western Europe.

We now turn to the question of how the military element of national power is developed to support these security objectives.

IV. CHINA'S NATIONAL MILITARY STRATEGY

The national military strategy that will now be discussed finds its roots in the late 1970s and early 1980s, when Deng Xiaoping reassessed the international security environment and made the decision to make economic modernization the central task for the coming decades. Deng's decisions not only sent the Chinese nation down a new and different path, but sent the PLA down a new road as well. It stands to

---

[15]"Political Report," op. cit., emphasis added.
[16]Ibid.

reason that a modernizing, outward-looking China seeking a leading role in the mainstream of the international order would require a new type of defense establishment and a new military strategic direction than that of a previously autarkic China seeking to lead the "Third World." What follows then is a description of what Chinese strategic planners might term the general thrust of the national military strategy "during the new historical period" and into "the cross-century period." It is current to the extent that the latest policy decisions coming out of the 15th Party Congress and 9th National People's Congress are incorporated as adjustments. But we should bear in mind that the general direction of China's NMS has been evolving for more than a decade. Consequently, there is both continuity and change to report.

V. NATIONAL MILITARY OBJECTIVES

China's national military strategy seeks to achieve three sets of national military objectives: *Protect the Party and Safeguard Stability, Defend Sovereignty and Defeat Aggression*, and *Modernize the Military and Build the Nation.*

These three objectives not only define what it is the PLA must achieve as a military force, but highlights the unique role it plays in the political economy of the PRC. Significantly, these three national military objectives are derived from and are mutually supportive of China's three national security objectives.

The formulation of these three national military objectives are strictly those of the author's and not authoritative PRC formulations. However, they should be familiar to students of the PLA because China's top military leaders often allude to them or aspects of them in public statements.

For example, in the April 1998 issue of *International Strategic Studies*, Deputy Chief of the General Staff Lieutenant General Xiong Guangkai states that the "basic objectives" of China's armed forces are to "consolidate national defense, resist aggression, defend the nation's sovereignty over its territorial land, sea, airspace as well as its maritime interests, and safeguard national unity and security."[17] In May 1998 Chief of the General Staff Department (GSD) General Fu Quanyou provided this iteration: "the PLA's mission is to strengthen the national defense, fend off aggression, safeguard territorial sovereignty and the rights and interests of territorial waters, and maintain national integrity and safety."[18] As yet another example, in 1996 Defense Minister Chi Haotian offered that the "basic objectives" of China's national defense are to "solidify the defensive capacity, resist foreign invasion, and safeguard the unification and security of the country."[19]

---

[17]Xiong Guangkai, "Gearing Towards the International Security Situation and the Building of the Chinese Armed Forces in the 21st Century," *International Strategic Studies*, No. 2., April 1998, pp. 1–8.

[18]"Fu Quanyou Stresses PRC's Defensive Policy," *Xinhua*, May 6, 1998, in FBIS-CHI-98-126, May 6, 1998.

[19]Chi Haotian, "Taking the Road of National Defense Modernization Which Conforms to China's National Conditions and Reflects the Characteristics of the Times—My Understanding Acquired From the Study of Comrade Jiang Zemin's Expositions on the Relationship Between Building the National Defense and Economic Development," *Qiushi*, No. 8, April 16, 1996, pp. 8–14, in FBIS-CHI-96-120l, April 16, 1996. Hereafter, "Taking the Road."

Although the three national military objectives that this paper posits are seemingly self-explanatory, it is worthwhile to briefly discuss each because through them we achieve a greater appreciation of just what it is the PLA is attempting to achieve, the challenges it faces, and a flavor for the organizational culture of this massive defense establishment.

Protect the Party and Safeguard Stability. The first and foremost mission entrusted to the PLA, and a national military objective, is to be the guardian of the CCP. The past two decades of professionalization and modernization have not altered the fundamental fact that the institutional loyalty of China's armed forces and the personal commitment of it top leaders is to the maintenance of the regime and the primacy of the CCP. The PLA remains the party's army.[20] China's "national command authorities" are the leaders of the Central Military Commission (CMC), which is a party organization under the Central Committee.[21] It is worth remembering that it has only been 71 years since the "Red Army" was founded. The revolutionary heritage of that army and its roots as a communist insurgent force that saved the fledgling CCP from annihilation in the 1920s and 1930s is not too distant a memory for some of the current PLA leaders. Minister of Defense and CMC Vice Chairman General Chi Haotian (66 years old) joined the PLA as a young man in the early 1940s during the "twin struggles" against the Nationalists and the Japanese and he spent a good part of his later career as a senior political commissar. The constant self-propagandizing within the PLA to remind the troops that the party is the focus of their loyalty has not abated as the Chinese armed forces have attempted to become a more proficient military force. Indeed, the first among the "Five General Requirements" issued by Jiang Zemin for all PLA soldiers is to be "politically qualified."[22]

It is often argued that today the party and ideology mean less and less in China. For the general populace and even some civil cadre that may be truer than not. However, the party and ideology still count for quite a lot in the PLA. The "new nationalism" that has been ascribed to the PLA is not necessarily at the expense of the Party and it may just be that in today's China the PLA is the only institution over which the CCP "center" in Beijing still has near total control from one end of the nation to the other. The constant public affirmations of the PLA's top leaders of the primacy of the Party is no more empty rhetoric than the statements of the top U.S.

---

[20]From late 1987 to early 1989 there were unconfirmed rumors in the Hong Kong press that the role of the PLA as a "state" or "national" army was going to be emphasized and analysts were watching to see if the personnel of the state and party military commissions would be "split out" to reflect this change. If there were in fact any plans to do this, they were quashed in the tense political atmosphere beginning with the September 1988 challenges to political and economic reform mounted by the left which were fueled by exceptional inflation and fears of instability, challenges to the party mounted by intellectuals (exemplified by the television series "*He Shang*," or "*River Elegy*"), and ultimately the student demonstrations that began in April 1989 and led to the showdown at Tiananmen in June. Since that time the PLA has reaffirmed its subordination to the party and continues to do so.

[21] The State Military Commission (MC) is identical in leadership to that CCP's CMC and its authority is invoked only in state or constitutional matters.

[22]The Five General Requirements are constantly cited in the PRC and PLA media. They call for the troops of the PLA to be politically qualified, militarily proficient, have a good "work style," practice strict discipline, and provide strong logistic support.

military leadership affirming the primacy of the Constitution and the principle of civilian control over the military.

In August 1997, PLA Academy of Military Sciences Commandant General Xu Huizi encapsulated many of the above thoughts in an article in *China Military Science*.

> After entering the new historical period, Deng Xiaoping repeatedly stressed that our military situation is that of the party commanding the gun, instead of the gun commanding the party; the military must follow the party's instructions, and must not at any time wave its own flag. . . . Comrade Jiang Zemin referred to the principle of the party commanding the gun as our military 'soul.'[23]

For these reasons, the PLA must be prepared to defend the CCP with military force against domestic challenges as well as external threats. Over the past few years much of the internal security mission of the PLA has passed to the People's Armed Police (PAP). However, the PAP is ultimately under the control of the CMC and the "regular" PLA has not been absolved from its requirement to provide for the defense of the party. One of the many reasons the PLA was called upon to converge on Beijing in the Spring of 1989 was because the PAP was incapable of handling a situation that was viewed by the CCP leadership to be burgeoning into a direct threat to the rule of the regime and the CCP.[24]

The PLA also pursues the related military objective of safeguarding internal stability. Some domestic challenges and threats may be aimed at the state, not the party per se. Examples would be acts of violence by separatist factions in non-Han China such as in Tibet or Xinjiang. Still other threats to stability with which the PLA must deal are natural disasters and manmade disasters. In these situations the PLA is often called upon to provide disaster relief, internal humanitarian assistance, as well as fight back the forces of nature such as floods or large-scale fires.

At the risk of being guilty of unsubstantiated generalization, I would offer that the top leadership of the PLA remains a very conservative group, which is virulently opposed to any domestic circumstance that could lead to instability (*luan*). The up-and-coming generation of leaders (colonel and above) are probably no less so. The chaos of the Cultural Revolution touched many of this latter generation personally and tragically and they fully understand the cost China paid in terms of development and modernization. And if in their hearts some dismiss as irrelevant Communist ideology, they may still see the CCP as the only political force in China that can successfully keep the nation together and lead it into the future.

Defend Sovereignty and Defeat Aggression. This national military objective brings us to the classical warfighting mission of the Chinese armed forces. Militaries exist to fight and so does the PLA.

---

[23] Xu Huizi, "Some Facts Concerning Our Historical Experiences in Building Quality in the PLA," *Zhongguo junshi kexue*, No. 3, August 1997.

[24] Once the student protests were labeled "counter-revolutionary" (meaning anti-CCP), the PLA was given the ideological justification for the use of force.

It is worth pointing out that the PLA considers itself a defensive force. Its leaders often declare that China since 1949 has never fought a war of aggression and has gone to war only when other nations have attacked China first or threatened its territory. Officially, Chinese will argue that their intervention in Korea (1950), the war against India (1962), the Sino-Soviet clashes of the late 1960s, and the incursion against Vietnam (1979) were all conflicts foisted upon China in defense of its sovereignty. Hence, they would couch their warfighting objective in terms of defense against aggression and the preservation of the nation's sovereignty.[25] Moreover, the PLA rejects the notion of fighting as part of a formal alliance, proudly claiming that no Chinese combat troops are stationed on foreign soil, nor does Beijing desire to do so.

However, as mentioned earlier, China's definition of "sovereignty" is much broader than just the sanctity of its borders. The PLA defense of China's sovereignty also includes being capable and prepared to employ force to achieve national unification and assert Beijing's maritime rights. Chi Haotian has commented directly on these points in the past. In a May 1996 interview he was reported to comment that China's national defense policy "is aimed at protecting China's territorial land, waters, and air space as well as China's maritime rights and interests against foreign aggression. It is also aimed at safeguarding China's unity. . . .[26] In the last decade, Beijing's use of the PLA to assert its claims to resources in the South China Sea has raised a great deal of concern within the region and beyond. As far as China's unification goes, only the issue of Taiwan remains unresolved. Although Beijing claims it prefers to settle this issue in a peaceful manner, China's leaders will not renounce the use of force.[27] This puts a tremendous amount of pressure on the PLA for obvious reasons. As we review China's analysis of its security environment later in this chapter, it will become clear just how much the PLA must accomplish as a professional military force.

Modernize the Military and Build the Nation. The modernization of the military deserves its own place among the PLA's national military objectives. The top civilian and uniformed leadership of the PRC consider China's overall modernization and the modernization of the military (often referred to as "army building") to be mutually dependent and supporting national objectives. "Strengthening national defense and army building," declared Jiang Zemin at the 15th Party Congress, "is the basic guarantee for national security and the modernization drive."[28]

---

[25]In March 1998, the president of a very prestigious Chinese think-tank commented to the author that China goes to war only over issues of sovereignty. He added that, "even when the correlation of military forces is obviously not in our favor China will still go to war over the issue of sovereignty."

[26]"Orientation of Chinese Army's Future Development—Exclusive Interview With Chinese Defense Minister General Chi Haotian," *Kuang chiao ching*, No. 284, May 16, 1996, in FBIS-CHI-96-106, May 16, 1996. Hereafter, "Orientation."

[27]In conversations on this point Chinese military officers and security analysts argue that if Beijing renounced the use of force "splittist elements" on Taiwan would attempt to declare independence and China would be forced to intervene militarily. Hence, according to their logic, a renunciation of the use of force by China would be destabilizing.

[28]"Political Report," op. cit.

The Chinese believe that without a modernizing and capable military the nation will not be able to enjoy the security from external threats it requires to continue to concentrate on economic reform. Without a capable military China cannot secure the internal stability it requires to modernize in the civil sector. Without a capable and modernizing military China cannot hope to secure any of its national security objectives in the realms of sovereignty or unification. Finally, without a modern PLA whose capabilities ultimately comport with China's (anticipated) economic, technological, and political strength, the PRC will not be able to take its seat at the table of world leaders in Beijing's much hoped for multipolar international order.

While both civil and military leaders subscribe to these views, the tension in the system revolves around the fact that military modernization is not proceeding at the same pace as economic modernization. Moreover, by long-standing political fiat going back to the beginning of the Dengist period, the decision *not* to invest state treasure in military modernization at the expense of economic reform still stands. Jiang Zemin reaffirmed this tenet at the 9th NPC in March 1998. Speaking to a full session of the PLA's delegation to the NPC, Jiang lectured that:

> It is also very important to correctly handle the relationship between economic development and the building of national defense. Building a modernized army and national defense is a guarantee for the country's safety and modernization drive. This is something the whole party and the whole nation always care very much about. The level of China's productive forces is still not high, and our economy is not that strong. Therefore, we must concentrate our energies on economic development. *Without a highly developed economy, it is also impossible to promote the modernization of national defense and the army. We must always insist on taking economic development as the central task while paying adequate attention to modernizing national defense.* . .[29]

Outwardly, at least, the top PLA leadership supports this basic line and does not question its validity. A serious challenge to the party leadership on this account should not be anticipated. Yet, as the men who are responsible for modernizing the military in a world (as we will later see) which they view as basically hostile, they do evince concern about how far behind economic development military modernization can lag. For example, General Xu Huizi has cautioned that "we must also understand that national wealth does not equate to military strength; there are many examples both past and present, in China and overseas, where a nation has been wealthy while its military was weak. We must try to keep our military quality development level at a level compatible with national defense security; keeping it at a level appropriate to China's international status as a great nation is a choice on which we must insist."[30] Defense Minister Chi Haotian made a more direct commentary on this point in 1996 in a lengthy article in *Qiushi*. "The building of

---

[29]"Chairman Jiang Zemin Stresses at PLA Delegation Meeting: Army Must Adapt Itself to the New Situation of Reform and Development and Subject Itself to and Serve the Country's Overall Situation with Enhanced Awareness," *Xinhua*, March 11, 1998, in FBIS-CHI-98-070, March 11, 1998. Emphasis added.

[30]Xu Huizi, "Some Facts Concerning Our Historical Experiences in Building Quality in the PLA," op. cit.

national defense," stated Chi, "cannot exceed the limitation of tolerance of economic construction, *nor can it be laid aside until the economy has totally prospered.*"[31]

As a result of this contradiction, the leadership of the PLA is going to have to be very focused in how it approaches the modernization of the military. The decisions they make (and are making) in this regard will to a large extent drive some of the major national military strategic concepts in their overall NMS.

The corollary to "modernizing the military" is "building the nation." Not only is the PLA's modernization expected not to be a drain on the state's coffers, but military modernization is also expected to enhance the economic, scientific and technological level of the state. The military, says Chi Haotian, is expected to do so by leading and participating in "key state projects," by bringing science and technology to the rural areas of China, by transferring superior military technologies to the civilian sector, by providing the nation with a pool of PLA veterans who are technically advanced relative to their civilian counterparts (the PLA as a "big school"), and providing infrastructure to the hinterlands.[32]

These, then, are the broad national military objectives the PLA seeks to achieve. To achieve these goals China's military planners need to articulate strategic concepts, or courses of action. Before they can do that, however, they must think through the environment in which they must operate, now and in the future. They must employ a "filter" that highlights both challenges and opportunities.

VI. THE STRATEGIC FILTER

At a minimum, the "strategic filter" for military planners at the national level must address three issues. First, they must account for political decisions that have been handed down to them or that have been brokered with civilian counterparts. These decisions usually encompass fiscal decisions and affect allocation of resources. Political decisions can also dictate or constrain military courses of action. Second, military planners must survey the security environment and assess what threats, current and future, must be addressed. Third, the nature of warfare must be examined to determine what kinds of engagements on the spectrum of military conflicts are most likely to occur. We attempt to use a "strategic filter" to look through Chinese eyes and attempt to see their strategic planning environment as they do. In some cases, we have definitive Chinese statements on these issues. In others, we can only try to transform ourselves into a PLA staff officer (*canmouguan*) sitting in the General Staff Department and speculate in an informed manner.

Political Decisions. Three key political decisions made by the top Chinese leadership affect what strategic military concepts are acceptable or possible for the PLA's

---

[31]Chi Haotian, "Taking The Road," op. cit. Emphasis added.

[32]Chi Haotian, "Orientation," op. cit. This is certainly not unique to China. In many parts of the developing world the military has often been on the leading edge of technology and built the national infrastructure. This was true of the United States in its early years after independence when the Army provided the nation with its engineers, its leaders of technology, and its captains of industry.

national planners. These political decisions are not recent developments but long-standing policies.

The first is that China does not participate in formal military alliances. This means the PLA must plan on being a "go-it-alone force" when it does go to war or when the military element of national power is employed. It also means that PLA planners do not have to concern themselves with questions of interoperability with potential allies or even temporary coalition partners. This goes for doctrinal issues as well as hardware and combat systems which, in any case, the Chinese would prefer to produce indigenously. It explains as well the deep reluctance of the PLA to accept invitations from foreign militaries to participate in combined training exercises even when these events are couched in terms of serving as confidence-building measures.

A second political decision in effect is that the PLA will not station combat forces abroad (trans., stationed on foreign soil). The reasons for this are a combination of geopolitical reality (who would invite the PLA to permanently station combat troops on its soil?) and budgetary considerations (it is expensive to station troops abroad).[33] Moreover, since China does not engage in formal military alliances, such an eventuality is difficult to imagine. Of greater significance, this policy also dictates that the defense of China must take place close in, on, or near China's actual land borders or off the Chinese littoral. These decisions make "forward presence" for the Chinese a relative concept. It also dictates that the PLA must maintain a certain scale or size, especially in its ground forces, to compensate for vast borders and the need to fight close to them.

The third political decision with which PLA planners must grapple has already been mentioned—that the funding the PLA requires to operate, train, maintain and especially to modernize the armed forces will be subordinated to other national priorities. We will not enter the great PLA budget debate here. The jury is still out on the real value of investments China makes in national defense, especially given the opaque nature of their system, the multifurcated streams of funding that find their way to the armed forces, the obviously low official figures, and the issues surrounding the "hidden budget" that is the result of PLA domestic and global enterprises. The numerical spread in foreign estimates of the real value of the Chinese defense budget continues to inspire an awe of its own. For 1994, the range of estimates ran from about $10 billion U.S. to $149 billion U.S.[34] For our purposes the trend is what matters. The recent "official" defense budget increase the PLA received during the 9th NPC (12.8 percent, for an official budget of 90.9 billion Yuan) indicates that while defense spending is going up yearly, no fundamental political decision has been made to reverse national priorities and provide the armed forces

---

[33]Even after the Korean War, the PLA pulled out of North Korea. Probably the last time China had forces stationed "abroad" when it was not at war itself was during the Second Indochina War when PLA air defense troops assisted in the defense of North Vietnam, especially Hanoi, against American air strikes. India has recently accused China of stationing elements of the PLA in Burma. This remains to be verified. But even if it turns out to be true, it is not on the scale of forward presence to which we are alluding.

[34]See Richard A. Bitzinger, "Military Spending and Foreign Military Acquisitions by the PRC and Taiwan," in James R. Lilley and Chuck Downs (eds.), *Crisis in the Taiwan Strait*, Ft. McNair, Washington, D.C.: National Defense University Press, 1997.

with a massive infusion of state funds. The PLA leadership undoubtedly does not receive the funding it would like (in this regard they have much in common with their brothers-in-arms across the globe). The question that we still have difficulty answering is: Are they getting what they need to fund their key current programs and underwrite their future requirements?

The PLA Analysis of the Security Environment. The basic and oft-repeated Chinese articulation of the nature of the world security environment is well known. Beijing sees the next two or three decades as a period in which the possibility of world war is negligible. The basic trend is toward "peace and development." China faces no immediate major direct military threats and the world is generally at peace. It is this assessment, originally made by Deng Xiaoping, that has impelled China to concentrate on economic reform as its primary national security objective. China continues to have, and continues to need, a window of relative peace to experiment and move forward in this endeavor. Jiang Zemin revalidated this assessment at the 15th Party Congress:

> The international situation at present as a whole continues to move toward relaxation, and peace and development are the main themes of the present era. . . . For a fairly long period of time to come, it is possible to avert a new world war and secure a favorable, peaceful international environment and maintain good relations with the surrounding countries.[35]

At the same time, Jiang also pointed out that there are serious security problems that require China's attention:

> However, the Cold War mentality still exists, and hegemonism and power politics continue to be the main source of threat to world peace and stability. Strengthening military alliance between various military blocs is not conducive to safeguarding peace and ensuring security. The unjust and irrational old international economic order is still infringing upon the interests of the developing countries, and the gap in wealth is widening. It is still serious that human rights and other issues are used to interfere in the internal affairs of other countries. Local conflicts due to ethnic, religious and territorial factors crop up from time to time. The world is not yet tranquil.[36]

Jiang's "caveat" is a relatively subdued echo of PLA concerns. For their part, Chinese military strategists seem to view the world as a place basically hostile to Beijing's national interests, especially China's sovereignty. It is a world where dangers to national security lurk everywhere. The strategists view competition between nations for advantage as the norm and as a zero-sum equation (*ni si wo huo*). Change in the global and regional security environment is viewed as constant and usually dangerous. The absence of war does not mean the absence of hostility toward China. And, over the horizon, today's much-needed trade partners can slowly transform into serious economic, political, and military rivals.

---

[35]Jiang Zemin, "Political Report," op. cit.

[36]Ibid.

China's Defense Minister General Chi Haotian is Beijing's top military spokesman on these issues and he continues to be quite straightforward and frank in publicly highlighting the dangers that PLA planners see bubbling beneath the surface of a relatively peaceful world. As recently as March 1998, he underscored that pockets of antipathy toward China continue to require military vigilance:

> Hostile international forces have never abandoned their strategic plot to westernize and split China, and the great cause of the motherland's ultimate reunification has yet to be accomplished. Under the long, peaceful environment and the situation centering on economic construction, we must be prepared for danger in times of peace and enhance our awareness of hardship. We must not become intoxicated by songs and dances in celebration of peace.[37]

Fair enough. It is the job of defense ministers around the world to remind the troops that preparedness is why they are paid. What is interesting is that in this particular talk to PLA delegates at the 9th NPC Chi couches the need for preparedness not just as a matter of general professional responsibility but as a result of clearly identified security challenges.[38]

In April 1996, General Chi spoke of the competitive nature of world relations when he wrote that "Major countries successively take contending for economic and technological superiority and enhancing comprehensive national strength as their development strategies and try by every possible means to gain the strategic initiative in the 21st century."[39] These types of official statements set the analytic tone for PLA strategic planners. But what are the specific concerns these planners see that might affect the formulation of national military strategic concepts, especially those that must be incorporated into their national military strategy? We will now survey them briefly. What we will find is that if one were a Chinese military planner contemplating the various possible contingencies for which military force might be required now or in the future, there would be very little reason to feel complacent. What does the PLA planner see when he looks around the map? What are the major near-term, mid-term, and long-term concerns and shaping factors?

Geography. Geography continues to be a critical factor for Chinese military planners. China shares land borders with fourteen other nations. All told, the Chinese claim more than 20,000 kilometers of land boundaries and 18,000 kilometers

---

[37] "Chi Haotian Warns Against Hostile International Forces," *Zhongguo tongxun she*, March 8, 1998, in FBIS-CHI-98-067, March 8, 1998.

[38] When one encounters these types of statements there are always the questions of how much is rhetoric, how much is deep-seated belief, how much is aimed at the PLA, how much is aimed at civilian leaders, and how much is intended for foreign consumption. I certainly have no definitive answer to offer on this count. But I do believe, and this is admittedly quite subjective, that this quote by General Chi is representative of the general feeling of apprehension and insecurity PLA military officials have when they survey their own security environment. The combination of their own historical baggage as "victims," their acknowledgment of their relative backwardness as a military force relative to the rest of the developed world and compared to some sectors within China itself, and what they believe is the heavy burden of defending China all combine somehow to make their top leaders prone to making statements positing extreme danger.

[39] Chi Haotian, "Taking the Road," op. cit.

of coastline that must be defended.[40] Moreover, these borders encompass almost every major type of climate and terrain known in the habitable world. Just being able to secure its land borders is an almost staggering proposition for the PLA. It should be pointed out that since 1949 all of China's wars or military campaigns (Korea, India, Indochina-Vietnam, Soviet Union) have been fought either to secure control over disputed portions of those borders (India, Soviet Union) or to preempt a potential threat to those borders (Korea, Indochina-Vietnam). But the new international security environment and proactive diplomacy by Beijing have guaranteed that threats to China's land borders remain minimal. Rapprochement with Russia has alleviated a tremendous pressure on the PLA. Moreover, the dissolution of the USSR broke up what was previously the longest land border in the world. Cautious but steadily improving relations with Hanoi continue to keep the Sino-Vietnamese border pacific. The 1996 five-nation agreement between China, Russia, Kazakhstan, Tajikistan, and Kyrgyzstan is addressing problems in the northwest. If China still has a major potential border security problem it is with North Korea, which we shall come to shortly. But for the most part, China's land borders have never been more secure than they are today.

Defense of the Economic Center of Gravity and the New Maritime Imperative. The same cannot be said of China's coast. In the last twenty years there have been two significant changes in China's security environment. The first is rapprochement with Russia. The second, ironically, is the result of two decades of successful Chinese economic modernization. China's economic center of gravity has shifted from deep in the interior, where strategic industries were sequestered in the 1960s to protect them from a possible Soviet invasion, to its current location on China's eastern coast, where Beijing's new market economy is strongest. This "gold coast," from Dalian in the north to Hainan in the south, defines the current economic *schwerpunkt* of China and its likely future location as well. And, arguably, Beijing's ongoing and future potential success in developing this coastal economy will be a major factor in defining China's importance in the future international order.

This shift has resulted in a profound change in the PLA's security calculus. Previously a large land force was required to protect the Chinese interior and its industries against a protracted land war with invading Soviets. Today, however, this pronounced shift in the economic center of gravity presents China with a littoral and maritime defense requirement that it probably has not had since the mid-Ch'ing Dynasty. For the PLA today (and more than likely tomorrow) *the essence of defending China will be defined by the PLA's ability to defend seaward from the coast in the surface, subsurface and aerospace battle-space dimensions.* This is precisely the type of warfare that the PLA is currently least well postured to conduct. Hence, the emphasis in military modernization over the past few years has been on naval forces, air forces, and missile forces—the three services whose force projection capabilities are required for and best suited to defend the new economic center of gravity. Because of this shift it could be argued that even if the PLA did not have a Taiwan

---

[40]"Chi Haotian on Defense Policy, Taiwan." Speech at U.S. National Defense University, December 10, 1996.

"problem" to consider, it would still emphasize force projection capabilities with a maritime (and maritime airspace) control and denial focus.

Defense of Maritime Sovereignty. In line with China's new coastal defense requirements, it is clear that the PLA has other maritime issues to contend with. Besides their 18,000 km of coast, the Chinese claim to have "more than 6,000 islands, and three million square km of territorial waters . . ." to defend.[41] The PLA must develop the capability to enforce Beijing's claims over disputed areas in the South China Sea and other maritime claims. While the waters have been relatively calm since the "Mischief Reef" incident, China could decide to employ the military element of national power once again if, from its perspective, it feels provoked by the other claimants in the region. Besides the Spratly Islands, China has competing claims with Vietnam over the demarcation of the Gulf of Tonkin and with Japan over the Diaoyu/Senkaku Islands. The issue of the South China Sea also highlights the importance China places on maritime resources (oil, fishing, and minerals) as it proceeds with economic modernization. And it is worth highlighting again that in his articulation of China's national military objectives earlier in this paper Defense Minister Chi Haotian specifically cites the need of the PLA to safeguard China's "waters and maritime rights and interests." These two requirements—defending China's maritime territorial claims and defending its maritime resources—are specifically included in the articulation of the objectives of the PLA Navy by its commander, Lieutenant General Shi Yunsheng.[42]

Close to Home: The Prospect of Instability. PLA and PAP military planners cannot discount the possibility that their classic internal defense mission will become more important in the next decade. As mentioned earlier, the social dislocations attendant to China's economic modernization are becoming greater and greater. Over the past few years worker protests, sit-ins, and physical attacks on local party officials and headquarters have been reported in the press with increasing frequency. As the social safety net continues to erode, by political mandate or by virtue of poor management, the possibility of increasing social instability cannot be discounted. China already has a "floating population" of some several millions which the state security apparati blame for increasing crime. If Jiang Zemin and Zhu Rongji actually implement the reforms of the state-owned enterprises as they have promised at the 15th Party Congress and 9th NPC, then over the next three years some three to five million additional Chinese could become unemployed and be without any social benefits at all. The "Great Amway Riots" of April 1998, as I like to call them, could be a harbinger of worse situations to come.[43] And should the Asian financial crisis eventually hit China hard, the social chaos could be totally unmanageable.

---

[41] "The Chinese Navy Moves From 'Coastal Defense' To 'Ocean Defense'," *Zhongguo tongxun she*, March 20, 1997, in FBIS-CHI-97-080, March 21, 1997.

[42] "Shouldering the Important Task of a Century-Straddling Voyage—Interviewing Newly Appointed Navy Commander Lieutenant General Shi Yunsheng," *Liaowang*, February 24, 1997, in FBIS-CHI-97-054, February 24, 1997. The "strategic mission of the Navy in the new period: Contain and resist foreign aggression from the seas, defend China's territory and sovereignty, and safeguard the motherland's unification and marine rights."

[43] I refer, of course, to the disturbances that reportedly occurred when Beijing unilaterally outlawed direct marketing in China.

But beyond the social unrest that could take place, China has other internal problems that must certainly keep PLA planners alert. In Xinjiang, the Uighur separatist movement has not been crushed. It is difficult to determine just how large or minuscule the problem is in fact. Last year several "terrorist" attacks in Beijing and other parts of Han China were reportedly tied to the East Turkistan freedom movement. We do not know if that linkage is true. But for the first time that one can remember (or that has been reported outside of the mainland) China now has a terrorist problem and the Chinese certainly believe the Pan-Turkic problem is potentially serious. This means that the PLA will need to have contingency plans on file if the PAP and other state security forces cannot handle future unrest in Xinjiang Province.

And always there is Tibet. The on-again, off-again unrest in that region is certainly a part of the PLA planning calculus. The issues of Xinjiang and Tibet directly attack Beijing's national security objectives of sovereignty and stability. Consequently, there is no reason to think that Beijing would not employ the PLA in either combat or military operations other than war (MOOTW) in either of these locations if its capabilities were needed to quell serious unrest.

Taiwan. Taiwan represents both a near-term contingency and a long-term readiness problem for PLA planners. Beijing will not renounce the right to use military force against Taiwan. Consequently, the PLA must develop a serious conventional deterrent capability vis-à-vis Taipei. The problem, of course, is that the only credible military deterrent is real conventional capabilities (we assume the political will is there). The next question must be, capability to do what? Whether the March-April 1996 mini-crisis in the Strait was a success or failure for Beijing is still being debated abroad. But one thing is certain; if one were a Chinese military planner, it should be painfully obvious that the PLA needs to develop a more diversified set of conventional military capabilities to employ as flexible deterrent options (FDOs). Relying heavily on missiles as a FDO runs the risk of being too provocative (witness the U.S. reaction). This is true whether the PLA must act against Taiwan or in defense of its claims in the South China Sea. In the 1996 affair, the PLA quickly escalated from "forward presence" (large military exercises on the coast) to "demonstration of force" (missile launches). It appears that either the PLA did not have many other options in between with which to *slowly* escalate or they totally miscalculated the external response. My own inclination is that it was both.

Russia. Russia should probably be viewed as a long-term potential problem for the PLA. In spite of increasingly friendly relations and their new "cooperative strategic partnership," there is no dearth of mutual distrust between Russia and China, especially in Siberia and the Russian Far East. China is still viewed by the Russians as a threat and these views have their political champions on the far right in Moscow. Beijing's economic strength at a time when Moscow is relatively supine is felt most sharply in Russian Asia, and the Maritime Provinces recoil at what they see as Moscow's kowtowing to Beijing at their expense. For their part, Chinese military planners likely view Russia as a possible "over the horizon" problem. They may worry about the day when the Russians recover their national strength and the imperatives of geography and the legacies of historical animosity act to unleash rivalries anew, not just on their common border but perhaps in Central Asia. These rivalries will probably be driven by competition for natural resources: lumber, oil, minerals, and food from the sea.

Korea. Chinese military planners discount the possibility of an all-out war in Korea á la 1950. Nevertheless, they should be worried about the security problems China could face if the economic and political situation in North Korea deteriorates to the point of a total meltdown. An implosion in the north has the potential to send tens of thousands of North Korean refugees streaming over the border into China's ethnically Korean provinces. This type of chaos on China's border is precisely the type of situation for which the PLA is historically deployed. But in this sort of scenario the PLA would not be fighting a classic war. China's armed forces would be engaged in MOOTW. They would have to secure their borders, deal with thousands of refugees, and possibly keep the Korean People's Army at bay simultaneously. The scenarios are endless. The point is, the PLA must be prepared to intervene unilaterally to secure China's own interests in the event of total chaos on its border with the Democratic People's Republic of Korea (DPRK).

Japan. The Chinese view Japan as the only Asian nation that has the political, economic, technological, and (potential) military strength to challenge Beijing for regional dominance in the next century. Japan is likely viewed by the PLA as the primary mid-term strategic concern. While it may seem irrational from an American perspective, Beijing believes that under the veneer of a generally pacifist Japan lurks an undercurrent of unrepentant Japanese militarism and a strong desire on the part of the Japan Self-Defense Forces to overturn constitutional limits on their roles and missions. Tokyo's reluctance to directly confront its role in World War II and Japanese military participation in United Nations Peace-Keeping Operations (PKOs) in Africa and the Middle East fuel these Chinese suspicions. However, *the* most worrisome development from the perspective of the PLA is the recent promulgation of the revised U.S-Japan Guidelines for Defense Cooperation. This instrument is viewed as a codicil under which the Japanese armed forces, especially the navy, will justify operating even further out from Japan proper while still claiming to be a force purely for the defense of the Japanese home islands. Even more damning, China considers Tokyo the "other black hand" behind Taiwan's drift toward independence. This is because of a variety of reasons that go back to Taiwan's status as an Imperial Colony (1895–1945)[44] and Lee Teng-hui's ties to Japan. These two factors (Chinese suspicions of Japanese military expansion and the Taiwan factor), when combined with maritime territorial disputes, makes for an imposing set of military planning requirements for PLA staff officers.

India. Up until the May 1998 nuclear detonations by India and their accompanying barrage of anti-China rhetoric, relations between Beijing and New Delhi seemed to be on the road to mending. The fact that the nuclear detonations occurred shortly

---

[44]For an overview of Taiwan's experience as a Japanese colony and the historical antecedents of the Taiwan independence movement (the "Formosan Home Rule Movement") during the colonial period see Chapter 2 in David M. Finkelstein, *Washington's Taiwan Dilemma, 1949–1950: From Abandonment to Salvation*, Virginia: George Mason University Press, 1993.

after the visit to India of GSD Chief General Fu Quanyou was a double slap in the face. It is still too early to say what impact these events will have on overall Sino-Indian relations or what impact the Indian detonations will have on China's nuclear policies. These questions notwithstanding, the Indians have handed PLA planners a set of potential problems that encompass both a readiness issue on land in the disputed border areas and possibly a mid- to long-term maritime-aerospace planning problem in the Indian Ocean. At a minimum, India will continue to be viewed as an enduring security concern.

The United States. The United States poses a special conundrum for the PLA. China needs good relations with the U.S. to achieve its overarching national security objective of modernization. But this should not be confused with a shared vision of how the post–Cold War security regime in Asia should unfold. For one thing, the Chinese do not subscribe to the U.S. argument that Washington's bilateral military alliances in the region are necessarily stabilizing.[45] Moreover, from the viewpoint of PLA planners, the forward presence of the American armed forces throughout the Pacific and Asia on land, sea and air cannot be a happy fact of life. The Pacific Ocean is still an American lake and the looming shadow of the U.S. Navy casts a pall over all others. It is in the interests of China and the PLA to see the U.S. military presence in Asia reduced at some point in the future. There is probably little debate in Beijing on this basic point. Where there may be significant debate is on the timing of a U.S. force withdrawal.

Some Chinese security analysts could argue that the American military presence has some utility for China by acting as a guarantor of the regional stability Beijing must have. They might argue that a quick drawdown by the U.S. would result in Tokyo filling the military vacuum quicker than the PLA could be prepared to credibly face down the Japanese. Others might see utility in the continued U.S. military presence in Korea as a check on instability close to home, although that argument will disappear after Korean unification or reconciliation.

Yet, there is a residual distrust and apprehension in some circles in the PLA about the true intentions of the United States in Asia and the role of its armed forces in the Pacific. The United States, through its forward military presence, has the *potential* to act as the great spoiler to two of Beijing's *core* security concerns: Taiwan and Japan. Because the United States underwrites the security of Taiwan, some Chinese security analysts argue that Taipei can continue to be recalcitrant in negotiating cross-Strait political issues and reckless to the point of provocation in its foreign and domestic policies. Moreover, it is the United States, some PLA planners would argue, that is goading the Japanese to rearm and pressuring Tokyo to expand its military role in the region under the false flag of increased host-nation burden sharing. Moreover, the "China Threat" theory that was in vogue in the U.S. a couple of years ago had its

---

[45]During his visit to Australia in February 1998, General Chi Haotian gave a significant speech at the Australian College of Defense and Strategic Studies in which he offered a Chinese vision of what the future Asia-Pacific security regime should be. The talk, which *Xinhua* hailed as a "New Security Concept," implicitly criticized the U.S. Asian security system of bilateral alliances (i.e., with Japan, Korea, Australia) and even the expansion of NATO as destabilizing. See "Chi Haotian Calls for 'New Security Concept'." *Xinhua*, February 17, 1998, in FBIS-CHI-98-048, February 17, 1998.

Chinese analogue. Some PLA analysts seriously believe that the U.S. is attempting to contain China militarily using Japan as the "northern anchor" and Australia as the "southern anchor." And Chinese opposition to NATO expansion is borne of a fear that the Partnership for Peace might spread to Central Asia and exponentially enhance U.S. influence on China's western doorstep.[46]

Consequently, PLA views of the U.S. are highly dichotomous. On the benign side, the U.S. is probably viewed by PLA strategists not so much as a *direct* military threat to Beijing, but as a lumbering but lethal giant that can wreak havoc on China's various national security interests because of (as viewed through Chinese lenses) vacillating American domestic political tacts and ignorance of the greater strategic implications of Washington's ever-changing policies. On the more cynical side, the U.S. is seen as capable of undermining core interests concerning Japan and especially Taiwan. More than likely, the PLA must constantly factor the potential reaction of the U.S. into almost every military contingency plan it may have. Assuming there are a variety of PLA contingency plans targeted against Taiwan, the "U.S. factor" is a significant unknown variable for the PLA. Beijing's suspicions about Tokyo's future path and the vagueness of the phrase "areas surrounding Japan" in the U.S.-Japan Guidelines for Defense Cooperation have probably only heightened concern about the U.S. role.

Nuclear Threats. Beijing's decision to develop nuclear weapons decades ago was in part a function of the desired status that capability conferred upon the PRC in the world order. However, Chinese security analysts are also quick to point out that China had no choice but to possess atomic weapons because China had been subjected to "nuclear blackmail" in the past—first by the United States during the Korean War and later on by the Soviets. For the past thirty years, Beijing has adhered to a nuclear strategy of minimum deterrence[47] and has publicly declared a no-first-use policy. It would appear that China's leadership has been satisfied with this approach. But whether this will remain the case indefinitely may be open to debate, certainly within China. One reason is the recent nuclear detonations by India. Second, over the past few years there has been growing concern among China's nuclear scientists and Beijing's security policy establishment over the possibility that the U.S. might in fact develop a national ballistic missile defense system and concerns about what China views as U.S. pressure on Russia to amend the Anti-Ballistic Missile Treaty to allow high-capability Theater Ballistic Missile Defense systems. The prospect of India having offensive nuclear capabilities, and the U.S. possessing highly capable high-altitude ballistic missile defenses cannot be a

---

[46]It was not surprising that the Chinese press attacked the U.S. Atlantic Command's combined U.S.-Russian-Kazakh exercise, CENTRAZBAT-97, in September 1997. The implications of U.S. 82nd Airborne Division troops jumping into Kazakhstan after what was hailed as the longest flight to a jump zone in the history of airborne operations was probably read much differently by the PLA than USACOM staff planners.

[47]Iain Johnston has posited that China's strategy of nuclear deterrence may be more correctly termed "limited deterrence." For a full reading of his hypothesis and its implications see, Alastair Iain Johnston, "China's New 'Old Thinking': The Concept of Limited Deterrence," *International Security*, Vol. 20, No. 3, Winter 1995/96, pp. 5–42.

comforting combination for those charged with responsibility for China's nuclear strategy.

China's Security Environment: A Recapitulation of the View From Beijing. In reviewing a PLA planner's view of China's security environment, we can come to the following generalizations:

- The PLA must be prepared to deal with internal unrest every day.

- The PLA must be prepared with military options for China's leaders to consider in dealing with Taiwan should the national leadership decide to employ the military element of national power to achieve its political ends. This is a current requirement and will endure.

- The PLA must develop a credible defense of its economic center of gravity: the coast. It must also be prepared to enforce Beijing's maritime claims.

- Any bilateral security concern that involves China with another country on its land border (India, Korea, Vietnam, Russia, etc.) should be considered an enduring security concern regardless of how pacific the situation is at the moment or promises to remain in the future.

- Russia is a long-term and enduring security concern for Beijing due to proximity, historical mistrust, and its potential to regain its great power status.

- For the foreseeable future, the United States remains an enduring security concern not because it is perceived as a direct military threat to China but because of its unpredictability, its power, the proximity of its military forces, its web of bilateral military alliances, and its potential role as "spoiler" for core Chinese security interests (Taiwan, Japan).

- Japan is probably the one country in the region which in the mid-term Beijing views with the most suspicion as a potential challenger in the military as well as political and economic realms.

- As for force structure and mix of arms, the PLA must enhance its maritime and aerospace capabilities. At the same time, because of the continuing possibility of internal unrest and current (India, Korea) and potential (Russia, Vietnam) security concerns along China's long borders, the PLA cannot totally neglect its ground forces.

- Finally, China must continue to field a credible nuclear deterrent, especially in light of India's recent actions and Chinese concerns about the potential for the U.S. or Russia to acquire credible ballistic missile defense systems. (See Table 1)

Analysis of the Operational Environment. The final element we will consider in the strategic filter is the PLA's assessment of the nature of warfare—what kinds of conflicts or wars PLA planners believe they have to be able to fight. This question is about the "what" of warfare, not necessarily the "who." For PLA planners—indeed, for all military planners around the world—this is a critical question because it

124

	NEAR-TERM CONTINGENCY	MID-TERM CONCERN	LONG-TERM CONCERN	ENDURING CONCERN	LAND FOCUS	AIR FOCUS	NAVAL FOCUS	NUCLEAR FACTOR
TAIWAN	X			X		X	X	*
INTERNAL UNREST	X			X	X			
DEFENSE OF ECONOMIC COG	X			X		X	X	
JAPAN		X		X		X	X	*
KOREA	X			X	X	X		
RUSSIA			X	X	X	X	X	X
INDIA	X			X	X	X	X	X
MARITIME CLAIMS	X			X		X	X	
VIETNAM				X	X	X	X	
UNITED STATES	**			X		X	X	X

*   Only to the extent that the United States might become involved.

**   Only to the extent that the United States might become involved in a Taiwan contingency.

This table is a shorthand way to review what this author posits are the likely military planning requirements confronting PLA staff officers who are tasked to develop what in the U.S. military planning system (JOPES) would be CONPLANS and OPLANS.

The first four columns place security concerns in a timeframe. While the "years" associated with each is purely subjective (see bullets below), the timeframes themselves reflect, I believe, the current PLA analysis. Thinking through the timeframes of potential threats usually shapes both short-term capabilities-based strategies and longer-term force developmental-based strategies. Moreover, thinking through timeframes also assists planners in making decisions about "out year" force structure and systems requirements.

- Near-Term Contingency: A military operation the PLA might very well have to perform today.
- Mid-Term Concern: A potential security problem 5 to 15 years out from today.
- Long-Term Concern: A potential security problem 15 to 20 years out from today.
- Enduring Concern: A potential security concern that is so fundamental that it persists over time.

The last four columns reflect which battle-space dimension(s) would likely be present in the case of each security concern. The "informational" dimension is not listed because it is always present in one form or another. The point of limiting the dimensions as I have is to highlight a way to think about which types of conventional forces the PLA must develop for each threat. The "Nuclear" dimension must be listed because of the number of potential scenarios in which China may come up against nations who possess nuclear weapons. This column does not imply China would use nuclear weapons; it means PLA planners have to consider this issue in their planning because it could constrain their freedom of action.

Table 1—PLA Threat Planning Matrix

affects the dedication of scarce resources, force structures, manning levels, equipment priorities and training methods. In most militaries, this is as much an internal and bureaucratic issue as it is a "warfighting" issue.

Among PLA defense intellectuals, the question of what kinds of wars the armed forces will have to fight and when generates a tremendous amount of study, research and publication. This question also generates a modicum of debate. As we know, "contending schools" on military issues is not new to the PLA. Different visions of what kind of wars the PLA must prepare for and how they will be fought have been going on since the days of the Red Army and the Long March, through the "Red versus Expert" debates of the 1950s and 1960s, and even through the 1980s when there was resistance to abandoning Mao's classic prescriptions for People's War (hence such politically correct but contradictory doctrines such as "People's War Under Modern Conditions"). In these previous debates the professional and personal stakes were extremely high, with losers often purged.

These days, however, the internal debates in the PLA about which kind of war(s) to prepare for are no longer live or die issues. Indeed, a benchmark of the level of professionalism that the PLA has achieved is that there are so many voices in the military journals and military press on this subject. Today the debates do not center about the political orthodoxy or heterodoxy of a point of view but around the questions of how much the PLA can accomplish, what it must accomplish, how quickly it can accomplish it, how much the PLA can afford, and whether one possible path should be funded at the expense of another.[48]

Today in the PLA there appears to be two major schools of thought on this question which live side by side. The first school of thought is that the PLA must prepare to fight "local wars under modern, high-tech conditions." The second school of thought looks out over the horizon and argues that the PLA must prepare to deal with the international "Revolution in Military Affairs" (RMA) of the 21st Century.[49]

While the PLA's top leadership is cognizant of both types of warfare and, presumably, would aspire to be able to deal with both, there is a profound difference between them. First, PLA strategists see local wars under modern, high-tech conditions as a form of warfare with which they must grapple now and well into the next century. This is because it is a form of warfare that many highly developed nations can conduct now, albeit with differing degrees of proficiency, while the PLA still cannot. Second, it is a form of warfare that is relatively well defined in both theory and practice. It has been seen in action in its incipient form (the Gulf War), and is intellectually digestible to a large portion of the top warfighting PLA leadership.[50]

---

[48]It is worth noting that for foreign analysts of the PLA this question has generated a bit of debate as well, for how one answers the question "what kind of war is the PLA preparing to fight?" seems to lead to other conclusions about the PLA.

[49]For an interesting and provocative essay on how key non-Western nations view the RMA, see Dr. Ahmed S. Hashim (Major, U.S. Army Reserve), my colleague at the Center for Naval Analyses, "The Revolution in Military Affairs Outside the West," *Journal of International Affairs*, Vol. 51, No. 2, Spring 1998.

[50]Nanjing Military Region Commander Lieutenant General Chen Bingde specifically refers to the Gulf War as an example of modern local war. See Chen Bingde, "Intensify Study of Military Theory To Ensure Quality Army Building: Learning From Thought and Practice of the Core of the Three Generations of Party

Third, there are discrete and finite measures the PLA can take in incremental steps to achieve some degree of ability to engage in this type of warfare. Put differently, there are metrics that can be used to peg achievable benchmarks of proficiency. And last but not least, Jiang Zemin, in his capacity as Chairman of the Central Military Commission "officially" coined this terminology in 1993 and instructed the PLA to use it as the guiding principle for military reform in the "new historical period."

This is not to say that the PLA leadership is not interested in the RMA or thinking about it or even experimenting with some of its concepts. The problem for the PLA is that the RMA is still a moving target. The concepts behind it are still evolving. Even in the United States, where it generates a great deal of thought among military futurists, it continues to be a subject hotly debated in various circles. This is precisely why the PLA's best military scholars and thinkers are paying close attention to the RMA debates in the U.S. and why they are even developing their own concepts of an "RMA with Chinese characteristics."[51] In some circles of military theorists in China there are even some who argue that it might be possible for the PLA to "leap-frog" the local wars stage of development and exploit technologies to become a mid- to late-21st century military. These are appealing ideas to a military organization such as the PLA that currently is so vast, so unevenly modernized, relatively underfunded, and, from their perspective, so perpetually threatened. Theoretical work on the RMA will continue.

The RMA, however, will not be the fundamental basis of PLA planning, training, professionalization, restructuring, or equipping for the first decades of the next century. Preparing for local wars under modern, high-tech conditions will serve as the foundation and focal point upon which and around which China will build a PLA capable of coping with the challenges of warfare in the first half of the next century. PLA views of local wars under modern, high-tech conditions is a subject too vast to do justice to in this paper. Nevertheless, a brief overview is in order.

As mentioned earlier, the decision to have the PLA focus on "local wars under modern, high-tech conditions" (hereafter LWUMHTC) as the basis for its future operational planning is officially credited to Jiang Zemin in a directive to the CMC in 1993.[52] The fact that this directive is credited to Jiang was but one of many moves he was making at the time to establish his *bona fides* as Chairman of the CMC (no doubt with Deng's approval). Jiang, as we know, had little to do with the military prior to his remarkable rise to national power in the wake of Tiananmen in June 1989.

The concept of LWUMHTC was the natural outgrowth of a previous major 1985 decision that revised the PLA's assessment of the kind of warfare they would have to conduct as major changes to the security environment unfolded. This was, of course,

---

Leadership in Studying Military Theory," *Zhongguo junshi kexue*, No. 3., August 1997, pp. 49–56, in FBIS-CHI-98-065, March 6, 1998.

[51] For an excellent sampling of Chinese military thinking on the RMA, see Michael Pillsbury (ed.), *Chinese Views of Future Warfare.*

[52] See Kuan Cha-chia, "Commander Jiang Speeds Up Army Reform, Structure of Three Armed Services to be Adjusted," *Kuang chiao ching*, No. 305, February 16, 1998, in FBIS-CHI-98-065, March 6, 1998. Hereafter "Commander Jiang."

Deng Xiaoping's "strategic decision," which declared that "early, major and nuclear war" (with either of the two superpowers) was unlikely and that the PLA now faced "local, limited war (*youxian jubu zhanzheng*)."[53] By the late 1980s the concept had a qualifier added to it, "under modern conditions" which began to take new technologies into account to a certain degree. But it was Jiang Zemin's iteration in 1993 that added "high-tech (*gao jishu*) conditions" to the formulation. So in fact, the PLA's concept of "local war" has been evolving for over ten years. It will undoubtedly continue to evolve, especially as PLA researchers continue to track and digest the "Military Technological Revolution."

When the PLA speaks of LWUMHTC they are describing two aspects of warfare. The first aspect, "local," means that this kind of war will be limited in geographic scope. It will be confined to one particular theater of operations and not a general war on all fronts. The second qualifier, "under modern, high-tech conditions," means that the weaponry the PLA expects a notional advanced enemy to employ against it will be of the most sophisticated technological types. Over the past five years there has been an explosion of PLA writing on the nature of LWUMHTC, and the list of the operational characteristics that define it gets longer all the time. For the sake of brevity, the list below captures the essence of the new operational environment PLA strategic theorists believe China now faces and will continue to face through the turn of the century.[54]

According to PLA military theorists, Local Wars Under Modern High-Tech Conditions are (will be) characterized by:

- Limited geographic scope

- Limited political objectives

- Short in duration

- High-intensity operational tempo

- High mobility and speed (war of maneuver)

- High lethality weapons and high destruction

- High in resource consumption and intensely dependent upon high speed logistics

---

[53]For a review of the events that led to this decision and the defining characteristics of local limited war, see Paul H.B. Godwin, "Force Projection and China's National Military Strategy," in C. Dennison Lane, Mark Weisenbloom, and Dimon Liu (eds.), *Chinese Military Modernization*, Washington, D.C.: The AEI Press, 1996, pp. 69–99.

[54]Chen Bingde, op. cit. According to Nanjing Military Region Commander LTG Chen Bingde, "Jiang Zemin formulated the military strategic guideline in the new period at the beginning of 1993 and made the major policy decision to lay the foundation of military preparedness on winning victories in local warfare under the conditions of modern technology, especially high technology." Chen also mentions in his article that in June 1991 Jiang attended a Gulf War symposium hosted by the Academy of Military Sciences. Although tentative, this allows us to speculate that observing cutting-edge U.S. technologies in action against Iraq (and a good amount of Chinese equipment purchased by Iraq) served as a wake-up call to those in the PLA leadership who still argued that the superior human qualities of the Chinese soldier ("will") could prevail, as in days of yore, against much more modern adversaries.

- Highly visible battlefield (near-total battlefield awareness)

- High speed C2 and information intensive

- Nonlinear battlefields

- Multidimensional combat (all battle space dimensions: land, aerospace, surface, subsurface, informational)

- Joint operations.

To prepare the PLA for the next major wave of reforms, GSD Chief General Fu Quanyou went to great lengths in a March 1998 article in *Qiushi* to provide his own view of the nature of future local wars under modern, high-tech conditions, warning that "meeting the challenge of world military developments is an historical responsibility that we cannot avoid." Going further, he argued:

> Along with the development of sophisticated technology, particularly the development of information technology and its widespread application in the military realm, the high-tech content of future wars will become greater and greater. With respect to the new forms of combat, it will primarily be new combat forms such as information warfare, air strike warfare, missile warfare, and electronic warfare. With regard to weapons and equipment, the focus will be on the development of digitized and smart equipment with new and sophisticated technology and long-range, precision-strike capabilities. With respect to structural organization, the trend is toward combined forces that are small and diversified. In the area of command and control, we will see the widespread application of C3I and C4I systems, holding administrative levels to a minimum and improving effectiveness. With regard to combat support, there will be an increased reliance on modern technical means to provide rapid, accurate, quality and complete support.[55]

Evolving Operational Doctrine: The Active Defense. Finally, an important concomitant of the evolving Chinese analysis of the operational environment over the years has been the evolution of the PLA's operational doctrine—the basic and fundamental principles that guide the employment of military forces. As Paul Godwin reminds us, the PLA has been aware for quite some time that modern military technologies developed abroad have been changing the nature of the operational level of war.[56] Indeed, the doctrine of "Active Defense" (*jiji fangyu*) predates the 1985 strategic decision by almost four years.[57] By 1981 it was already apparent to Deng and the PLA leadership that luring the enemy deep and fighting a protracted war of attrition in the Chinese interior was no longer acceptable and that the PLA would have to stop an invading army as far forward as possible. "Active Defense" has been the official operational doctrine of the PLA ever since.[58] But

---

[55]Fu Quanyou, "Make Active Explorations, Deepen Reform, Advance Military Work in an All-Round Way," *Qiushi*, No. 6, March 16, 1998, in FBIS-CHI-98-093, April 3, 1998.

[56]Godwin, op. cit., p. 72.

[57]Hwang Byong-moo, "Changing Military Doctrine of the PRC: The Interaction Between People's War and Technology," *The Journal of East Asian Affairs*, Vol. XI, Number 1, Winter/Spring 1997, pp. 221–266.

[58]Two points about the term "active defense." First, the Chinese refer to "active defense" as their *military strategy*. I have chosen to refer to it and identify it as a *doctrine* to maintain parallelism with U.S. terminology for comparative purposes. Were you to ask a PLA officer to identify China's "military strategy"

again, the interaction of the current security environment (no threat of a Soviet invasion, but many potential peripheral conflicts) and the new operational environment (highly lethal local wars under modern, high-tech conditions) is changing the fundamentals of how the PLA is thinking about employing force on the battlefield (which is meant to include all the battle space dimensions).

The characteristics of the "Active Defense" doctrine as adapted to LWUMHTC are still evolving and, like the subject of LWUMHTC itself, there is no dearth of Chinese military writing on what the doctrine should actually entail. What has been officially adopted is still an open question. Nevertheless a representative list of common characteristics is warranted and I believe it is useful to also indicate the previous maxims to appreciate the magnitude of the change in operational thinking that is going on inside the PLA:

- From luring deep to fighting forward[59]

- From a war of annihilation to a campaign against key points

- From a war of attrition to a decisive campaign with a decisive first battle

- From waiting for the first blow to deterring the first blow by force

- From a defensive campaign to an "offensive defense" campaign

- From "advance and retreat boldly" to checking the initial enemy advance

- From a "front army campaign" to a "war zone" campaign

- From the principle of mass to the principle of concentration of firepower

- From four single service campaigns to joint campaigns.

These are the attributes of a proactive doctrine much more forward-leaning than in the past.

VII. NATIONAL STRATEGIC MILITARY CONCEPTS

By way of review, National Military Strategic Concepts are broad, overarching statements which provide the general outline of *how* a military plans to achieve its National Military Objectives.[60] NMSCs, therefore, should be linked to NMOs. They should be articulated at a level of broad generalization to allow military planners a high degree of flexibility in developing the specific plans, programs and capabilities that are required to enable the National Military Strategy. They are building blocks

---

he would certainly say "active defense." Second point, The PLA Navy (PLAN) usually refers to this doctrine as "Active Offshore Defense" (*jijide jinhai fangyu*).

[59]Major General Gao Guozhen and Senior Colonel Ye Zhen, "Gao Guozhen and Ye Zhan View Operational Doctrine Since 1980's," *Zhongguo junshi kexue*, November 20, 1996, in FBIS-CHI-97-066, November 20, 1996. These authors, from the Academy of Military Science, provide an informative overview of doctrinal evolution in China. This list is adopted from their article.

[60]I have added the adjective "National" to "National Military Strategic Concept" as a shorthand reminder that we are dealing at the level of grand strategy as opposed to the operational level of war. In the U.S. national military strategy, the Joint Staff simply uses the term "strategic concepts."

and a useful shorthand that can provide a quick conceptual overview of what the armed forces deem important and how they basically plan to fight, deter, and develop capabilities.

Depending on the priorities of a military organization, NMSCs can be used to describe not only operational (warfighting) matters but critical programs of a more systemic or administrative nature. Moreover, it is the synergistic relationship among NMSCs, doctrine, and security imperatives that drives the decisions for the development of military capabilities (hardware, software, personnel), which in turn generate the myriad of tasks that a military trains for to attain certain levels of professional competency. The process of analysis, conceptualization, program development, and evaluation is iterative. Therefore, while one can capture a "snapshot" of how NMSCs fit into a larger NMS, we keep in mind that the image is in motion and subject to change based on either radical alterations in the security environment or fundamental political decisions.

Having developed the PLA's National Military Objectives, thinking through the Strategic Filter, and reviewing the PLA's key doctrinal principles, we can now state what should be the four basic NMSCs that drive practically all the PLA programs we read about. The four NMSCs are: *nuclear deterrence, political work, forward defense, and army building.*[61] Each of these NMSCs is linked to NMOs (see Figure 2) and the combination of the four enable the overall NMS. Moreover, each of the four NMSCs has a component that is directed internally as well as the outward orientation that militaries usually take.

Nuclear Deterrence. Nuclear weapons are, and will continue to be, the mainstay of Chinese strategic deterrence. This NMSC directly supports the NMO, "Defend Sovereignty & Defeat Aggression." It does so by deterring nuclear threats against China proper and, indirectly, injecting a "nuclear factor" into the resolution of outstanding sovereignty issues, such as Taiwan and border problems with other nuclear states. Moreover, China's nuclear force and the scientific establishment that supports it indirectly support the NMO, "Modernize the Military & Build the Nation," thanks to the technological crossover effect that nuclear weapons technology and know-how has for the overall scientific development of China. One could also argue that China's possession of nuclear weapons—its only real *global* power projection capability—serves as the ultimate antidote to a conventional military force whose *sustainable* reach is barely regional.

As Iain Johnston has pointed out, the Chinese view the possession of nuclear weapons not just as a military necessity, but a prerequisite for international stature.[62] As a member of the "Nuclear Club," Beijing accrues tremendous prestige and some degree of political leverage in the international order. When coupled with

---

[61]These four NMSCs are not authoritatively listed by the PLA as such. We are using a foreign model to analyze a Chinese system. They do, I believe, capture the essence of what drives the PLA.

[62]Iain Johnston, "Prospects for Chinese Nuclear Force Modernization: Limited Deterrence versus Multilateral Arms Control," *The China Quarterly,* June 1996, p. 550. "China's decision-makers have generally accorded a great deal of status and military value to nuclear weapons."

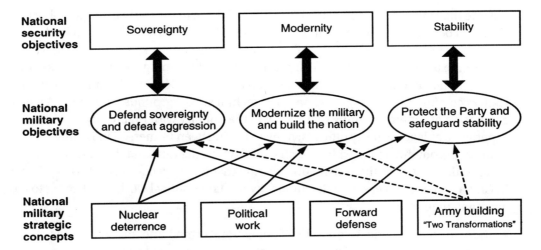

Figure 2a

the PRC's permanent seat on the U.N. Security Council, China is able to claim "major power" status despite its extremely low level of domestic development relative to the other permanent UNSC members. Moreover, China's nuclear status has an internal component: the development and possession of a nuclear arsenal are touted by the regime as a credit to the scientific and technological prowess of the Chinese nation *and the leadership of the CCP in particular.*

Even though the conventional PLA is making great strides toward becoming a more competent military force, the Chinese nuclear arsenal has the potential to become even more important to Beijing. China's nuclear neighborhood changed dramatically when India coupled its recent nuclear detonations with incantations of "China as threat number one."[63] Additionally, the Chinese are carefully watching developments in the U.S. theater missile defense (TMD) program, on-going debates about a future U.S. national missile defense program, and U.S.-Russian negotiations for adjustments to the Anti-Ballistic Missile Treaty. Insecurity on the part of the PLA about the future credibility of its nuclear deterrent (capabilities and doctrine) should be of great concern; greater, perhaps, than the concern often evinced about enhancements to the Chinese conventional forces.

Political Work. Military Political Work (*Junduide zhengzhi gongzuo*) remains an important tool of the PLA leadership. It deserves to be rated as a NMSC by virtue of the fact that it supports the two NMOs, "Protect the Party & Safeguard Stability," and "Modernize the Military & Build the Nation."

Political work is the key link between the PLA and the CCP. As mentioned earlier, the first mission of the PLA is to protect the CCP, and "absolute loyalty to the party" (*dangde juedui lingdao*) is non-negotiable. Although the PLA is becoming more "professional" and less "political" (meaning, less involved in internal political issues

---

[63]While one could argue that the Chinese always suspected India had a nuclear capability and that therefore nothing has changed, I would argue that from the perspective of a PLA military planner the reality of that nuclear capability coupled with anti-China rhetoric is not to be taken lightly.

outside the realm of security issues), the PLA's linkage to the party through political work will become more important because the "Third Generation" civilian leaders of the CCP (Jiang Zemin, Li Peng, Zhu Rongji, *et al.*, have no background in military affairs or experience in the PLA. The passing of Deng Xiaoping signaled the end of an era when the top party leaders had also been top PLA leaders. With China's civilian leaders coming from purely state, party, or technical backgrounds, and the PLA's new generation of generals coming to primacy in the military with practically no experience as leading party officials, or even having been immersed in Beijing politics, the potential for a widening chasm between the CCP and the PLA is increased. Add to this the general observation that ideology counts for less and less in China and one can easily understand why military political work is not going to go away any time soon.

At the same time, political work is itself being transformed into an asset for the modernization of the PLA. The General Political Department is working to rationalize the PLA personnel system to meet the needs of a more professional and technically competent officer corps. Also, under the rubric of political work comes the adjustment of the PLA's military justice system and working the critical and increasingly problematic issue of civil-military relations (what I refer to as the problems of "host-region support").

Political work is going to become increasingly critical for another reason directly related to modernization and in support of the NMO "Modernize the Military & Build the Nation." The next wave of PLA structural reforms (which will be discussed subsequently) are promising to generate some resistance from the top PLA leadership. Political work will be necessary to explain what these reforms are, why they are necessary, and that the personal and professional sacrifices involved must be endured. It is of no little significance that in March 1998 Chief of the GSD General Fu Quanyou warned that "At present, there is still quite a bit of old thinking and concepts in our army building and military work," and that:

> Implementing a new size and structure will inevitably involve immediate interests of numerous units and individual officers and men. Leading organizations must take the lead in unifying our thinking, in paying attention to the overall situation, in implementing the new size and structure, in enforcing various disciplinary items, in *resolutely overcoming selfish departmentalism and overemphasis of local interests*, by no means permitting discussions of prices or bending the rules.[64]

Political work is the vehicle that will be used to smooth the way, overcome resistance, and rally support for the changes coming down the road.

Forward Defense. "Forward Defense" is more in line with traditional Western notions of what NMSCs are supposed to describe, e.g., "warfighting" concepts. "Forward Defense" directly supports the NMO "Defend Sovereignty & Defeat Aggression." It also supports "Protect the Party & Safeguard Stability" when placed in the context of internal challenges to the regime. This particular NMSC is an

---

[64]Fu Quanyou, *Qiushi*, op. cit. Emphasis added.

invention on the author's part, not an authoritative or official PLA articulation.[65] The concept of "Forward Defense" takes into account the previous analyses of how the PLA views its security environment, its operational environment, and the imperatives of the "Active Defense" doctrine. It is also descriptive of what we have been able to read in PLA professional journals and newspapers.

"Forward Defense" means that the PLA prefers to fight a military conflict as far away from China's borders and coastline as is possible. No enemy will consciously be permitted to move onto mainland China, control Chinese airspace, or dominate the Chinese littoral and the adjacent seas if they can be stopped. But this is currently problematic for the PLA due to limited force projection capabilities, weak logistics, an uncertain degree of combat sustainability, and no overseas basing. Therefore, while the "Active Defense" doctrine calls for a new emphasis on offensive operations, the PLA, by virtue of the four factors above, is almost always on the defense and tied to interior lines of communication. In other words, while the offense is being given new emphasis at the operational and tactical levels of war, strategically the PLA is on the defense. Consequently, the next best alternative to "power projection" and "overseas presence" (two U.S. strategic concepts) is "defending" (engaging the enemy) as far forward on land, on sea, and in the air as is possible.

For the ground forces, this means concentrating the best PLA forces in critical Military Regions and being able to rush key units forward to engage an enemy as quickly as possible. Hence, the doctrinal importance of the PLA's "rapid reaction units" (*kuaisu fanying budui*). An example of the former imperative is the emergence over the past few years of the Nanjing Military Region (MR) as critical because of the PLA's requirement to maintain military pressure on Taiwan, a near-term contingency and enduring security problem.

For the PLA Navy this means more green water and blue water training and presence as well as extending the navy's reach as far out to sea as possible. As Paul Godwin informs us, PLAN strategists and planners have been engaged in working out the details of a "forward deployed navy" (my term) since the late 1980s, when they were directed to shift their planning from fighting naval engagements close to the Chinese littoral (coastal defense, *jinhai fangyu*) to fighting further out at sea (offshore defense, *jinyang fangyu*).[66]

The implications for the PLA Air Force (PLAAF) and PLAN Air Forces (PLANAF) are equally clear and require no detailed explanation: get out further and linger in the combat zone longer. But the NMSC of "Forward Defense" also serves to highlight why the Strategic Rocket Force (SRF), also known as the Second Artillery, is an increasingly important service arm for the PLA. It may just be that the PLA missile force is currently the only service arm which can actually enable the NMSC of

---

[65] I have chosen the word "Defense" because the PLA would probably prefer it as a politically correct term given official Chinese pronouncements that China never initiates hostilities and only uses military force to defend against aggression. The fact that the word "Defense" is used does not mean that the PLA will not consider pre-emptive strikes after the commencement of hostilities. The new characteristics of the "Active Defense," specifically "offensive defense," legitimizes such an action doctrinally.

[66] Paul Godwin, "From Continent To Periphery," op. cit.

"Forward Defense" in any credible manner today. This, of course, creates a different set of problems for the PLA in MOOTW. Using the SRF for a "flexible deterrent option," such as "demonstration of force" (as was the case in March 1996), could provoke foreign conventional military responses which the rest of the PLA is not yet ready to meet. Nevertheless, the SRF will continue to serve as the point of the spear for "Forward Defense" while the remainder of the PLA continues to enhance its capabilities to enable this NMSC.

There are two other points to make about "Forward Defense" as it pertains to engaging a foreign force. First, this NMSC is consistent with the recent spate of foreign weapon systems purchases made by the PLA (KILO-class submarines, SU-27 fighter aircraft, and *Sovremenny-class* destroyers). These acquisitions are clearly intended to serve as near-term, "quick-fix" capabilities to shore up the currently constrained ability of the PLAN and PLAAF to meet the requirements of "Forward Defense," which is a *strategic* imperative.[67] A second point to make is that rapprochement with Russia has the potential to free up a good deal of the Chinese armed forces from their previous dispositions close to the northern border. Consequently, analysts should be sensitive to major shifts in basing to other more insecure areas on the Chinese periphery in support of this NMSC.

The internal component of "Forward Defense" also applies to those regions in China where unrest, separatist activities, or porous and unstable border situations exist—primarily in Tibet, Xinjiang, and the border with North Korea. Consequently, the Shenyang, Chengdu, and Lanzhou Military Regions will probably continue to be important commands, but they may not require state-of-the-art military capabilities to deal with internal unrest. Just being there *en masse* to support the PAP if needed will be sufficient.

Army Building. The fourth NMSC is "Army Building" (*jundui jianshe*, sometimes translated as "Military Construction"). "Army Building" is a general term, but it is a critical NMSC for the PLA and for foreign students of the PLA. "Army Building" refers to the sum total of the policies, programs, and directives that guide the current reform (*gaige*) as well as the ongoing and future modernization (*xiandaihua*) of the armed forces of the PRC. "Army Building" cuts across all facets of military issues to include:

- Doctrine

- Organization and force structure

- Personnel matters

- Individual, small unit, and large unit training

---

[67]One often hears arguments that these purchases were made with the Taiwan problem in mind. That is probably true. But these types of purchases are equally applicable to other potential near-term contingencies, such as in the South China Sea. Nevertheless, trying to explain the purchase or development of military capabilities by looking solely at the tactical or even the operational levels of war can only take one so far. There must be a strategic context as well and it appears that the generic NMSC of "Forward Defense" provides such a context.

- Political work

- Resource allocations

- Weapons (hardware) requirements

- The development of specific military capabilities

- Reserve and militia utilization

- Mobilization and demobilization issues

- Defense industrial policies

- Civil-military relations.

"Army Building" is where the strategic vision for the PLA is articulated. That strategic vision is then translated into concrete programs, directives and policies under the rubric of "Army Building." Therefore, "Army Building" directly supports all three NMOs. Moreover, "Army Building" has a direct impact on the other three NMSCs. Studying this NMSC is the key to understanding *how* the PLA plans to modernize. It outlines what kinds of *National Military Resources* the PLA must develop to enable the rest of the current National Military Strategy and anticipated changes to it in the future.

To a certain degree, then, "Army Building" represents a "strategy" within China's National Military Strategy. "Army Building" is often confused with China's larger National Military Strategy (the focus of this paper), because it cuts across so many facets of PLA and PLA-related issues. It does represent a substrategy, because it deals directly with military modernization (a subject most PLA-watchers focus on), and certainly because it is the most discussed PLA issue in the Chinese military press and journals.

With these general comments as background, we can now be more specific about the current "line" for "Army Building."

The "Two Transformations." Since late 1995, "Army Building" has been guided by a remarkably ambitious line, "the Two Transformations" (*liang ge zhuan bian*). This guidance is attributed directly to Jiang Zemin in a speech made to an enlarged meeting of the CMC in December 1995. It is currently considered "the military strategic guideline" for "Army Building," in the "new period" and a "cross-century task."[68]

The "Two Transformations" call for the PLA to transform itself, (1) from an army preparing to fight local wars under ordinary conditions to an army preparing to fight and win local wars under modern, high-tech conditions, and (2) from an army based on quantity to an army based on quality. A corollary which usually accompanies

---

[68]Chen Bingde, op. cit., and Zhang Qinsheng and Li Bingyan, "Complete New Historical Transformations—Understanding Gained From Studying CMC Strategic Thinking on 'Two Transformations'," *Jiefangjun bao*, January 14, 1997, in FBIS-CHI-97-025, January 14, 1997. One often sees this term as "the two basic transformations" or the "two basic conversions" (both being given as *liang ge jiben xingzhuanbian*). For the sake of brevity I shall use "two transformations."

136

these two imperatives in PLA literature is that the PLA must also transform itself from an army that is *personnel intensive* to one which is *science and technology intensive* (S&T) (emphasis added).[69] It is readily apparent that the specific plans and programs that the PLA develops to enable this "new line" will cut across every facet of military affairs listed above.

While the linkage of the "Two Transformations" to LWUMHTC and the new military technologies emerging abroad is clear and direct, it is not clear how or if the "Two Transformations" are expected to change PLA doctrine (e.g., Active Defense). PLA writings are confusing on this point. Some articles link the "Two Transformations" directly to the Active Defense doctrine for the here and now and see them as enablers. Others talk about the "Two Transformations" as the way the PLA will be able to cope with the Military Technological Revolution in the West in the near to mid term. Still other PLA military theorists invoke the RMA when discussing the long-term durability of this program. In fact, all may be at work. The passage below in *Jiefangjun Bao* (January 1997) is representative of the various planes to which the "Two Transformations" are often linked in theoretical discussions.

> The military strategic principle for the new period (i.e., the "Two Transformations") has epoch-making significance in the history of our army's development. There have been several strategic transformations in our army's history; *various strategic principles of active defense with different content were set forth based on the new tasks for military struggles.* Their focuses were mostly a selection of operational guidance and means of conducting comprehensive war under ordinary technical conditions. *The strategic principle for the new period proceeds from the high-tech condition of the information era; it guides the preparations for military struggles in the new period and the employment of military force in future wars, while guiding the building and development of military force.* The former requires converting to a foothold to winning high-tech local wars, while the latter requires proceeding from improving quality and efficiency, and increasing the content of science and technology (emphasis added).[70]

It is also unclear how long the PLA leadership believes it will take to accomplish these "Two Transformations." The PLA literature is clear that this is a process that will go on for quite a few years. But for how long? In PLA articles discussing the "Two Transformations" one runs across vague terms such as "cross-century task" and "long-term historical process." My own sense from reading the literature is that the PLA has linked this metamorphosis to Jiang Zemin's goal of China becoming a "moderately developed country" by 2049, the centenary of the People's Republic of China.

What we can state with certainty is that the "Two Transformations" principle is now a politically mandated "line" accompanied by the usual rallying support of the senior

---

[69]This latter point, of course, also parallels Jiang's recent calls during the 9th NPC for the entire Chinese people to raise their general level of S&T education. It should be pointed out that while Jiang is linked with the formulation of the term "Two Transformations," the essence of this program may in fact belong to Liu Huaqing, who had called for quality and science and technology since the early 1990s.

[70] Zhang Qinsheng and Li Bingyan, op. cit.

PLA leadership (the NMSC, "Political Work"). Once again, Lieutenant General Chen Bingde:

> The formulation of the military strategic guideline in the new period and the decision to effect the "two transformations" in army building are explicit characteristics of the times. They reflect the objective need of army building in the new period. They are not only a major strategic choice, but also a theoretical achievement of significant importance. They indicate that our army will enter a brand new period of development.[71]

In theory, then, the "Two Transformations" is the PLA's shorthand for the multitude of national military reforms, programs, plans and policies that will enable the PLA to accrue the **National Military Resources** required to successfully fight and win on the high-tech battlefield as they have analyzed it. Ultimately, the programs under the "Two Transformations" must turn the *operational requirements* of the high-tech battlefield into *operational capabilities*.[72]

A military force does not have an operational capability until it can actually perform the operational requirement to standard. The process of turning a *requirement* into a *capability* is a complicated process at the strategic and operational (theater and campaign) levels of warfare. It is the result of the synergy that accrues when the critical elements of a military system are *developed* and then *integrated* holistically to focus on a particular warfighting requirement or set of requirements. The possession of weapons systems or "hardware" is necessary but by no means sufficient to develop an operational capability. At a minimum, systems (hardware and software), doctrine, personnel, and force structures must all be developed and then integrated and synthesized through fielding, testing and training before an operational capability results. These four major elements as well as the budgets that enable them are the basic NMRs of any military force (See Figure 3).[73]

At this point, we have arrived at our final destination; articulating the major components of a notional "National Military Strategy" for the PLA and explaining the rationale for them. As a "cross-century" NMS, the PLA will strive to achieve three national military objectives through the implementation of four national military strategic concepts. For comparative purposes, Table 2 and Table 3 juxtapose the

---

[71]Chen Bingde, op. cit. It is also worth pointing out that the "Two Transformations" is, I believe, another vehicle being used to validate Jiang Zemin's military *bona fides* by crediting him with the credentials of a military theorist. Because the "Two Transformations" are sometimes juxtaposed next to Deng's 1985 "strategic decision" as the two most important theoretical analyses in military strategy during the "new period," we infer that Jiang is being elevated to near-Deng status. But the political utility of this "line" for Jiang should not cloud our thinking about the momentous consequences for the modernization of the PLA *if* this program, or even parts of it, can actually be implemented over time.

[72]At the national level of military planning, operational requirements are couched in broad terms. Generic examples would be: sea denial, strategic lift, amphibious warfare, special operations, information warfare, joint operations, or long-range air interdiction, to name just a few.

[73]This model is based on the work of Kenneth Kennedy at the Center for Naval Analyses who originally developed the basic structure. I have modified his model by adapting it to the particular circumstances of the PLA and adding the need for "Force Structure." Additionally, I have added to the general model the linkage to NMSCs (in this case, the "Two Transformations" for the PLA), and a top block underscoring that, for the PLA, "Operational Requirements" are the result of Chinese analysis of LWUMHTC. Finally, I have added "testing" and "fielding" to what was originally a block containing only "Training."

elements of this notional PLA NMS with stated U.S. national security objectives (1997 and 1994) and elements of the published U.S. NMS for both 1997 and 1995.

As we continue to track the progress of the PLA in attaining its goals, we should focus on the critical elements shown in Figure 3. Some elements will be easier to track and analyze than others. For example, major weapons purchases eventually become known. Major personnel policies are usually publicized thanks to political work. But other critical variables will remain unknown to analysts because of the closed nature of the Chinese defense establishment. For example, we may become cognizant of major training exercises but we cannot easily measure their success in integrating all the elements "to standard." Nor can we be certain of the efficacy of new command and control policies ("software") or even command and control systems ("hardware"). Nevertheless, being aware of major policy decisions and their relationship to this process will provide key benchmarks for analysis. This brings us to the most recent military policy decisions emanating from the 15th Party Congress and the 9th NPC.

VIII. THE 15th PARTY CONGRESS AND THE 9th NPC

This paper began by asserting that the statements on defense issues emanating from the 15th Party Congress and the 9th NPC may well indicate that the PLA is about to embark on its next great wave of reforms. If the various statements by Jiang Zemin and other PLA leaders at these two events are to be taken at face value, and their exhortations are not mere rhetoric, then over the next few years we may witness significant changes in the Chinese defense establishment, especially in the realm of force structure and organization. In the notional NMS we have constructed, the strategic concept "Army Building" may be given increased emphasis and specific policies and programs under the rubric of the "Two Transformations" may be launched.

The central theme of the 15th Party Congress was the need for dramatic structural reform of the Chinese economy, especially in State Owned Enterprises (SOEs). Yet, for PLA watchers there were three potentially significant outcomes in the relatively small section of Jiang's political report dealing with defense issues. The first, of course, was the most obvious: the announcement of a half-million-man reduction over three years. The significance of the announcement is as much in the realm of the "potential" as the "actual." Specifically, the last major reduction of personnel in the PLA (one million men, announced in 1985) was part of a larger package of significant structural changes. These included the reduction of the number of Military Regions from eleven to its current seven (announced in 1985) and the creation of the Group Army system (first announced in 1983 but continuing in implementation through 1985).[74] Consequently, if history is any indicator, all antennae should be on alert for other major structural reforms down the road to accompany the 500,000-man reduction.

---

[74]Godwin, op. cit.

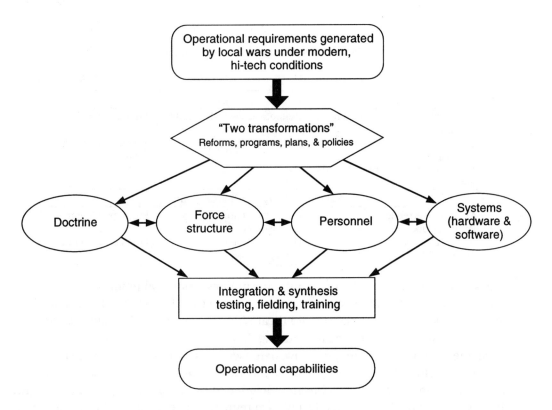

This diagram depicts in very simplistic form the four basic elements of a military system that are required to translate an operational requirement into an operational capability. The PLA analysis of the operational environment under Local Wars Under Modern, Hi-Tech Conditions identifies the operational requirements that must be developed. The NMSC, the "Two Transformations," articulates the reform programs that will be instituted to develop the National Military Resources required. The minimum required NMRs are Doctrine, Force Structure, Personnel, & Systems. An operational capability cannot be realized until they are integrated to be able to perform to standard. At the operational level of warfare this process requires inter-service coordination (jointness). At the tactical level of warfare this process must be repeated hundreds of times for a myriad of tasks and skills.

Figure 3

Table 2

U.S. and PRC National Security Objectives

Notional PRC National Security Strategy	May 1997 U.S. National Security Strategy	July 1994 U.S. National Security Strategy
Sovereignty	Enhance security	Enhance security
Modernity	Promote prosperity	Promote prosperity at home
Stability	Promote democracy	Promote democracy

Table 3

PRC and U.S. National Military Strategies

Notional PRC NMS	1997 U.S. NMS	1995 U.S. NMS
National Military Objectives		
Protect the Party and safeguard stability	Promote peace and stability	Promote stability
Defend sovereignty and defeat aggression	Defeat adversaries	Thwart aggression
Modernize the military and build the nation		
National Military Strategic Concepts		
Nuclear deterrence	Power projection	Power projection
Political work	Decisive force	
Forward Defense	Overseas presence	Overseas presence
Army-Building	Strategic agility	

The second potentially significant outcome in Jiang's political report was also relatively stark. After the necessary exhortations to the PLA to "uphold the party's absolute leadership" (the NMSC, "Political Work"), Jiang went through a list of tasks the PLA must implement. The first among them was that the PLA "should implement a military strategy of active defense."[75] As mentioned earlier in this paper, the Active Defense doctrine has been around for some time. Yet, specific mention of the Active Defense was not included in Jiang's political report to the 14th Party Congress in 1992.[76] Usually, political reports refer to army building issues and seldom comment on operational issues. Having the CMC chairman list it as a requirement for the PLA in a Political Report gives it the aura of an uncontestable political decision, not merely a necessary military imperative, thereby ending any debate there may have been within the PLA over this doctrine and, by extension, validating the NMSC I have termed "Forward Defense." Moreover, it implies that the PLA will have to become serious in developing the operational capabilities that this doctrine demands under LWUMHTC.

The third significant outcome of Jiang's comments on national defense is probably less obvious. When reviewing the rest of his list of PLA tasks, one can see that he is listing the key requirements to enable the "Two Transformations" (the NMSC, "Army Building"), although Jiang does not use the specific term. Since this formulation had not been put forward until late 1995, the 15th Party Congress provided the first opportunity for Jiang to place his imprimatur on the program outside the PLA. Encapsulating the gist of the list, Jiang calls upon the PLA to stress quality over quantity ("take the road to building fewer but better troops"); upgrade defense and "combat capability in a high-technology environment"; adapt defense industries to new market mechanisms; strengthen education and training; "gradually upgrade weapons and equipment"; and improve the reserve and militia system.[77]

---

[75] Jiang Zemin, "Political Report," op. cit.

[76] See, "Jiang Zemin Delivers Political Report," October 12, 1992, in FBIS-CHI-92-198-S.

[77] Jiang Zemin, "Political Report," op. cit. The points are paraphrased from Jiang's Political Report to the 15th Party Congress. It is worth pointing out that at the 14th Party Congress in 1992 Jiang was still

The upshot of these three outcomes from the 15th Party Congress was that it was clear that major changes were in the wind. It was not until the 9th NPC in March 1998, however, that one started to get a sense of where Jiang and his generals are headed in the near-term and where their decisions fit into the notional NMS we have generated.

In a major speech to PLA delegates to the 9th NPC, Jiang Zemin provided the rationale for new military reforms. He stated that "global military reform keeps gaining momentum . . ." and that "military powers in the world have one after another readjusted their development strategies for their armies in a bid to speed up army modernization and overtake other countries in army quality." He stressed that developments in science and technology are having a "profound effect" on warfare. Jiang declared that the Central Committee and the CMC "have set forth new strategic principles for military development . . ." that are aimed at the PLA being able to fight LWUMHTC, "emphasizing army strength with science and technology rather than the quantity of troops, and bringing about a shift from a personnel-intensive to a technology-intensive armed force." In effect, Jiang reiterated that the "Two Transformations" concept is what the PLA must seek and that by moving forward with these military reforms China will be on the road to "significantly narrowing our differences with the world's advanced levels and laying a solid foundation for future development."[78]

What Jiang's comments do not illuminate is the specific decisions that the Central Committee and CMC have made along these lines. Neither do the comments of senior PLA leaders at the NPC. However, their general comments do give us some insight. The most important theme to come out from statements of senior PLA officers at the 9th NPC, I would argue, is that the PLA must now use the occasion of the 500,000-man reduction to make significant and much-needed organizational and structural changes to the armed forces. Comments along these lines made by Generals Fu Quanyou, Chi Haotian, and Yu Yongbo are worth highlighting because, in the typical sweeping speeches men such as these make at such events, it is often difficult to sort out the "signals" from the "static." GSD Chief General Fu Quanyou stated: "We should take the favorable chance of reducing 500,000 soldiers to conscientiously implement the policy decisions of the party Central Committee and Central Military Commission; *make a breakthrough in optimizing structures*, balancing relations, and improving work efficiency. . . ."[79] Defense Minister General Chi Haotian declared: "While readjusting and streamlining the army, we should stress not only a smaller army but also an *optimal structure* . . . The PLA structural

standing in the shadow of Deng Xiaoping, Yang Shangkun, Yang Baibing, Liu Huaqing and Zhang Zhen. By the 15th Party Congress Deng had died, Liu and Zhang had retired, the two Yangs were not much heard from, and a new generation of general officers was coming into its own with Jiang at the helm. Consequently, one can posit that the 15th Party Congress is the point at which Jiang Zemin finally consolidated his position as titular leader of the PLA (both as State President and Chairman of the CMC) and was in a position to call for bold reform within the PLA.

[78]Wu Hengquan and Luo Yuwen, "Chairman Jiang Zemin Stresses at PLA Delegation Meeting."

[79]Ma Xiaochun and Zhang Yanzhong, "While Relaying the Guidelines of the First Session of the Ninth National People's Congress at the General Staff Headquarters, Fu Quanyou Calls for Conscientiously Implementing the Guidelines of the NPC Session and Achieving Military Work in a Down-to-Earth Manner," *Jiefangjun bao*, March 21, 1998, in FBIS-CHI-98-091, April 1, 1998. Emphasis added.

and establishment readjustment and reform is a project of complex system engineering. . . . Charged with a sacred mission of safeguarding China's national security and social stability, the *PLA is faced with an arduous task of reorgranizing itself at present*" (emphasis added).[80] GPD Chief General Yu Yongbo opined: "The CPC Central Committee and the Central Military Commission have decided to *readjust and streamline the army establishment*, this being a major strategic measure intended to further army quality-building and an important step toward cross-century army modernization-building. The army should actively, assuredly, and successfully fulfill this important task . . ." (emphasis added).[81]

What we take from all of this is that in the near-term (the next three to five years) we can expect the PLA to institute significant changes in force structure and organization. All the "indications and warnings" are out on the table. While the specifics are not yet clear, we can probably engage in some informed speculation by carefully studying the PLA literature relating to the current and future operational environment, the Chinese analysis of LWUMHTC, and works of military science dealing with the operational level of warfare.

The independent Hong Kong newspapers and other international news agencies have already picked up on the subject of impending PLA structural reform. Various articles have touted PLA plans to reduce the seven Military Regions to "five major theaters."[82] others have reported rumors of suggestions that the Ministry of National Defense should be civilianized.[83] Still other articles allege that the PLA is studying the U.S. model and considering revamping the CMC along the lines of the Joint Chiefs of Staff and the Joint Staff.[84] Even more intriguing, the PRC-affiliated Hong Kong media are heralding impending changes as well. In February 1998 a lengthy and detailed article in *Kuang chiao ching* provided what it claimed were the details of a three-year plan for a major structural readjustment of the PLA.[85] In April 1998 *Ta gong pao* echoed, in abbreviated form, the *Kuang chiao ching* article (without directly citing it) just in case interested observers had missed it the first time around.[86]

Whether or not the particular initiatives in any of these articles come to pass remains to be seen. What is important is that these articles may in fact be indicative of an

---

[80]Gao Aisu, "Deputy Chi Haotian calls at NPC Panel Discussion for Earnestly Pushing Forward Army Building, Reform, Taking the Road of Building Small But Better Trained Army with Chinese Characteristics," *Jiefangjun bao*, March 7, 1998, pp. 1, 4, in FBIS-CHI-98-090, March 31, 1998.

[81]Tan Jian, "Deputy Yu Yongbo Calls at NPC Panel Discussion for Submitting to Overall Interests; Supporting Reform; Earnestly Pushing Forward Modernization Building," *Jiefangjun bao*, March 31, 1998, pp. 1, 2, in FBIS-CHI-98-090, March 31, 1998.

[82]Hsiao Peng, "Seven Major Military Regions to Be Changed into Five Major Theaters—A Great Change in PLA Commanding System Is Under Deliberation," *Sing tao jih pao*, April 15, 1998, p. A4, in FBIS-CHI-98-105, April 15, 1998.

[83]Willy Wo-Lap Lam, "Proposal to Give Civilian Top PLA Jobs," *South China Morning Post*, June 1, 1998, Internet edition.

[84]Fong Tak-ho, "PLA Turns to US Role Model for Modernization Role Model," *Hong Kong Standard*, April 3, 1998, Internet edition.

[85] Kuan Cha-chia, "Commander Jiang Speeds Up Army Reform."

[86]"Special Article" by staff reporter Kung Shuang-yin: "The Curtain Has Been Raised on Army Restructuring," *Ta gong pao*, April 7, 1998, p. A1, in FBIS-CHI-98-097, April 7, 1998.

environment of military reform in Beijing in which truly significant changes to the way the PLA is organized and commanded are being considered.[87] We realize the weighty constraints the PLA faces in making major adjustments, but we must not discount the possibility that a serious series of changes is about to unfold.

IX. CONCLUDING COMMENTS

The purpose of this paper has been to offer a notional "National Military Strategy" for the People's Liberation Army of China. It remains my argument that it is difficult to make analytic sense out of the myriad of changes going on within the PLA, or past policies for that matter, without a stated conceptual overview of what the PLA intends to achieve as a national military force and how it plans to go about achieving it.

An American template has been used to build this strategy for seven reasons. First, the model has general utility in organizing one's thinking about national-level military issues. Second, it provides a useful baseline for comparative use by imposing rigor of definitions. Third, it affords ease of cognizance ("recognition value") for those who carefully follow U.S. military issues. Fourth, working through the model requires analytic justification. Fifth, it forces one to think about linkages. Sixth, once the detailed rationale is ingested the essence of the strategy is captured in a short list of National Military Objectives and National Military Strategic Concepts. Finally, it highlights where PLA strategic issues have a universal quality and where issues unique to the Chinese armed forces stand out.

We should keep in mind that a national military strategy is a template and road map. It is not written in stone. It is dynamic and subject to change. This is true for all armed forces and it is true for the PLA. Significant changes or minor adjustments are driven by changes within the "Strategic Filter." Testing and challenging the assumptions upon which an NMS is based is an iterative requirement. New political decisions (or altered domestic environment), significant developments in the security environment, or changes in the operational environment determine the mix of continuity and change.

It is important to understand that it is one thing to articulate a national military strategy (in this case by proxy) and quite another to accomplish it. In the realm of "Army Building" especially, Jiang Zemin and the CMC leadership have levied some ambitious requirements on the PLA without necessarily providing all the resources required. The PLA starts from the premise that the funding it requires from the state will not be adequate. The PLA leadership is painfully cognizant that the quality of the personnel it requires is not readily available across the breadth of the armed forces. Neither is the defense industrial base adequate to produce indigenously *all* the

---

[87]My own inclination is that the first significant structural reforms will occur at the Military Region and Group Army levels on an experimental basis. This is because the changes can be dictated from above and the whole point of the structural reform program is first and foremost to enhance real combat capabilities. I am also prepared to envision significant near-term and major adjustments within the staffs of the (now) four General Departments and the command relationships between the General Departments and the Military Regions.

weapons systems the PLA would like to have in the near term. The list of impediments and challenges is endless. This is not meant to suggest that the situation is hopeless for the PLA. To the contrary, we may be surprised by what unfolds over the next few years. What this means is that tradeoffs must be made, economies sought, and careful choices made in expending resources to enable the critical components of this NMS.

Given the totality of the resource constraints under which it will continue to labor, the PLA will remain too large and too diversified in its missions to expect that it can reform and enhance the capabilities of the entire force any time soon, even with the impending 500,000-man reduction. Consequently, it would make eminent sense if the PLA leadership focused its army building efforts in particular on a small, well-chosen segment of the PLA.

Well into the first decades of the next century three distinct PLAs will continue to coexist.[88] The vast majority of the PLA will probably remain a low-tech force, ground-oriented, and charged mainly with guarding borders or watching vigilantly over those regions of China where internal unrest is possible or where humanitarian relief operations are required. This PLA will remain, to borrow a PLA phrase, a "millet and muskets" army. This PLA will receive little in resources but will be important for accomplishing the internally directed elements of this notional NMS.

A second PLA will be composed of an extremely small (but very vocal) group within the Chinese armed forces. This PLA will consist of the strategic thinkers at the Academy of Military Sciences (AMS) and other security-related think-tanks who deal with doctrine, the theoretical scientists and applied technology specialists at COSTIND's research institutes, and perhaps even PLA C4I specialists. This PLA will be thinking way over the horizon and experimenting with military technical revolution (MTR) and RMA concepts. We shall call this small group of thinkers, technicians, and potential innovators the "MTR/RMA PLA." They will be a "blackboard PLA" that will probably be able to accrue some resources for their work through channels beyond the declared defense budget, perhaps by teaming with civilian technology-oriented counterparts. Their impact on the PLA is open to speculation at this point.

The third PLA is the PLA to which the external elements of this notional NMS will apply. We can best term this PLA the "warfighting PLA." This is the military force that will be required to employ the doctrine of "Active Defense" and enable the NMSC "Forward Defense," today and into the foreseeable future. It is within this PLA that a subset of the relatively best combat units in the Army, PLAN, PLAAF, and Second Artillery can be found. This subset is where I suspect the CMC will concentrate its efforts in "Army Building" and which will likely undergo the most interesting and potentially significant experiments in reorganization, restructuring,

---

[88]The idea of "many PLAs" and what they represent is a result of extended conversations with my colleague Lieutenant Colonel Dennis Blasko (USA, Ret.), formerly Assistant U.S. Army Attaché to China and subsequently Assistant Army Liaison Officer in Hong Kong. For Blasko's own thinking on this issue, see his later chapter, "A New PLA Force Structure."

and C2 modernization. It is this PLA which will be the focus of the "Two Transformations" and a sizeable investment of resources over time.

Even within this small subset it is difficult to predict how successful the PLA will be in crafting a smaller but exponentially more capable force. There are a host of factors beyond the control of the PLA that could impel or impede success: a domestic economic meltdown, a radical change in internal politics, domestic unrest, or a radically altered security or operational environment, to name just a few. With a reform program of such complexity as the one we suspect the PLA is about to undergo taking place in the context of a complicated and ever-changing domestic and international environment, linear projections are almost meaningless.[89]

Moreover, the PLA will accrue inevitable advances in *tactical* capabilities through the acquisition or development of highly sophisticated weapons systems on and under the sea, in the air, or on the ground, but these will not necessarily bear any relationship to an enhanced capability to conduct *sustained combat* at the *operational* level of warfare, or the PLA's ability to achieve key national military objectives. The closed nature of the PLA will ensure we see the "glitz" but little of the "guts" of their armed forces. Consequently, it will be relatively difficult to measure how well the PLA is doing in closing the large gap between its current conventional capabilities and the strategic concepts posited in this NMS (especially the NMSC "Forward Defense").

As a concluding comment, I would offer that models, by their very nature, are artificial representations of reality. As I mentioned early on, the PLA certainly would not present its NMS as I have using an American template. However, I am confident that the concepts in this NMS, the policies outlined, and the analysis and intentions of this NMS are constructs that would be familiar to Chinese officers serving on the staffs of the major military organs in Beijing and in the field on the headquarters staffs of the MRs.

It is my hope that students of the PLA will leave this study with some new ways to think about where their own particular studies fit into a larger strategic context. As military technologies around the world continue to become more advanced, the traditional lines separating the strategic, operational, and tactical levels of warfare are becoming more blurred. All levels of military analysis must now take the strategic level of analysis as its point of departure.

---

[89]The best way to get at this question is probably to utilize scenario-based analysis as developed and applied by the Royal Dutch/Shell Oil Company. This is the soundest way to develop a manageable set of plausible, alternative "PLA futures" as a result of a serious analysis of drivers, trends, predetermined factors, and key uncertainties. For details on this methodology, see Peter Schwartz, *The Art of the Long View: Planning for the Future in an Uncertain World*, New York: Doubleday, 1996.

8. THE PLA'S EVOLVING CAMPAIGN DOCTRINE AND STRATEGIES

Nan Li[1]

China's military planners like to claim that in *overall* terms, the PLA is technologically inferior if compared with its potential adversaries. Such inferiority will continue in the foreseeable future, even though the People's Liberation Army (PLA) has made some progress in modernizing its weaponry in recent years. On the other hand, the strategic principles of the PLA have undergone a major shift since the mid-1980s. Eschewing the traditional emphasis on "victory through inferiority over superiority (*yilue shengyou*)," one new principle stresses "victory through superiority over inferiority (*yiyou shenglue*)."[2] The contradiction between the technological inferiority of the PLA and the requirement of the PLA to achieve superiority over potential adversaries begs an important question: What are the PLA's strategies that may turn PLA absolute inferiority to local and temporary superiority?

Traditionally, on the premise of a major invasion of China by a technologically superior adversary such as one of the superpowers, the PLA would compensate for its technological inferiority with its abundance in space, manpower and time. The vast, familiar territory of China, coupled with a protracted, manpower-intensive people's war of tactical dispersion, mobility, harassment, and attrition, would eventually gain China sufficient time. This in turn would allow China to gradually weaken the overextended invading forces, identify their weaknesses, reconstitute domestic military forces, and finally win the war through more decisive, strategic offensives.

The decline and final end of the Cold War, however, have denied the PLA the opportunity to fight a protracted, manpower-based total war with deep depth. This is because the prospects for massive foreign invasion and total conquest of China have largely diminished. On the other hand, local, limited wars involving national unification and disputes over maritime and land territories are assumed to be more likely to take place. Unlike total war, however, these local wars presumably are shorter in duration, fought in remote border regions or on the high seas, most of which are sparsely populated and have less depth for maneuverability. Such wars

---

[1]Nan Li has been a post-doctoral fellow at Olin Institute of Harvard University and a senior fellow at the U.S. Institute of Peace. He has taught political science at Dartmouth College and the University of Massachusetts at Amherst. He is currently researching ways to mitigate the potential for possible military conflicts in Asia.

[2]Hu Guangzheng *et al.*, *Yingxiangdao ershiyi shiji de zhengming* (*Contention Affecting the 21st Century*), Beijing: Liberation Army Press, 1989, p. 113.

also usually require technology-based forces and arms capable of forward deployment. These in turn make it difficult to accommodate a drawn-out, mass mobilization-oriented, heartland-based total war. Finally, potential adversaries in these local wars may not be as powerful as either of the two superpowers during the Cold War.[3] All these have somewhat reduced the relevance of the PLA's old comparative advantage in space, manpower and time to PLA's war planning. Also, the two-decade-long defense modernization has produced some "pockets of technological excellence" within a generally backward PLA. As a result, a range of new strategies has been articulated for turning the absolute PLA inferiority to local and temporary superiority under the new doctrine of "local war under high-tech conditions." This essay investigates these strategies.

The study is significant for both empirical and policy reasons. Empirically, some recent studies have mentioned the new strategic principles, but since they focus on illustrating the major differences between the old and the new, they fall short of fleshing out the specific components of these new principles.[4] This essay intends to fill this void by examining a body of PLA literature that has emerged following the PLA war games over the Taiwan Strait in early 1996. More important, since the PLA is becoming increasingly involved in China's security policy process, an in-depth examination and interpretation of the PLA's new campaign doctrine and strategies may help to gain a better understanding of the nature and direction of China's security posture, which may have significant policy implications for Asian security.

This essay shows how China's military planners have articulated the doctrine of war zone campaign (WZC, or *zhanqu zhanyi*) as a major type of local war that may enhance local and temporary PLA superiority. It then shows how they have fleshed out three major campaign strategies that may enhance the probability of such superiority. The strategies include "elite forces and sharp arms" (*jingbing liqi*), "gaining initiative by striking first" (*xianji zhidi*), and "fighting a quick battle to force a quick resolution" (*suzhan sujue*).

Three caveats are in order. First, this study is more concerned with the theoretical dimension of the PLA's war planning than the extent to which theory translates into reality. Theory may or may not become reality. What is important is that China's military planners now believe that China's defense modernization will be more theory-driven than situation-driven, and they give some good reasons. First, they believe that the PLA's relinquishment of the smothering tasks of large-scale domestic class struggle and preparing for an external total war since 1985 provides an element of breathing space. Such a space provides an opportunity for PLA planners to think systematically about the most likely kind of security threat that China may face, and the optimal strategies to cope with such a threat. This in turn may lead to more efficient use of scarce resources for defense modernization. Also, arming the still technologically backward forces with advanced theory may quicken the formation of combat effectiveness once the advanced arms are available. Indeed, the prospect for

---

[3]See Nan Li, "The PLA's Evolving Warfighting Doctrine, Strategy and Tactics, 1985–95: A Chinese Perspective," *China Quarterly*, No. 146, June 1996, pp. 445–448, 451–453.

[4]*Ibid.*, pp. 443–463; Paul Godwin, "From Continent to Periphery," pp. 464–487.

such arms to be available to the PLA is not that bleak, as the defense budget grows in proportion with economic growth. Furthermore, a technologically advanced military that lags behind in theory may mean higher costs in war, as the French and the British militaries discovered during the early phase of World War II. Finally, simply because the PLA is technologically inferior, conceptual innovation is all the more indispensable to compensate for such inferiority if the PLA intends to fight and win wars in the future.[5] All these reasons are sound enough to justify this study, if we are serious about understanding the nature and direction of China's security posture.

Second, this study examines the PLA's "hard-kill" strategies. "Soft-kill" types, or the so-called asymmetrical strategies such as cyber warfare, are not the central concern here, except where they are articulated and integrated into the overall PLA campaign strategies. While there has been extensive discussion of concepts associated with the revolution in military affairs (RMA) in recent PLA literature, how to operationalize these concepts to PLA conditions is still being debated. This topic deserves separate analysis.[6]

Moreover, this study relies heavily on materials whose circulation is confined to the PLA (*junnei faxing*). Because they are limited to a smaller group involved with actual military planning, they should deal with genuine policy issues. They therefore are deemed more credible and less propagandistic compared with materials that cater to general Chinese and foreign readers. The official bias of such literature, however, will be counterbalanced by author's critical concluding evaluation. The essay's first four sections discuss the concept of WZC and three campaign strategies. The conclusion summarizes the findings and evaluates the practical implications of these doctrines and strategies.

WAR ZONE CAMPAIGN

Since the 1985 "strategic transition" of the PLA from preparing for "early, total, and nuclear" war to local and limited war, there have been debates on what type of local war the PLA should be prepared to fight.[7] The 1991 Gulf War and the 1996 Taiwan Straits crisis have apparently convinced the PLA planners that a likely war scenario for which the PLA should be prepared to deter or fight is a medium-sized local war comparable to a PLA WZC. "In terms of scale and nature, possible future local wars that involve large-scale sea-crossing and amphibious landing operations, counter-

[5]Major General Zhang Yuqing, "*Zhai 'gaojisu tianjianxia jubu zhanzheng zhanyi lilun yantao hui' kaimushi shang de jianghua*" (Speech at the Opening Ceremony of the 'Symposium on the Theory of Campaign in Local War under High-tech Conditions'); Lt. General Hu Changfa, "*Zhai 'gaojisu tiaojianxia jubu zhanzheng zhanyi lilun yantao hui' jiesu shi de jianghua*" (Speech at the Conclusion of the 'Symposium on the Theory of Campaign in Local War Under High-tech Conditions'); Senior Colonel Chen Youyuan, "*Junshi jisu geming yu zhanyi lilun de fazhan*" (Military Technological Revolution and the Development of Campaign Theory), all in NDU (National Defense University) Scientific Research Department and Campaign Research and Teaching Section (eds.), *Gaojishu tiaojianxia zhanyi lilun yanjiu* (*Research on the Theory of Campaign under High-tech Conditions*), Beijing: National Defense University Press, 1997, pp. 2, 14–15, 21–22. See also Hu, *Contention Affecting the 21st Century*, pp. 202–204.

[6]For the most recent analysis of the PLA's information warfare strategies, see the chapter by James Mulvenon in this volume.

[7]Hu, *Contention Affecting the 21st Century*, pp. 173–177.

offensive operations in the border regions, and repelling local foreign invasion all belong to this category."[8] If WZC is a primary mode of operations that may shape PLA war planning, the central issues that need to be addressed become: What are the defining characteristics of WZC? How can such campaigns increase the chances to turn PLA absolute inferiority to local and temporary superiority?

Defining Characteristics

Several major features have been advanced to define WZCs. First, the WZC is an intermediate campaign mode between combined arms group army (CAGA) campaign, CAGA group (*jituanjun qun*) campaign, and war zone front (*zhanqu fangxiang*) campaign on the one hand, and a major or total war involving more than one war zone, and partial or total national mobilization on the other. Unlike the former three types of campaign where the ground forces dominate and other service branches play only a supportive role, a WZC is a *joint service* campaign where each service conducts relatively independent subcampaigns (*zhi zhanyi*). Since a war zone usually has one strategic direction, several campaign fronts, and multidimensional space, subcampaigns may include electronic warfare operations, conventional strategic and campaign missile operations, air operations, sea operations, and front army or CAGA operations. Therefore, unlike CAGA-level campaigns that emphasize ground forces, the WZC gives equal weight to all four services (ground, navy, air, and conventional strategic missile forces) in the war zone, which may be reinforced through the national supreme command by forces outside the war zone. They "have a system of unity between military region [MR, which encompasses several adjacent provinces] and war zone. [This means that] the peacetime MR becomes a WZ during the war time . . . and has jurisdiction over the ground, navy, and air forces within it." Generally speaking, a WZC "is the total sum of several service-based subcampaigns, while a CAGA-level campaign is the total sum of ground battles."[9]

Furthermore, a WZC may last from several weeks to several months and involve several phases. Even though "the trend is toward shorter duration, it is still longer than CAGA, CAGA group and war zone front campaigns." On the other hand, unlike a major war where a single campaign may affect but not directly decide the strategic outcome, a WZC is limited and local in the sense that its outcome directly determines whether national strategic objective is realized. Since WZC itself constitutes local war

---

[8]Senior Colonel Huang Bin, "*Shenhua zhanqu zhanyi yanjiu de jidian sikao*" (Several Reflections on Deepening Research on War Zone Campaign), in NDU Scientific Research Department, *Research on the Theory of Campaign*, p. 43.

[9]*Ibid.*, pp. 43, 45; Yu Shusheng *et al.* (eds.), *Lun lianhe zhanyi* (*On Joint Campaign*), Beijing: National Defense University, 1997, p. 16; Xu Guoxin *et al.* (eds.), *Zhanqu zhanyi houqin baozhang* (*Logistics Support for War Zone Campaign*), Beijing: National Defense University, 1997, p. 2; Major General Yu Chenghai, "*Dui zhanqu lianhe zuozhan zhihui konzhi wenti de jidian renshi*" (Several Understandings Regarding the Question of Command and Control Concerning War Zone Joint Operations), in NDU Scientific Research Department and Force Command Research and Teaching Section (eds.), *Gaojishu tiaojianxia zuozhan zhihui yanjiu* (*Research on the Command of Operations under High-tech Conditions*), Beijing: National Defense University, 1997, pp. 20–21.

but not an element in a larger war, the "campaign objective overlaps war objective."[10]

Finally, a WZC is conducted by the unified, joint service command at the war zone level under the guidance of the national supreme command. Also, some such campaigns may be carried out under the threat and deterrence of nuclear, chemical and biological warfare. Moreover, since the political stake of such campaign is high, both sides may utilize their best forces, high-tech arms and advanced C4I, leading to fierce battlespace competition that fuses defense with offense.[11]

Advantages of WZC

How then can WZC help to transform the absolute PLA inferiority into local and temporary superiority? First, a CAGA-level campaign is too short in time, too limited in scope, and too dominated by regular ground forces to give full play to the PLA's newly developed "pockets of excellence," particularly in naval, air, conventional strategic missile, rapid reaction, and special operations capabilities. On the other hand, total war may overwhelm and diminish the relevance of these "pockets of excellence," since the technologically superior superpower(s) is likely to employ its most advanced weapons simultaneously on all fronts and throughout the war process, thus dwarfing PLA's "pockets of excellence." Compared to a CAGA-level campaign, however, a WZC is sufficiently big and long for the PLA to concentrate its "pockets of excellence" to a local and temporary situation to reverse its absolute inferiority. On the other hand, a WZC is more limited and shorter than a total war, because the PLA is more likely to deal with an adversary that is much less powerful than a superpower in such a campaign, as long as *the powerful adversary* (*qiangdi*, referring to the superpower) is denied sufficient reasons to intervene, or is deterred from doing so. In this way, the WZC may work to the advantage of the PLA.[12]

Moreover, unlike CAGA-level campaigns that have limited space, the joint service campaign associated with the multidimensional space and deeper depth of a war zone "allows sufficient leeway for asymmetrical strikes (*buduideng daji*) through flexible assembling of diverse means and innovative combination of versatile styles." The joint service campaign may provide the conditions for the PLA to "use its strength against enemy's weakness (*yiqiang jiruo*)," and avoid matching PLA's weakness with the enemy's strength. It, for instance, may lead to situations where "we can use our air power to strike enemy ground and naval targets, use our ground forces to deal with enemy air and naval operations, use our navy to fight enemy ground forces, and use our combat forces to strike enemy non-combat aspects such as C4I and logistics." If "it is inevitable that we fight a service-matching war of symmetry with the enemy, it is still necessary that in comparative capabilities we

---

[10]Huang Bin, "Several Reflections," pp. 44, 45; Yu, *On Joint Campaign*, p. 16; Xu, *Logistics Support*, p. 2.

[11]Huang Bin, "Several Reflections," pp. 43, 45.

[12]Colonel Yu Guohua, "*Shilun wojun lianhe zhanyi ying zunxun de jige jiben yuanzhe*" (An Exploratory Comment on Several Basic Principles That the Joint Campaign of Our Army Should Follow), in NDU Scientific Research Department, *Research on the Theory of Campaign*, pp. 96–97.

constitute superiority at the sub-campaign and battle levels."[13] This articulation of the WZC illustrates conditions that may increase the chances for the PLA to achieve a measure of local and temporary superiority. But for the probability of such chances to be enhanced, more concrete strategies have to be systematically articulated.

ELITE FORCES AND SHARP ARMS

One new PLA campaign strategy is "elite forces and sharp arms (EFSA)." Several major reasons have been advanced to justify EFSA. First, unlike total war, the limited nature of local war makes it possible to achieve local and temporary superiority through the concentrated use of EFSA. According to one PLA author, "after many years of army building, we have acquired a certain number of high-tech elite forces and sharp arms, and they are capable of competing with a powerful enemy." Finally, operations in areas close to the homeland provide the favorable base, logistics, and battlefield conditions for utilizing EFSA.[14] If the PLA is endowed with the necessary "material conditions" (the PLA expression for "pockets of excellence"), the key question to be addressed is: how does the PLA optimize the use of such forces and arms to achieve local and temporary superiority? The answer to the question lies in the PLA articulation of several major concepts associated with *deployment, coordination* and *command.*

Deployment (*bushu*)

On deployment, a new concept is *transregional support operations* (TRSO, or *kuaqu zhiyuan zuozhan*), which refers to deploying the best forces and arms from other MRs to reinforce the war zone where local war may occur. Several rationales have been proposed for TRSO. One is that it enhances political and diplomatic initiatives. Rather than sustained forward deployment of a large number of forces in the disputed area as during the Cold War, TRSO focuses on developing rapid reaction units (RRU) and capabilities in the rear while maintaining a moderate level of forwardly deployed forces. This in turn reduces tension in the disputed area. On the other hand, the highly mobile and effective RRU constitutes "indirect forward presence" (*jianjie qianyan cunzai*). This itself may contribute to local and temporary superiority in psychological terms, which may help to deter provocation, thus preventing escalation.[15]

Also, TRSO provides a strong incentive for military modernization, since it stresses technology-based mobility and effectiveness. In the longer run, this enhances PLA

---

[13] *Ibid.*

[14] Major General Zheng Shenxia (PLAAF), "*Lianhe zhanyi kongjun yunyong ying zhaozhong bawo de jige wenti*" (Several Questions That Should Be Emphatically Addressed in Applying Air Force in Joint Campaign); Major General Liu Yinchao, "*Gaojishu tiaojianxia jidongzhan zhanyi de jige wenti*" (Several Questions on Mobile Warfare Campaign under High-tech Conditions), both in NDU Scientific Research Department, *Research on the Theory of Campaign,* pp. 102, 321.

[15] General Liu Jinsong, "*Guanyu shishi zhiyuan zuozhan de jige wenti*" (Several Questions on Implementing Support Operations), in NDU Scientific Research Department, *Research on the Theory of Campaign,* pp. 32–33.

overall superiority. Moreover, the development of infrastructure in the civilian sector and the progress that has been made in PLA reconnaissance and force projection capabilities all facilitate material conditions for TRSO. Furthermore, unlike the Cold War, the lack of clear and imminent danger from different directions simultaneously allows for TRSO of short duration and concentrated direction with reduced risk of escalating tension on other fronts. But more important, there are many potential flash points (*redian*) along China's long and complex land and maritime borders. On the other hand, "our forces are sparsely deployed . . . if tension escalates into conflict, particularly if a local war of medium or bigger size breaks out, forces and arms of one war zone may not be enough." This is particularly so since "our equipment is still relatively backward and our comprehensive operational capabilities are still rather weak." All these render TRSO indispensable for the PLA to achieve local and temporary superiority.[16]

While TRSO deals with concentration of EFSA at the strategic level, how to deploy EFSA at the campaign and battle levels is crucial to whether local and temporary superiority can be achieved. While concentrated use of forces and arms has always been a PLA warfighting principle, it is now being reinterpreted to adapt to new conditions. A central new condition is that "on the high-tech battlefield, annihilating enemy vital forces and arms can no longer be achieved by simply adding numbers of forces, planes, tanks, and artillery pieces." This is because rapid advances in arms, surveillance and communications technology lead to longer range, higher mobility, more stealth, faster speed, higher lethality of arms, and a more transparent battlefield. If "we concentrate a large number of forces and numerous average arms in a designated area, it may not only lead to waste but also cause huge losses."[17] Therefore, several major principles on deployment have been formulated that may contribute to local and temporary superiority.

First, rather than average forces and arms, it is necessary to use carefully selected forces and arms. The selection may be based on unified and comprehensive planning, to constitute "comprehensive strike effects" (*zhonghe daji gongneng*). This planning is based on careful analysis of the nature, types, and number of targets, to determine the nature, types, and quantity of strike arms. The old practice of "using single service, one type of arms to strike all types of targets" may be replaced by "selecting different types of arms to strike different types of targets."[18]

Within this general selection of forces and arms, however, a flexible "component assembly" (*bankuai goujian zhuhe*) method may be applied to the combination of forces and arms. At a crucial place and moment, for instance, it may be necessary for the "trans-organizational system" (*kua jianzhi*) and "supra-normal" (*chaochang*) reinforcement of key categories such as electronic warfare, air assault/defense,

---

[16] *Ibid.*, pp. 33–34; Major General Zhang Qihong, "*Guanyu dongyuan he zhuzhi difang liliang baozhang zhanyi juntuan jidong wenti*" (The Question of Mobilizing and Organizing Civilian Strength to Support Mobility of Campaign Forces), in NDU Scientific Research Department, *Research on the Theory of Campaign*, pp. 351–352.

[17] Yang Yi, *Gaojishu tiaojianxia zuozhan fangshi, fangfa yanjiu yu sikao* (*Research and Reflection on the Styles and Methods of Operations under High-tech Conditions*), Beijing: Military Sciences Press, 1997, p. 97.

[18] *Ibid.*, p. 98.

precision-guided munitions, and attacking forces to be implemented. The objective is the optimal use of forces and arms so that "sufficient munitions, arms, and forces—neither too little nor too much—are used to destroy selected targets in the shortest possible time."[19]

Moreover, instead of the old practice of concentrating forces and arms in a point area, the new principle requires "dispersed deployment" (*sushan bushu*) of forces and arms to deny the enemy a clear "window of vulnerability" that may trigger an effective preemptive strike. This in turn may increase the chances of survival, the basic condition leading to local and temporary superiority. The degree of dispersion may be determined by the shooting or operational range and degree of mobility of the forces and arms. This dispersion, on the other hand, is accompanied by *concentration of effects*. This means that the destructive effects of these forces and arms may be concentrated on primary direction, right timing, major targets and key nodes. Once objectives are realized, both forces and arms may swiftly change position through high mobility. The purpose, again, is to constitute local and temporary superiority based on minimizing casualties caused by enemy strikes and swiftly destroying enemy targets.[20]

Coordination (*xietong*)

EFSA are largely associated with the technology-intensive services such as the Navy, Air Force, and the Second Artillery (China's strategic missile force). Also, as discussed earlier, future local war is more likely to be a joint force-based WZC than a ground-force-dominated CAGA-level campaign. Traditionally, in CAGA-level campaigns, technology-intensive services play only a supportive role. Therefore, the central challenge facing a CAGA campaign is coordination among ground-based functional arms such as infantry, armor, artillery and engineering (hence combined arms, or

---

[19] *Ibid.*; Yu Guohua, "An Exploratory Comment," pp. 94–95; Liu Yinchao, "Several Questions," pp. 321–322; Lt. Colonel Wang Wei *et al.*, "*Gaojisu tiaojianxia jubu zhanzheng zhanyi zhidao de jige wenti*" (Several Questions on Directing Campaign in Local War under High-tech Conditions); Lt. General Zheng Baoshen (PLAAF), "*Lukong lianhe zhanyi yuanzhe qiantan*" (An Elementary Investigation of Principles regarding Land-Air Joint Campaign); Major Li Bo *et al.*, "*Shilun zhanyi zhanfa de duikang xing ji duikang xingshi*" (An Exploratory Comment on the Confrontational Nature and Style of the Fighting Methods in Campaign); Nanjing MR Command Department, "*Denglu zhanyi de tedian yu yaoqiu*" (Characteristics and Requirements regarding Amphibious Landing Campaign); Major Yu Guangxuan, "*Denglu zhanyi zhidao de jige wenti*" (Several Questions on Directing Amphibious Landing Campaign); Vice Admiral Yang Yushu, "*Shilun lianhe denglu zhanyi zhong duoqu he baochi zhihaiquan wenti*" (An Exploratory Comment on the Question of Capturing and Sustaining Sea Superiority in Joint Amphibious Landing Campaign); Vice Admiral Wang Yongguo, "*Haishang fengsuo zuozhan wenti tantao*" (An Investigation of the Question of Sea Denial Operations); Senior Colonel Sun Xiaohe *et al.*, "*Shilun denglu zhanyi zhong huolizhan de zhuzhi yu xietiao*" (An Exploratory Comment on Organizing and Coordinating Fire Power Support in Amphibious Landing Campaign); Lt. General Fu Binyao, "*Bianjing diqu fanji zhanyi zhidao de jige wenti*" (Several Questions on Directing Counter-offensive Campaign in Border Region); Major General Peng Cuifeng, "*Lianhe zhanyi juntuan bianjing fanji zuozhan chutan*" (A Preliminary Investigation of Border Counter-offensive Operations by Joint Campaign Forces); Major General Wang Guiqin, "*Qiantan gaojishu tiaojianxia jubu zhanzheng zhanyi zhong de fei zhenggui zuozhan*" (An Elementary Discussion of Non-Regular operations in Campaign in Local War under High-tech Conditions), all in NDU Scientific Research Department, *Research on the Theory of Campaign*, pp. 60–61, 89–90, 137, 155, 159–160, 191, 196, 221, 271–272, 275–276, 428–429.

[20] Yang, *Research and Reflections*, p. 98; Liu Yinchao, "Several Questions," p. 322.

hecheng), even though the role of other services in such a campaign has become an increasingly important subject for investigation.

A joint (*lianhe*) service campaign, however, is where each service branch plays the leading role in each of several relatively autonomous subcampaigns in separate time and (sometimes) space. In such a campaign, it is possible for the Air Force to conduct an air subcampaign where the ground forces, Navy electronic warfare and special operations elements, and long-range surface-surface missile forces play a supportive role. Similarly, the Navy may become the dominant player in a sea denial and crossing campaign where the air and ground forces play a supportive role. As a result, rather than the traditional dominant role of ground forces and a subordinate and dependent role for others in a CAGA-level campaign, the relationship among all service branches is now defined by *equality* (*pingdeng*) and *partnership* (*huoban*) in the joint service campaign.[21] On the one hand, this development may contribute to local and temporary superiority if effective interservice coordination leads to complementarity of interservice effects toward fulfilling the campaign's goals. But poor coordination may also mean interservice friction and internal attrition, or diversion of energy and resources, thus causing waste and even defeat. Therefore, how to coordinate joint-service campaigns becomes another crucial issue that determines whether the PLA may achieve local and temporary superiority.

Several principles for coordinating joint campaigns have been advanced. One is that it is now necessary to cultivate the consciousness of "equality" and "partnership" among commanders and the rank and file of all services. This, in the long run, may contribute to equality-based interservice cooperation and reduce the chances for interservice bickering and friction, especially conflicts caused by a sense of superiority by some services and inferiority by others.[22] Moreover, macro-level coordination, such as clarifying and explaining campaign goals and the relationship between goals, specific targets, and tasks assigned to each service, are deemed crucial in providing a common understanding and purpose to which diverse service-based operations can contribute.[23]

Also, unlike CAGA-level campaigns, where coordination has always favored ground forces, the new principle requires coordination to be centered around whichever service conducts the separate subcampaigns. "For ground operations, the ground forces commander coordinates and controls air force elements, paratroopers, and navy marines that participate in such operations. For naval operations, the naval fleet or base commander coordinates and controls all the elements. The air campaign forces commander coordinates and controls air, air defense, and air shipment operations." This should "lead to a heightened sense of responsibility and

---

[21]Yu Guohua, "An Exploratory Comment," p. 94; Yang Yushu, "An Exploratory Comment," p. 191; Senior Colonel Xue Yanxu, "*Gaojishu tiaojianxia lianhe zhanyi xietong chutan*" (A Preliminary Investigation of Coordination regarding Joint Campaign under High-tech Conditions), in NDU Scientific Research Department, *Research on the Theory of Campaign*, pp. 72–73.

[22]Xue Yanxu, "A Preliminary Investigation," p. 75; Zheng Shenxia, "Several Questions," p. 108.

[23] Xue Yanxu, "A Preliminary Investigation," pp. 73–74.

initiatives by service commanding officers." This in turn may reduce interservice friction, which again may contribute to local and temporary superiority.[24]

Other crucial joint operations requiring careful coordination have also been specified. These include the timing and manner of transition from one phase of the campaign to another (such as from electronic and air subcampaign, to sea-denial and crossing operations, then to amphibious landing and ground operations) to minimize confusion or neglect associated with such transitions, and coordination between positions of various services to reduce internal chaos and friendly-fire casualties. Furthermore, coordination in stratagem, such as outflanking troop movements to divert enemy forces to secondary fronts, is also deemed crucial for determining whether local and temporary superiority may be achieved on the primary front.[25] Finally, several coordination modes have been specified, including strike zone–based, target-based, and timing-based coordination.[26]

On how to coordinate, one major method of interservice coordination is mutual dispatch (*shuangxiang paichu*) of service representatives to establish air-land, air-sea, or sea-land operations. These groups inform each other of timing, methods, requirements and targets of operations, and formulate and implement coordinating plans.[27] But the most important coordinating mechanism at the higher level is the newly conceived joint force command.

Command (*zhihui*)

The traditional MR commanding organs are ground-force-dominated, since they reflect the ground force nature of the CAGA or war zone front campaigns. Such domination may cause interservice friction and attrition if it is applied to the joint campaign where all services are equal partners. Also, the lack of professional and technical experience of some MR commanding officers in non-ground force services may not contribute to an optimal combination of various service components, the basic condition for achieving local and temporary superiority. Finally, in operations (but not in administrative matters), non-ground force service forces stationed in an MR do not answer to the MR commanding officers, but rather to service

---

[24]Yu, *On Joint Campaign*, pp. 136–137.

[25]*Ibid.*, pp. 137–138, 139–140; Xue Yanxu, "A Preliminary Investigation," p. 76; Yu Guohua, "An Exploratory Comment," p. 98.

[26]Strike zone–based coordination means establishing two adjacent strike zones: a tactical depth zone where firepower can be coordinated among ground forces artillery, attack helicopters, and air force attack aircraft, as well as a campaign depth zone where coordination is conducted among ground forces campaign/tactical missiles, air force fighter-bombers and attack aircraft, and Second Artillery conventional missiles. Target-based coordination means that firepower is coordinated around two types of targets. Air firepower, for instance, may be responsible for targets that are mobile and easier to detect, and have weaker air defense, while ground firepower deals with targets that are hidden and fixed, and have stronger air defense. Timing-based coordination refers to ordering the sequence of strikes. Between ground and air, for instance, ground strikes will go first. Among mobile targets of varying speed, the most mobile and threatening targets will be dealt with first. See Senior Colonel Cui Changqi *et al.*, "*Lushang jidong zhanyi lianhe huoli zhiyuan wenti chutan*" (A Preliminary Investigation of the Question of Joint Fire Power Support in Ground Mobile Campaign), in NDU Scientific Research Department, *Research on the Theory of Campaign*, p. 340.

[27]*Ibid.*, p. 339; Yu, *On Joint Campaign*, p. 109.

headquarters in Beijing. This means that for MR headquarters to have operational commanding power (*zuozhan zhihui quan*) over service forces, a complex procedure such as "war zone command—supreme command—service command—war zone service command—war zone the service forces" has to be dealt with.[28] This certainly impedes the swift assembling of joint forces, a basic condition for realizing local and temporary superiority.

The potential configurations of the WZC joint command attempt to tackle these problems. One conception proposes "introducing a system where command, political, and logistics departments of the joint campaign are established under the campaign commanding officers." This concept is practical since it requires only slight modification of the current MR three-department system by adding service-based personnel to these departments. The drawback of such a system, however, is obvious. "By giving equal weight to the three departments, it fails to demonstrate the leading role of the command department. It may also bloat the joint campaign commanding organs."[29]

As an alternative, a new concept has been advanced. Using the existing MR command department as the foundation, the joint command may consist of intelligence, decision control, communications and electronic warfare, and fire control and coordination components. Also, rather than the ground force domination as in the old MR headquarters, this command may be truly joint, with a higher proportion of both commanding and staff officers from non-ground force services. It is even possible that "the joint forces commander and chief of staff come from services other than the ground forces." Under usual circumstances, however, the commanders of WZ Air Force, Navy, and Second Artillery forces serve as the deputy commanding officers of the joint command. Moreover, the *decision control component* may be composed of staff teams headed by a deputy chief of staff from all service commands. This component assists decisionmaking by the joint force commander, formulates joint service operations plans, and conducts interservice coordination. Other components may also be multiservice-based rather than single-service dominated. But most important, this command may be relegated the operational commanding power over all service commands and forces within the war zone.[30] By expanding representation of services in the joint command proportional to their role in the joint campaign, it may alleviate interservice friction. This expansion also provides the necessary expertise that may optimize assembling of joint forces. Also, as the joint command gains complete operational control of service commands and forces within the war zone, swiftly organizing the joint forces

---

[28]Yu, *On Joint Campaign*, p. 105.

[29]Senior Colonel Wang Jianghuai, "*Dui wojun lianhe zhanyi zhihui tizhi wenti de tantao*" (An Exploration into the Question of the Joint Campaign Command System of Our Army), in NDU Scientific Research Department, *Research on the Theory of Campaign*, p. 130.

[30]*Ibid.*; Yu, *On Joint Campaign*, pp. 99–101, 105, 108. Major General Liu Chaoming, "*Lun gaojishu tiaojianxia zhanyi silingbu gongzuo gaige*" (On Reforming the Command Department Work in Campaign under High-tech Conditions), in NDU Scientific Research Department, *Research on the Theory of Campaign*, p. 435. Wang Jianghuai, however, proposes to separate communications from electronic warfare and add a special operations component. He also suggests the establishment of a reserve (*yubei*) command and a rear-area (*houfang*) command.

becomes possible. All these may increase the chances for achieving a measure of local and temporary superiority.

Besides command structure, the degree of (de)centralization of command is another central issue that is intensively scrutinized. In joint operations, expansion of the battlespace, the new equal relationship among services, and the increasing technical complexity in both arms and specialization of personnel may cause centrifugal tendencies. This in turn requires a stronger measure of centralized command to organize and channel the otherwise disparate elements, energy and attention to the campaign goal. Also, unified control is necessary to optimally manage and use intelligence, diverse weapon platforms, electronic warfare capabilities, and radio frequencies.[31]

Sometimes, however, "joint force commanders not only issue specific tasks to each service command and forces, but also specify concrete steps, methods and means to accomplish these tasks, and discretely change the timing and methods of implementing these tasks chosen by the service commanders." This excessive centralization may impede the achievement of local and temporary superiority in several ways. It may restrict local initiatives and foster a mentality of dependence, and cause delay due to requirement for frequent and detailed vertical debriefing, incurring higher cost for loss of opportunities. It also increases the workload of the higher joint command, lengthens campaign preparation, and may create a situation where one small error at the top causes big losses on the battlefield. Also, its heavy reliance on communications may mean loss of control if the communications network is severely damaged.[32]

Also, the autonomous, service-based nature of subcampaigns and rapidly shifting situations on the battlefield due to intensified competition caused by the application of high-tech weapons may all require a higher degree of decentralization and local initiatives for reducing cost and maximizing gains. On the other hand, excessive decentralization may have its own price. The lack of updated knowledge about campaign progress by the joint command may mean weaker coordination and adjustment, which may even lead to loss of control of the whole campaign.[33] Therefore, the extent to which, when, and how to centralize or decentralize are crucial issues that determine whether local and temporary superiority may be realized. As a result, several principles have been formulated to guide command in joint campaigns.

The extent of centralized command is deemed crucial in determining and deciding the general goal, scale, and time of the campaign, as well as the objectives, tasks, timing, and space of operations for each service force. It is also indispensable in arbitrating and mediating the relationship between the general campaign goal and subcampaign tasks, the transition from one subcampaign to another, coordination

---

[31] Zheng Baoshen, "An Elementary Investigation," p. 88; Yu Guohua, "An Exploratory Comment," p. 99; Yang Yushu, "An Exploratory Comment," p. 192; Yang, *Research and Reflection*, pp. 132–133.

[32] Yu, *On Joint Campaign*, p. 114.

[33] *Ibid.*, p. 115; Yu Guohua, "An Exploratory Comment," p. 99.

between services within a subcampaign, and in adjusting the strategic direction of the campaign. Other than these key areas, "decentralized command and service autonomy are desirable, particularly in intraservice matters regarding manner of deployment and specific methods for implementing subcampaigns."[34]

In timing, centralized command is absolutely necessary in the planning and preparation phase prior to the campaign. Central intervention is also necessary for strategic direction and the use of key arms and forces that may have imminent and important impact on the campaign outcome. "To avoid loss of opportunities and ensure rapid response at key moments, it is even necessary to conduct 'bypassing-level command' ('yueji zhihui')," allowing the campaign commander to be "squad leader." But otherwise, decentralized command is desirable "after a campaign is initiated, during the normal progress of the campaign, and at the secondary fronts."[35]

In execution, unity may be achieved between centralized control and service-based initiatives. Accountability, for instance, may be established between the joint command and service commands through "target control and management." This means that the joint command clarifies and assigns targets to services, and treats and monitors the extent of target appropriation as the central criteria in determining the degree of success for each service-based subcampaign. "As for how subcampaign targets are realized," however, "it is up to service commanders to determine. The joint command should not interfere."[36] All these principles, again, may help to reduce internal confusion, friction, and attrition, thus increasing the probability of local and temporary superiority.

GAINING INITIATIVE BY STRIKING FIRST

Whether local and temporary initiative (zhudong quan) can be achieved on the battlefield is another central variable in determining whether superiority can be realized. On this issue, the new principle of *Gaining Initiative by Striking First (GISF)* has been introduced for several reasons. First, the increasing precision and lethality of high-tech weapons of modern war means unprecedented destructiveness. Rather than gaining initiative, the side that "strikes only after the enemy has struck" (*houfa zhiren*) may lose momentum and face the prospect of decisive defeat. For similar reasons, the premise of a clear line between offense and defense which underlies the notion of first and second strike may become less relevant, since opposing sides on the modern battlefield shift and disperse their forces frequently to reduce casualties and create opportunities for preemption (*xianzhi*), the basic conditions leading to

[34]Zheng Baoshen, "An Elementary Investigation," pp. 88–89; Yu Guohua, "An Exploratory Comment," p. 100; Yu Guangxuan, "Several Questions," p. 160; Yu, *On Joint Campaign*, p. 115.

[35]Yu Guohua, "An Exploratory Comment," p. 100; Yu Guangxuan, "Several Questions," p. 161; Yu, *On Joint Campaign*, pp. 115–116; Lt. Colonel Fu Binzhong, "*Lianhe zuozhan tigao zhengti weili wenti chutan*" (A Preliminary Investigation of the Question on Enhancing the Comprehensive Power of Joint Operations), in NDU Scientific Research Department, *Research on the Theory of Campaign*, p. 185; Yang, *Research and Reflections*, p. 133.

[36]Yu Guohua, "An Exploratory Comment," p. 100; Yu, *On Joint Campaign*, pp. 115–116.

local and temporary initiative and superiority. Also, by causing substantial losses to enemy forces and arms through a first strike, an unfavorable balance of forces and arms may be reversed and chances for local and temporary superiority are enhanced. Finally, in some future local wars, "our adversary is on the defensive, and in a passive and reactive (*beidong yingfu*) position." This may create a situation where "we not only have the initiative to launch the campaign . . . but also the initiative to choose the transition between phases of the campaign, and the primary direction, location, and pace of the campaign."[37] But since first strike does not automatically ensure local and temporary initiative and superiority, the central question is: how to create conditions that enhance the chance of success for the first strike? The answer to this question lies in the examination of two key issues: the element of surprise, and initiation and initial battle of the campaign.

Element of Surprise

First strikes may not lead to local and temporary initiative and superiority if the adversary is well prepared and anticipates the strike. Therefore, whether an element of surprise (*turanxing*) can be achieved so that the enemy can be caught unprepared by the first strike is of crucial significance. Since the basic conditions of *turanxing* are transparency of enemy intentions and capabilities to the PLA on the one hand, and successful concealment of the PLA's intentions and capabilities to the enemy on the other, how to realize these two conditions becomes the central focus of the examination.

For the first condition, to become familiar with the warfighting doctrine and styles, command and organization, and weapons systems of potential adversaries at the strategic level requires long-term, systematic, and institutionalized study and analysis of the target military. At the campaign preparation level, however, it is necessary to extensively acquire and *affirm* intelligence on and continuously monitor enemy movements, weapons, and targets through secured and effective reconnaissance means such as spy satellites, surveillance ships and planes, ground monitoring radar and radios, and human intelligence.[38]

For the second condition, it is recognized that rapid advances in reconnaissance technology lead to real-time, multispatial, multidirectional, and more effective surveillance. Therefore, the traditional practice of "covering objects with tree twigs and bed sheets" is no longer effective in concealing PLA campaign preparation. On the other hand, there were many cases of successful concealment through nontechnological means on both sides during the 1991 Gulf War, and they all helped to achieve local and temporary superiority through reducing losses and maximizing gains. Also, campaigns may be fought close to the familiar environment of the

---

[37]Hu, *Contention Affecting the 21st Century*, p. 112; Yang Yushu, "An Exploratory Comment," p. 188.

[38]Zheng Shenxia, "Several Questions," p. 107; Nanjing MR Command Department, "Characteristics and Requirements," p. 153; Sun Xiaohe, "An Exploratory Comment," p. 224; Hu Changfa, "Speech at the Conclusion," p. 13; Major General Zhou Wenbi, "*Dui Gaoyuan diqu bianjing zhanyi de sikao*" (Reflection on Border Campaign on High Plateau), in NDU Scientific Research Department, *Research on the Theory of Campaign*, pp. 289–290.

homeland, which may increase the chances of successful concealment. Therefore, several major methods of concealment have been formulated. First, it is necessary to create a situation of "internal intensity and external relaxation" ("*neijin waisong*"). This means that while war preparation is intensified within the war zone, political and diplomatic measures may be employed to conceal strategic intentions, to dispel the atmosphere of war preparation, and thereby reduce enemy vigilance. Such measures may also gain time for war preparation, thus improving the odds of local and temporary initiative through first strike.[39]

Besides disinformation campaigns, other methods that may cause enemy misjudgment include "concealing the real" (*yinzhen*), "demonstrating the false" (*shijia*), "blending the real with the false, and the illusory with the substantial" (*zhenjia jiaozhi, xushi bingju*), and electronic measures during the campaign preparation. "Concealing the real" means hiding real forces and arms through campaign camouflage capable of stealth effects such as ambivalent color (*michai*), distorted form (*bianxing*), sound reduction (*xiaosheng*), and defilading (*zheyan*). "Demonstrating the false" refers to installing dummy planes, ships, tanks, vehicles, artillery pieces, missiles and other arms, and using reserve and militia units to conduct false deployment in directions where no operations are planned.[40]

"Blending the real and the substantial with false and illusory" means first of all mixing the real and substantial forces and arms with false and illusory ones. Second, since China's land and maritime borders are wide in direction and complex in geography, and they have numerous islands, harbors, towns and cities, and well-developed infrastructure in some regions, they also provide the conditions for meshing forces and arms with civilian facilities (*yujun yumin*). Both may increase the difficulty for the enemy to differentiate the real from the false. Moreover, night darkness, bad weather, and difficult terrain may provide favorable concealment conditions for campaign preparation and initiation. An asymmetry of transparency favors daytime, normal weather, and smooth landscape in spite of advances in surveillance technologies. Furthermore, times of relaxation provide favorable opportunities for the first strike to achieve surprise. Finally, electronic measures may be employed to conceal campaign preparation and initiation. These measures may include electronic interference or jamming to disrupt enemy surveillance, electronic flanking movements (*dianzhi yangdong*), deception (through transforming a multifighter formation into an electronic signature of one civilian plane, or one electronic warfare plane into a large attacking formation) to mislead the enemy, and

---

[39]Wang Wei, "Several Questions," p. 59; Fu Binzhong, "A Preliminary Investigation," pp. 181–182; Nanjing MR Command Department, "Characteristics and Requirements," p. 153.

[40]Yang, *Research and Reflection*, pp. 86–87; Nanjing MR Command Department, "Characteristics and Requirements," p. 153; Liu Yinchao, "Several Questions," p. 323; Major General Zheng Shouzeng, "*Shilun jituanjun bianjing diqu jidong fanji zuozhan de zhanyi jidong*" (An Exploratory Comment on Group Army Campaign Mobility in Mobile Counter-Offensive Operations in Border Region); Major General Zhang Xianglin, "*Binhai shandi jingong zuozhan yuanzhe*" (Principles for Offensive Operations in Maritime Mountainous Region); Major General Yuan Xinhua, "*Yilu zhihai, yidi zhikong*" (Use Land to Control the Sea, Use Ground to Control the Air), all in NDU Scientific Research Department, *Research on the Theory of Campaign*, pp. 295, 381, 385.

electronic silence during campaign initiation.[41] All these measures may confuse the adversary about the timing, place, scale, nature, and direction of the campaign initiation. This in turn enhances the probability of surprise, a basic condition to catch the enemy unprepared through first strike, thus achieving local and temporary initiative and superiority.

Initiation and Initial Battle

PLA planners may claim in public that "active defense (AD)" is a central principle that guides its initiation of military operations. But the PLA's internal writings have yet to demonstrate a serious concern for the contradiction between AD and first strike. On the contrary, the PLA interpretation of the principle is flexible to the extent that Gaining Initiative by Striking First (GISF) becomes a central component of AD. For example, one account stresses that "at the time when we are to restore islands and territories now occupied by the enemy, after adopting political and diplomatic initiatives that have failed to produce results and being faced with enemy aggression, we are forced to initiate offensive operations. At the strategic level, this belongs to the realm of 'active defense' . . . It is not only unscientific in theory but also lands us in a vulnerable position (*beidong diwei*) in practice if we separate GISF from AD, or establish a dichotomy between the former and the latter."[42] But for local initiative and superiority to be realized through preemption at the campaign level, the element of surprise, optimal timing, and proportion of forces devoted to the initial battle must become the central focus of attention.

According to PLA planners, the "window of opportunity," or the optimal timing for a first strike, is the brief period between the failure of political and diplomatic initiatives at the strategic level and the constitution of enemy comprehensive strike capabilities through completed deployment. "In the circumstance of enemy loss at the strategic level campaign commanders should grasp the favorable opportunity when the enemy's campaign deployment is still incomplete, launch a preemptive offensive within the scope permitted by the strategic objective, throw enemy campaign deployment into confusion, and force the enemy to fight us under the conditions of insufficient preparation and unfavorable posture."[43]

---

[41]Yang, *Research and Reflection*, p. 89–90; Nanjing MR Command Department, "Characteristics and Requirements," p. 153; Zheng Shouzeng, "An Exploratory Comment," p. 295; Liu Yinchao, "Several Questions," p. 324; Zhang Xianglin, "Principles for Offensive Operations," p. 381; Yuan *Xinhua*, "Use Land," p. 385; Lt. General Qiao Qincheng (PLAAF), "*Kongjun zhai lianhe zhanyi zhong de yunyong*" (Use of Air Force in Joint Campaign); Major Chao Kuofa (PLAAF), "*Denglu zhanyi zhong kongjun zuozhan yunyong de jige wenti*" (Several Questions on Air Force Operations in Amphibious Landing Campaign); Major General Wen Yuzhu, "*Gaojishu tiaojianxia yanhai diqu zhanyi zuozhan de zhanchang jianshe*" (Battlefield Construction in Coastal Regions For Campaign Operations Under High-tech Conditions), in NDU Scientific Research Department, *Research on the Theory of Campaign*, pp. 112, 116, 147–148.

[42]Hu, *Contention Affecting the 21st Century*, p. 113.

[43]Wang Wei, "Several Questions," p. 60. See also Major General Zhang Erwang, "*Lianhe zhanyi zhong changgui daodan budui zuozhan yunyong de jige wenti*" (Several Questions on the Use of Conventional Missile Forces in Joint Campaign), Colonel Wang Xiaodong *et al.*, "*Daodan budui zhai jingong zhanyi zhong de yunyong wenti*" (The Question of Using missile forces in Offensive Campaign), both in NDU Scientific Research Department, *Research on the Theory of Campaign*, pp. 228, 234.

Once the campaign is initiated, the outcome of the first battle is deemed crucial in determining whether local initiative is realized. Therefore, it is necessary to "throw a powerful and superior initial strike force (*shoutu bingli*) into the initial battle." For the air force, for instance, "as high as 80% of the campaign air force should be used in the initial battle in coordination with surface-surface missiles, ship artillery, ground force aviation, electronic warfare, and special operations capabilities." The objective is "temporary suppression of enemy capabilities" by striking key targets that are "relatively vulnerable but crucial in constituting enemy comprehensive operational capabilities." A substantial initial strike force is necessary because it increases the chances of expanding the positive effects of initial strike. "Under high-tech conditions, local and temporary superiority achieved through initial battle can be easily and quickly lost . . . it is therefore necessary to engage in resolute, active and continuous offensive to deny enemy breathing space."[44]

FIGHTING A QUICK BATTLE TO FORCE A QUICK RESOLUTION

While GISF is associated with the beginning of the campaign, "quick battle and quick resolution" (QBQR) deals with its prosecution and conclusion. Several reasons are offered to explain why QBQR is more desirable than the old notion of protracted war. First, border wars are not fought for total conquest of a country but rather for disputed territories in a confined area. It is therefore necessary to contain the conflict, and "swiftly smash enemy occupation or diplomatic blackmail, and achieve a final resolution at one stroke (*yicixing jiejue*)." Also, "some border regions are remote, have bad and limited roads and harsh natural environment. The material basis in these regions for war is very weak." As a result, "arms may be negatively affected by harsh conditions, and it is difficult to organize logistics to sustain the war effort for long." Moreover, the high speed, precision, and lethality of high-tech weapons may quicken the war process. But the high cost of high-tech weapons also means that war can be too costly if it is allowed to drag on or escalate. Also, prolonging the war may not only mean higher expense to sustain the war, but also severely damage the civilian infrastructure, thereby negatively affecting the national economy. Finally, simply because the PLA is inferior in technology, it is all the more necessary for it to fully exploit the local and temporary superiority it could achieve through preemption to resolve the war in its favor in the shortest possible time. If the war is allowed to become protracted to the extent that the temporarily suppressed enemy regains its superiority, the PLA may face the prospect of being defeated.[45] If QBQR is both desirable and necessary, the central question is: how will the PLA

[44]Wang Wei, "Several Questions," p. 60; Zheng Shenxia, "Several Questions," p. 103; Qiao Qingcheng, "Use of Air Force," p. 112; Chao Kuofa, "Several Questions," pp. 115–116; Liu Yinchao, "Several Questions," p. 322; Major General Ma Diansheng, "*Guanyu Kongjiang zhanyi de jige wenti*" (Several Questions regarding Airborne Campaign); Major General Gui Quanzhi, "*Gaohan shandi bianjing fanji zuozhan de zhidao yuanzhe yu zhanfa*" (Guiding Principles and Fighting Methods Regarding Counter-Offensive Campaign on High and Cold Plateau), both in NDU Scientific Research Department, *Research on the Theory of Campaign*, pp. 250–251, 307–308.

[45]Fu Binyao, "Several Questions," pp. 267–268; Yang, *Research and Reflection*, p. 93; Major General Zhu Wenquan, "*Shandi jingong zhanyi de jige wenti*" (Several Questions regarding Offensive Campaign in Mountainous Region), in NDU Scientific Research Department, *Research on Theory of Campaign*, p. 363.

realize QBQR in its favor? The answer to this question is associated with the articulation of two key concepts: mobility (*jidong*) and offensive operations (*jingong zuozhan*).

Mobility

Because of the highly destructive nature of high-tech weapons, static positional defense may not lead to temporary and local superiority, the basic condition to achieve QBQR in PLA's favor, but may instead increase the chances for the PLA to be quickly defeated. Therefore, continuous, dynamic offensives through high mobility are now deemed absolutely necessary. Successful mobility may contribute to local superiority for several reasons. It may reduce inferiority through swiftly deploying superior forces and arms at crucial times and places. It may reduce one's own vulnerability by denying the enemy fixed targets. It also generates initiative and momentum through shifting and separating enemy forces, confusing the enemy about intentions, and thus creating opportunities to annihilate separated enemy forces by taking advantage of a favorable time and place of one's choosing.[46]

While movement warfare (*yundong zhan*) has always been an important element of the PLA's fighting style, it is now recognized that the means of mobility have changed. Modern mobile warfare (*jidong zhan*), for instance, involves motorized transport, air, and rail, while the old warfare relied on nonmechanized transport. Even though the latter may reduce losses because it provides smaller targets that are easier to conceal through camouflage and in difficult terrain and night conditions, it is too slow, and therefore is no longer applicable to the mobility of PLA campaign forces. While faster, however, mobile warfare is more dependent on good roads, rails, and sometimes complex and cumbersome logistics; it is noisier; and it offers more visible profiles to the enemy. Also, reconnaissance technology has developed to the extent that it is now possible to identify subtle signs of mobility using detection of light, heat, magnetic, and acoustic waves. The marriage of technology-based reconnaissance of the battlespace with long-range, precision-guided munitions from the air may neutralize the central objective of PLA mobility: reducing one's own losses and creating local and temporary momentum through preemption at unexpected times, places and directions and with unexpected intensity and styles.[47] It is clear that whether local information and air superiority can be achieved may determine whether mobility leads to local and temporary initiative, so that QBQR can be realized in PLA's favor.

For information superiority, the concept of "counter-reconnaissance" (*fanzhencha*) is advanced. This concept deals primarily with the "intelligence" aspect of the enemy

---

[46]Hu Changfa, "Speech at the Conclusion," p. 8; Zheng Baoshen, "An Elementary Investigation," p. 90; Peng Cuifeng, "Preliminary Investigation," p. 275; Senior Colonel Zhan Xuexi, "*Xiandai zhanyi tedian*" (Characteristics of Modern Campaign), Lt. General Han Reijie (PLAAF), "*Zhanyi jidong zhong de kozhong shusong wenti*" (The Question of Air Transportation in Campaign Mobility), both in NDU Scientific Research Department, *Research on the Theory of Campaign*, pp. 53–54, 344; Yang, *Research and Reflection*, p. 73.

[47]Yang, *Research and Reflection*, pp. 84–85.

C4I and intends to neutralize enemy "eyes," "ears," and "noses" through centrally organized, comprehensive, integrated countermeasures. Most of the measures that may help to gain an "element of surprise" in campaign initiation (as discussed earlier) are also applicable to concealed mobility. These include disseminating disinformation, camouflage and concealment, electronic interference and deception, applying anti-photoelectric, anti-infrared, anti-radar, anti-sonar, and other stealth technology to arms, introducing weight-reduction technology to arms, and exploiting favorable natural and social conditions. More proactive countermeasures are also specified, including developing and deploying anti-surveillance satellite measures; damaging enemy ground and air surveillance radars, and unmanned drones; monitoring radios and other major sensors; and strikes by long-range missiles, attack airplanes, ship artillery, and special forces; and employing laser, kinetic energy (*dongneng*), and particle beam (*lizhi shu*) weapons during early operations (*xianqi zuozhan*). Finally, the security and effectiveness of the PLA's own "eyes," "ears," and "noses" may be enhanced through both preventive and protective measures.[48]

For threats to mobility from the air, it is acknowledged that the traditional close air raid may be replaced by the over-the-horizon (*chao shiju*) surface-to-surface and air-launched missiles and smart bombs as the dominant threat, and the launching platforms may stand-off beyond the scope and reach of the PLA campaign air defense network. On the other hand, stealthy tactical aircraft and attack helicopters mean that close air attack may still be necessary to strike tactical targets.[49] Therefore, achieving local and temporary air superiority may depend on comprehensive planning and integrated air defense measures.

It is, for instance, necessary to establish one or several inter-connected "mobility corridors" (*jidong zoulang*) or a "local security umbrella" (*diyu anquan shan*) at the locality and time when campaign forces are to move. Such an umbrella is sustained by several major measures. One is using surface-to-surface missiles, bombers, and other long-range arms to strike and destroy as many enemy stand-off platforms as possible in early operations. This is then supplemented by Air Force interceptors, as well as mobile and fixed long-range SAMs and large-caliber AAA batteries deployed adjacent to the mobility zone. In the meantime, mobile air defense units equipped with portable SAMs and mobile small-caliber AAA batteries deal with attack helicopters and other close-in flying targets. Such an umbrella may be connected with the strategic air defense for early warning and early interception. Furthermore, the umbrella may be reinforced by continuous electronic measures to reduce or confuse the enemy's awareness and enhance its transparency to the PLA. The survivability of the air defense forces themselves needs to be enhanced through mobility and concealment. Finally, forces and arms mobility must not be conducted under daylight conditions without surveillance and air superiority.[50]

---

[48] *Ibid.*, pp. 86–88.

[49] *Ibid.*, p. 88.

[50] *Ibid.*, p. 88–89, Major Gong Qing, "*Lianhe zhanyi zhong fangkong liliang zhengti shiyong wenti tantao*" (An Investigation of the Question Regarding Comprehensive Use of Air Defense Forces in Joint Campaign), NDU Scientific Research Department, *Research on the Theory of Campaign*, p. 263.

In addition to surveillance and air threats, several major modes of mobility may enhance deception or reduce exposure and casualty, thus improving the odds of local and temporary initiatives. These include exterior-line mobility (ELM, or *waixian jidong*), interior-line mobility (ILM, or *neixian jidong*), and leap-forward mobility (LFM, *yuejinshi jidong*).

ELM refers to mobility outside but close to the exterior flank of the enemy's deployment. It is usually conducted when enemy deployment is relatively concentrated, the battlefield is large in scope, and time is sufficient. Since the purpose of ELM is to create momentum (*zhaoshi*) that may shift enemy forces and create favorable fighting opportunities (*zhanji*) and vulnerable targets, it may not have to be always concealed. Small-scale mobility may divert attention from large-scale mobility (*xiaodong yan dadong*), or a posture of encircling and capturing key targets or positions may intend to trigger enemy reinforcement, thus creating opportunities for annihilating enemy vital forces through mobility.[51]

Unlike ELM, ILM is mobility through gaps between forces within the enemy's deployment. ILM is conducted when favorable opportunities have been created with concrete targets. Finally, LFM means that forces are dispersed and organized into smaller, more agile tactical units that take many routes (*xiaoqun duolu*), and move swiftly from one point to the next simultaneously and in darkness. It is assumed that LFM may lead to lower force density, less exposure time, and shortened length of mobile columns, which in turn may enhance both survivability and speed of mobility.[52] All these modes may be employed in a combined way or separately, depending on how they contribute to local and temporary initiatives under specific circumstances. The central objective of mobility, however, is to create favorable conditions for offensive operations, the crucial step leading to QBQR in PLA's favor.

Offensive Operations

Surprise attack (SA, or *xiji*) is defined as the central mode of PLA offensive operations. "Information offensive" (*xinxi jingong*) is a new mode of PLA offensive operations. Three major types of SA have been explicated, including firepower SA (*huoli xiji*), ambush SA (*fujixing xiji*), and mobile SA (*jidong xiji*).

Firepower SA is "a new type of SA for the PLA," and is based on the concentrated effects of conventional strategic, campaign, or tactical missiles; bombers or attack aircraft; and campaign or tactical artillery. It can also be solely based on missile or air firepower. The attack can be sudden, concentrated, and brief, or continuous within a limited period of time. The advantage of firepower SA is that it is less restricted by distance, easy to organize and command, and easy to control. It is, however, limited since it cannot capture territories, positions, and materials. Because of this, it may be primarily conducted at the strategic and campaign levels for objectives such as battlefield denial, cutting off sea and air transportation, striking political and

---

[51] Yang, *Research and Reflection*, p. 90; Wang Yongguo, "An Investigation," p. 195.

[52] Yang, *Research and Reflection*, pp. 90–91; Liu Yinchao, "Several Questions," p. 324.

economic targets, and destroying military targets. For similar reasons, it is more desirable to use missiles than aircraft to avoid the complex preparation for gaining conditional air superiority, and more desirable to be sudden, concentrated, and brief than continuous to achieve QBQR.[53]

The purpose of ambush SA, however, is to deal with mobile enemy forces. Through concealed predeployment of forces close to the zone where possible enemy operations may be conducted, comprehensive capabilities based on air-land firepower, force mobility and assault are applied suddenly to enemy forces entering the zone. While ambush has always been a PLA fighting method, there are several major differences between the new and the old. First, since enemy mobility may be swift using multiple routes and many tactical groupings, the new ambush zone will be reasonably large so that ambush forces may have sufficient space to move and choose favorable positions while having better chances of avoiding detection. Also, unlike the old ambush where forces are deployed very close to the strike zone, the mobile ambush forces will maintain an optimal distance from the strike zone to conceal intentions, and close in swiftly through a "sudden, fierce" assault on the enemy forces that enter the strike zone. Moreover, unlike the old single-service-based ambush, the new ambush combines air-land firepower with ground forces assault. Finally, to ensure local superiority, rigorous preventive measures will be implemented against enemy mobility, air raid, and other counter-ambush measures.[54]

Mobile SA attracts the most attention. Unlike a firepower SA or ambush SA, the purpose of a mobile SA is to fight "temporarily stationed enemy forces and a hastily constituted enemy defense." The essence of a mobile SA is "suddenly annihilating (*jianmie*) the enemy through concealed and swift mobility," so that QBQR in PLA's favor can be realized.[55] Specifically, three types of mobile SAs have been articulated, including continuous assault (*lianxu tuji*), deep strike (*zhongshen gongji*), and vertical strike (*liti gongji*, or three-dimensional strike).

Continuous assault is meant to be directed against temporarily stationed enemy forces. The central premise is that since the enemy forces and arms are precise, lethal, fast, and responsive, PLA forces may exploit the brief opportunity when the temporarily stationed enemy forces are unprepared or not fully prepared. In this circumstance, PLA forces may launch a multi-direction, multi-route, multi-echelon, continuous assault on key enemy positions at both the *front* and in the *depth* of enemy deployment. First, several frontal and deep operational groupings (*zhengmian zuozhan qun* and *zhongshen zuozhan qun*) may be organized. Each grouping is capable of air-land operations, and consists of several echelons (*tidui*) that may conduct continuous assault alternately. Continuous assault may begin with "sudden, heavy, and brief firepower assault" on key targets, such the enemy C4I, and against missile, plane, and artillery platforms "to reduce enemy coordination and fire

---

[53]Yang, *Research and Reflection*, pp. 78, 95; Zhang Erwang, "Several Questions," p. 230.

[54]Yang, *Research and Reflection*, pp. 95–96.

[55]*Ibid.*, p. 96.

power." This is to be immediately followed by simultaneous assault by all operational groupings to cut apart (*fenge*) enemy forces; to prevent coordination between enemy firepower support, counter-attack forces, and divided enemy forces; and to encircle (*baowei*) each slice of enemy forces. At the same time, deep operational groupings may strive to intertwine with (*jiaozhi*) enemy depth forces to reduce enemy superiority in long-range firepower support and mobile reinforcing forces, and to give full play to PLA's advantages in close combat. To prevent the divided enemy forces from either being diverted (*zhuanyi*, the PLA euphemism for relief from encirclement) or being consolidated, thus running the risks of fighting a battle of routing (*jikuizhan*) or a battle of attrition (*xiaohaozhan*) rather than an ideal battle of annihilation (*jianmiezhan*), all groupings must conduct assault continuously to preclude breathing space for the enemy, and to achieve QBQR in PLA's favor at one stroke.[56]

Deep strike is directed mainly against fortified enemy defenses. Its underlying premise is that since enemy defenses are more likely to be mobile than fixed in order to reduce losses caused by the adversary's preemption, its firepower support system, main counter-attack forces, and reserve forces (*yubeidui*, referring to regular backup forces but not the reserves) are likely to be deployed in the close depth (*qian zongshen*) of defense. They are also likely to be connected with C4I that coordinates firepower, force mobility, and counter-attack, which are likely to be deployed in the deep depth (*quan zongshen*) of defense. They generally constitute the "belly" that is crucial to successful defense. On the other hand, this "belly" may be soft since various components may be loosely connected and secured. As a result, it may provide a window of vulnerability for deep strike. Still, "it is not realistic for our army to conduct the kind of simultaneous deep strike as conducted by Western armies, since in overall terms, we will remain inferior in high technology for a long time." On the other hand, "it is absolutely necessary to conduct deep strike. The focus of our deep strike, however, is on decomposing enemy defense, restricting enemy mobility in firepower support, forces, and electronic capabilities, weakening and sabotaging enemy comprehensive defense capabilities, and creating conditions for annihilating enemy forces separately."[57]

Specifically, deep strike forces may be composed of a carefully selected armor-based assault component, a raid and sabotage component, and sometimes an airborne and heliborne component. Each component is to be capable of ground and air defense and well armed with instruments and techniques of destroying or sabotaging various types of enemy targets. While taking into consideration alternative plans in the face of possible enemy interception, blocking, or delay through air-land firepower strike, mobile obstacles, and charges (*chongji*) by enemy counter-deep strike forces, PLA deep strike forces may either force their entry into the enemy's defense perimeter with the firepower support of air, missile, artillery and frontal attacking forces, or infiltrate through darkness, bad weather, and favorable terrain. Once reaching their targets, a swift, fierce, and coordinated attack is to be launched, aimed at destroying

---

[56] *Ibid.*, pp. 98–99.
[57] *Ibid.*, pp. 99–100.

the crucial parts of the enemy's C4I system, weapons platforms, and backup forces in the shortest time possible. Depending on circumstances, strikes can be concentrated on one target system after another, or dispersed against several separate systems simultaneously.[58] The central objective is to weaken and paralyze the enemy's comprehensive defense capabilities, so that QBQR in PLA's favor may be hastened.

Vertical strike may be applied in both continuous assault and deep strike, or conducted independently. Its basic premise is that since enemy defenses have become more effective, the traditional PLA practice of "using the ground to control the air" may also be supplemented by vertical strike, making offensive operations more effective. But since "the PLA aviation elements are still rather limited in both quantity and quality, it is not possible to wage vertical strike throughout the whole process of offensive operations." Therefore, vertical strike is to be applied to key directions, at key times, and against key targets in order to enhance local and temporary superiority. The basic elements of vertical strike include:

- dropping paratroopers into the enemy campaign and tactical depth to attack enemy forces;

- using transportation planes and helicopters to ship deep strike forces into the enemy's defensive depth for striking predetermined targets, and for outflanking (*yuhui*), encircling, and cutting apart enemy forces;

- landing forces in the enemy's rear to surround (*hewei*) and encircle enemy forces;

- using helicopters to transport forces to favorable positions, to block and fight incoming enemy reinforcing units, thus ensuring complete annihilation of encircled enemy forces;

- using attack helicopters to provide support for outflanking, encircling, penetrating, and cutting up enemy forces;

- using concealed helicopter shipments of special forces to raid vital enemy targets; and

- using fighters to deny airspace over encircled enemy forces, and using attack planes and bombers to provide airborne fire support for attacking ground forces.[59]

Application of vertical strike may be concealed and sudden, though only under local air superiority. Vertical strike may be assembled in such a way that it is highly streamlined, capable of independent operations, and well equipped with anti-

---

[58]*Ibid.*, pp. 100–101; Zhu Wenquan, "Several Questions," pp. 364–365; Major General Ding Shouyue, "*Lun gaojishu tiaojianxia jituanjun chengshi jingong zhanyi zhanfa*" (A Comment on the Fighting Methods of the Group Army in Urban Offensive Campaign under High-tech Conditions), in NDU Scientific Research Department, *Research on the Theory of Campaign*, p. 394.

[59]Yang, *Research and Reflection*, p. 101; Ma Diansheng, "Several Questions," p. 252; Zhu Wenquan, "Several Questions," p. 364; Ding Shouyue, "A Comment," p. 394.

air/anti-tank capabilities and electronic warfare capabilities that can sabotage and suppress enemy communications and fire control systems. Furthermore, it may be used in such a way that it coordinates well with ground fire support and attacking forces. Finally, technical and operational maintenance needs to be strictly organized. The basic purpose of vertical strike is to create conditions for QBQR to be realized in PLA's favor.[60]

Finally, the notion of an "information offensive," particularly relating to the communications and computer aspects of C4I, have been explicated under two underlying premises. One is that sensor-based information is transmitted to command and control through channels such as communications satellites, laser-based communications facilities, and wired or wireless radios and televisions. Also, information is analyzed, processed, stored, and disseminated to commanders for decisions and weapons platform control through digitally-linked computer networks.[61] Therefore, neutralizing or destroying communications channels and computers through a combination of "soft" and "hard" means may cause severe disruption and even disintegration of enemy decisionmaking and coordination, the basis for enemy comprehensive fighting capabilities.

Several methods have been proposed for an information offensive. On the soft side, if enemy communications are immune to monitoring, various types of electronic interference may be applied to disrupt transmissions. Similarly, "special computer reconnaissance equipment" may be used to collect "weak electromagnetic signals from enemy operating computers" that may be translated into legible information. Otherwise, the method of "touching the vital point" (*dianxue*, a derivative from the martial arts mythology of paralyzing or killing a person by touching a vital point of the body) may be applied. For example, "computer virus assault" to damage the software of enemy computers, hacking into an enemy computer network to acquire vital information, or hacking into enemy computers to replace real information with disinformation. Moreover, microwave "bombs" (referring to electronic magnetic pulse, or EMP) may be used to burn the circuits and other vital parts of enemy computers. Finally, power supplies may be sabotaged to paralyze enemy C4I.[62]

But since "soft kill" methods "can only temporarily paralyze the enemy information systems," "hard kill" methods are indispensable in "fundamentally destroying the whole system." This requires reconnaissance satellites, ground and air surveillance radar and radios, and other means to systematically and continuously collect and analyze information on the status and position of enemy information facilities and equipment. This lays the foundation for employing missiles, attack planes, ground/ship artillery, laser and kinetic energy weapons, and special forces to conduct precision strikes against or sabotage of enemy C4I. In the meantime, an

---

[60] Yang, *Research and Reflection*, p. 102.

[61] Senior Colonel Wang Jun *et al.*, "*Xinxizhan zhanyi qianjian*" (An Elementary Perspective on Information Warfare Campaign), in NDU Scientific Research Department, *Research on the Theory of Campaign*, p. 455.

[62] *Ibid.*, p. 455–456; Major General Jia Fukun, "*Xinxizhan—weilai zhanzheng de zhongyao zuozhan fangshi*" (Information Warfare—An Important Operational Style of Future War), in NDU Scientific Research Department, *Research on the Theory of Campaign*, p. 443.

information offensive may be accompanied by measures to enhance protection and security of the PLA's C4I, particularly in reinforcing its anti-interference (to ensure smooth and secure flow of communications), anti-virus (to ensure normal functioning of computers), and anti-destruction (through enhancing mobility or hardening bunkers) capabilities.[63] The paralysis of enemy C4I and the survival of the PLA's C4I may greatly enhance PLA local and temporary superiority, thus contributing to a QBQR in PLA's favor.

CONCLUSION

This essay has outlined new campaign doctrine and strategies that may enhance the chances for the PLA to turn its absolute technological inferiority to local and temporary superiority. The newly articulated notion of *WZC*, for instance, may contribute to such superiority, since it may enable the PLA to fully exploit its new joint forces-based "pockets of technological excellence." Such "excellence," however, may be either underutilized or overwhelmed in the other two major types of war: the CAGA-level campaign and total war. Similarly, the chances of such superiority may be better if the joint forces–based *EFSA* can be optimally deployed, commanded, and coordinated, so that casualties can be reduced and internal friction and attrition minimized on the one hand, and positive, interservice complementarity for common goals may be enhanced on the other. Moreover, the odds of such superiority may be improved by a strategy of *GISF*, if such a strike catches the enemy by surprise with optimal timing and optimal use of initial strike forces and arms. Finally, the chances for the PLA's local and temporary superiority may be enhanced by successful mobility, based on conditional information superiority, air superiority, and surprise attacks that may divide and disintegrate enemy coordination.

A CRITIQUE

What then are the likely implications of the new doctrine and strategies for China's security posture? Through a reading of this essay, one may easily conclude that the PLA has adopted a more forward-deployment-based, offensive posture. Such a conclusion may need to be qualified in two major ways. First, by adopting a military mission-focused offensive doctrine, such discourse may serve a domestic political purpose: to call for reducing the PLA's nonmilitary departments and functions, such as intervention in politics and business, and demand more military budget from the civilian leadership. For instance, Senior Colonel Wang Jianghuai's article on the joint forces command, which was first published in *Guofang daxue xuebao* (*NDU Journal*) and later summarized in the *Liberation Army Daily*, openly criticized the political departments and the nonmilitary aspects of the logistics departments for contributing to the bloating of the PLA command structure.

It is now apparent that those who advocate more focused military professionalism have won half of their political battle: they have convinced the top PLA and civilian

---

[63]Jia, "Information Warfare," p. 447; Wang, "An Elementary Perspective," p. 456.

leadership to ban the PLA's business activities, to increase the defense budget, and to downsize the PLA so that more money can be saved for improving its quality. On the other hand, they have not been successful in separating the PLA from the party. The recent State Council Defense White Paper has reiterated the party's absolute leadership over the PLA and the need to maintain political departments within the PLA.[64] If articulation of a more offensive doctrine and strategies is part of a domestic political agenda that seeks to create an apolitical national army (*guojun*),[65] then one should not expect its implementation anytime soon.

Second, some major points made by PLA's planners in this essay may reflect wishful thinking and may not be realizable. An important external condition that may enhance the chances of a successful WZC without escalation, for instance, is the absence of intervention by a "powerful" third party, or the absence of a second front if or when the campaign takes place. But the PLA may not have direct influence or control over either of these two conditions. Also, newly conceived joint services operations may require more fundamental reinvention of the current PLA command structure. Such reinvention may require establishing a genuine *state-based* Ministry of National Defense that absorbs all the functions that are essential but not immediately related to military operations. This in turn may enable the introduction of a streamlined, relatively autonomous joint forces coordinating mechanism and command system that can be highly focused on joint operational matters. This is not likely to happen in the short run, since it may mean the demise of the current nonjoint, highly administrative system of four general departments answering to a *party-based* central military commission. Both vested interests and habit may work against such change.

Third, in spite of all measures to increase the enemy's transparency to the PLA and to reduce the PLA's visibility to the enemy, successfully concealing a WZC-sized force for a preemptive strike that may catch the enemy by surprise is an immensely difficult deed to accomplish. Finally, quick battle and quick resolution in PLA's favor may not be easy to achieve in border wars over disputed territories. China fought a border war against Vietnam in 1979. The war may have ended quickly, but the outcome was apparently not in PLA's favor, and the border skirmishes continued even after the war ended. This happened largely because the Vietnamese used a "People's War" strategy against the PLA, thus triggering a security dilemma in the border region. Indeed, recent PLA writings on campaign doctrine and strategies do not contain alternative plans if or when the PLA cannot gain the initiative by striking first or if quick battle and quick resolution become difficult to obtain.

---

[64]See *China Daily*, July 28, 1998.

[65]For adopting an offensive ideology to promote the military's institutional interests in the domestic political arena, see Jack Snyder, *The Ideology of the Offensive*, Ithaca, NY: Cornell University Press, 1984; and Barry Posen, *The Sources of Military Doctrine*, Ithaca, NY: Cornell University Press, 1984.

A CRITIQUE OF THE CRITIQUE

In spite of these qualifications, the new campaign doctrine and strategies deserve serious attention and further analysis for several major reasons. First, major weapons systems that are being acquired or developed by the PLA, including guided missile destroyers and improved IRBMs and ICBMs, may be intended to deter "powerful" third party intervention by making the cost unacceptably high. Nuclear and missile technology transfer and improved IRBMs may also be designed to deflect and deter pressure for a possible second front. To conclude that the PLA has little influence over the external environment that may enhance or reduce the chances of a successful WZC may be premature.

Furthermore, if articulation of a new doctrine serves a domestic political purpose, it may also translate into genuine policy programs, especially if civilian leadership is successfully persuaded by the PLA about the benefits of the offensive doctrine and the domestic political struggle is resolved in PLA's favor. Also, the current PLA four-department system may not be so detrimental to successful joint services operations. It is, for instance, possible to simulate and practice a more streamlined wartime joint forces command structure separately. In the meantime, the current four-department system may continue to function for the purpose of peacetime PLA administration.

The conventional wisdom that the Party's political control may decimate military professionalism is also inconclusive. Recent history has shown that one-party authoritarianism sometimes produces effective militaries. A new book by the MIT/Harvard historian Loren R. Graham, *What Have We Learned about Science and Technology from the Russian Experience?*, shows that science and technology flourished during the most repressive years of the Soviet period. This happened largely because scientists were bestowed with abundant research money and privileges, they tended to be intensely loyal to the homeland, and they were focused on their work to escape the horrible social reality around them.[66] By the same logic, it is plausible that as long as the PLA remains loyal to a party that still makes most important decisions, and a tacit agreement is reached for both not to meddle too much in each other's affairs, the PLA in return may gain more resources for its institutional, professional, and technological development.

Furthermore, the emerging irredentist nationalism in a rapidly modernizing China may provide a latent motivation for defense modernization. Among China's irredentist claims, Taiwan probably provides a focus upon which the otherwise disparate resources and energies in an increasingly commercial society may be brought to bear. Taiwan is neither an overly easy nor a totally insurmountable objective to achieve. It is sufficiently "Chinese" but separate enough to justify the irredentist claim. It is located in the center of China's maritime-oriented economic development and security design, and may need to be overcome if it becomes an obstacle in realizing this design. All these factors may provide the PLA with a perfect rationale to request and acquire sufficient resources, and an ideal incentive to optimize the use of these resources. This may in turn transform the PLA from a labor-

---

[66]Cited in *Washington Post*, August 12, 1998, p. A21.

intensive, ground-force-dominated, defense-oriented force to a technological, joint operations–based, and forward deployment–oriented military.

Moreover, it is certainly true that a glaring gap has existed between theory and practice in PLA's past. This explains why many (including this author) argue that some major points articulated by PLA planners in this essay may not become reality. But it is also plausible that after the substantial technocratic development of the PLA for the past 20 or so years, the old Maoist-ideology-based factional bickering has been declining. This means that the old factional balancing behavior, which usually causes policy paralysis (hence the gap between theory and reality), may be gradually supplanted by more conciliatory, bandwagoning behavior. Tentative evidence shows that the gap between theory and practice may be somewhat narrowed as central policy capacity is being enhanced. The relative success of the 1985–1987 PLA reorganization, for instance, contrasted sharply with the abortion of the 1975 PLA downsizing, when Deng and his followers were severely criticized by the Maoist faction as "reversing the verdict of the Cultural Revolution." One may argue that the post-1985 PLA business activities provide a counter example. But such activities had occurred with the full consent of top leaders such as Deng and Zhao Ziyang. Therefore, there is little reason to believe that the recent central decision to ban such activities would fail, particularly if the top leadership has promised to compensate for ban-related financial shortfalls with defense budget increases.[67]

Similarly, a careful reading of this essay may show that the PLA theorists are sufficiently sophisticated to discuss *conditions* that may lead to certain desirable outcomes. This means that if these conditions (concealment of intentions and capabilities, for instance) cannot be met, certain desirable outcomes (e.g., catching the enemy by surprise) may not be realizable. Therefore, certain actions (e.g., preemptive strike) should not be taken, since the cost may be too high. This implies that discussions of alternative plans do exist, but may not be easily accessible, perhaps because they are protected by higher levels of classification. More careful analysis may be necessary. Indeed, one runs the risk of self-denial if one discards the new PLA doctrines and strategies as simplistic, chest-beating bravado and propaganda.

Finally, while the PLA may be cautious enough not to threaten or fight wars unless the benefit is perceived to be high, the cost low, and the prospect of winning good, it does sometimes take a hawkish position within China's foreign policy establishment. It has become almost routine for the PLA to confront the more dovish Ministry of Foreign Affairs on issues such as missile technology transfer, the Nansha (Spratly) Islands crisis in the late 1980s, and, most recently, Taiwan.[68] As shown in this essay, the PLA openly demands diplomatic deception for achieving military objectives. In the end, the inter-departmental conflict is arbitrated by the top party leadership. What is disturbing is that on many of these issues, the Party's arbitration has favored

---

[67] For the promise made by Hu Jingtao, a member of the Politburo Standing Committee, see *Qiao Bao* (*The China Press*), July 29, 1998.

[68] What matters here is the dominant institutional pattern. This, however, does not exclude the possibility that a minority of doves exists on the military side, as does a minority of hawks on the State Council side.

the PLA. If domestic civilian control of the military has a tendency to be compromised by collusion between the Party and the PLA, some type of external control may be inevitably necessary in the future, particularly when the PLA becomes stronger and more assertive. For this reason, further analysis of the PLA's intentions and capabilities may be highly desirable.

9. THE PLA AND INFORMATION WARFARE

James Mulvenon

INTRODUCTION

Among recent discussions of the evolution of Chinese military doctrine, few subjects have received as much attention as information warfare (IW).[1] China is arguably only one of three countries pushing the envelope on IW strategy development, behind the United States and Russia.[2] It has an active offensive IW program and has devoted significant resources to the study of IW. Chinese military journals are replete with articles that either directly or indirectly address the subject, and a significant number of full books by PLA authors have been published in the past few years.[3] Granted, IW's current cachet in both China and the United States can be partly explained by the hip, futuristic, attractively ill-defined nature of the subject, which invites the frenetic pace at which some of the nation's most forward thinkers are attempting to coin the permanent neologisms and concepts of this new type of combat.[4] At the same time, however, I would argue that behind all the rhetoric and hype, IW presents

---

[1]Two good introductions to the subject are Bates Gill, *China and the Revolution in Military Affairs*, Carlisle, PA: Strategic Studies Institute, 1996; and John Arquilla and Solomon Karmel, "Welcome to the Revolution . . . in Chinese Military Affairs," *Defense Analysis*, 13:3, December 1997, pp. 255–269.

[2]For an excellent cross-section of U.S. writings on the subject, see John Arquilla and David Ronfeldt (eds.), *In Athena's Camp: Preparing for Conflict in the Information Age*, Santa Monica, Calif.: RAND, MR-880-OSD/RC, 1998. See also Martin Libicki, *What Is Information Warfare?* Washington, DC: National Defense University Press, 1996; John Arquilla and David Ronfeldt, "Cyberwar Is Coming!" *Comparative Strategy* 12, No. 2, Spring 1993, pp. 141–165; Richard Szafranski, "A Theory of Information Warfare," *Airpower Journal*, Spring 1995, pp. 56–65; Alan Campen *et al.*, *Cyberwar*, Washington, DC: AFCEA Press, 1996; Norman Davis, "An Information-Based RMA," *Strategic Review*, Winter 1996; C. Kenneth Allard, "The Future of Command and Control Warfare: Toward a Paradigm of Information Warfare," in L. Benjamin Ederington and Michael Mazarr (eds.), *Turning Point: The Gulf War and U.S. Military Strategy*, Boulder, Colo.: Westview Press, 1995. For the best summary of Russian IW writings, see the work of Timothy Thomas.

[3]Among books, the most notable are Wang Pufeng, *Xinxi zhanzheng yu junshi geming* (Information Warfare and the Revolution in Military Affairs), Beijing: Junshi kexueyuan, 1995; Shen Weiguang, *Xin zhanzheng lun* (On New War), Beijing: Renmin chubanshe, 1997; Wang Qingsong, *Xiandai junyong gaojishu* (Modern Military-Use High Technology), Beijing: AMS Press, 1993; Li Qingshan, *Xin junshi geming yu gaojishu zhanzheng* (New Military Revolution and High Tech War), 1995; Zhu Youwen, Feng Yi, and Xu Dechi, *Gaojishu tiaojianxia de xinxizhan* (Information War Under High Tech Conditions), Beijing: AMS Press, 1994; Zhu Xiaoli and Zhao Xiaozhuo, *Mei'E xin junshi geming* (The United States and Russia in the New Military Revolution), Beijing: AMS Press, 1996; Dai Shenglong and Shen Fuzhen, *Xinxizhan yu xinxi anquan zhanlue* (Information Warfare and Information Security Strategy), Beijing: Jincheng Publishing House, 1996.

[4]One can identify a similar dynamic in the early years of the literature on nuclear strategy. See Fred Kaplan, *The Wizards of Armageddon*, New York: Simon and Schuster, 1983.

the Chinese with a potentially potent, if circumscribed, asymmetric weapon. Defined carefully, it could give the PLA a longer-range power projection capability against U.S. forces that its conventional forces cannot currently hope to match. In particular, I would argue that these weapons give the PLA a possible way to attack the Achilles' Heel of the advanced, informatized U.S. military: its information systems, especially those related to command and control and transportation. By attacking these targets, the Chinese could possibly degrade or delay U.S. force mobilization in a time-dependent scenario, such as Taiwan, and do so with a measure of plausible deniability.

This paper seeks to outline the current debate within the PLA over information warfare, emphasizing its remarkably heterogeneous character. It draws upon a sizable number of full-length books and journal articles. What this paper does *not* do, however, is assess PLA capabilities in information warfare, since nearly all of the relevant data resides in the classified realm. Nonetheless, this literature analysis is an important first step in understanding the role of information warfare in the 21st century PLA.

DEFINING TERMS

Before proceeding further, it is necessary to define terms, although this exercise is fraught with terminological, political, and ideological peril. In some ways, however, the Chinese themselves have made the job a little easier. Chinese writings clearly suggest that IW is a solely military subject, and as such, they draw inspiration primarily from U.S. military writings. The net result of this "borrowing" is that many PLA authors' definitions of IW and IW concepts sound eerily familiar. For our purposes, therefore, we shall use the definition of information warfare found in Joint Pub 3-13, *Joint Doctrine for Information Operations (IO)*:

> Information operations conducted during time of crisis or conflict to achieve or promote specific objectives over a specific adversary or adversaries.[5]

"Information operations" are defined in Joint Pub 3-13.1, *Joint Doctrine for Command and Control Warfare (C2W)* as:

> actions taken to achieve information superiority by affecting adversary information, information-based processes, information systems, and computer-based networks, while defending one's own information, information-based processes, information systems, and computer-based networks.[6]

More concretely, the Army in FM-100-6 *Information Operations* defines "information operations" as

> continuous military operations within the military environment that enable, enhance, and protect the friendly force's ability to collect, process, and act on

---

[5] Joint Chiefs of Staff (JCS) Pub 3-13, *Joint Doctrine for Information Operations (IO)*, October 9, 1998, p. I9.

[6] Joint Chiefs of Staff (JCS) Pub 3-13.1, *Joint Command and Control Warfare (C2W) Operations*, February 7, 1996, p. I3.

information to achieve an advantage across the full range of military operations; information operations include interacting with the global information environment and exploiting or denying an adversary's information and decision capabilities.[7]

The goal of these operations is "information dominance," or

> The degree of information superiority that allows the possessor to use information systems and capabilities to achieve an operational advantage in a conflict or to control the situation in operations short of war, while denying those capabilities to the adversary.[8]

By introducing these definitions, I am not precluding that the Chinese may eventually develop an indigenous IW strategy, and there is limited evidence of movement in this direction. Instead, these U.S. definitions provide a baseline by which to judge PLA writings.

CHINESE INFORMATION WARFARE STRATEGY: HETEROGENEITY AND INNOVATION

This section examines the early-stage Chinese IW literature, offering the following preliminary conclusions.

The literature:

- focuses on disrupting logistics and communications

- understands the U.S. threat and admits their own technical weaknesses, including poor reliability, survivability, and security

- reveals a surprising grasp of U.S. IW doctrine, but borrows concepts inappropriate for the PLA's technological level

- correctly identifies the important lessons of DESERT STORM, but in some cases draws the wrong conclusions

- overestimates Chinese capabilities to develop effective defensive countermeasures.

Evolution of Chinese IW Strategy

In the mythology of PLA IW study, Shen Weiguang, a soldier in a field unit, began writing about information warfare in 1985, publishing a book entitled *Information Warfare* that was later excerpted as an article in *Liberation Army Daily*.[9] Chinese IW doctrine did not achieve an analytical focus, however, until the Gulf War in 1991. As has been documented in many other places, the Chinese military leadership was very

---

[7]Field Manual 100-6 *Information Operations*, August 1996.

[8]Ibid., p. 8.

[9]Shen Weiguang, "Focus of Contemporary World Military Revolution—Introduction to Information Warfare," *Jiefangjun bao*, November 7, 1995, p. 6.

impressed by the performance of U.S. forces in DESERT STORM, especially the ease with which they destroyed the Iraqi's largely Soviet and Chinese equipment. From their writings, it seems clear that PLA theorists believe that IW played a significant role in the U.S. victory. A commonly held belief, for example, is that the U.S. military used computer viruses to disrupt and destroy Iraqi information systems.[10] In their descriptions of DESERT STORM, these authors point to other allied operations and technologies as examples of information war. First, Wang Pufeng singles out superior satellite reconnaissance of strategic sites and Iraqi positions, as well as attacks on Iraqi command and control systems, as key elements of the rapid allied victory against Saddam's forces.[11]

On the lessons of DESERT STORM for the PLA, however, there is some divergence between those who believe the next war will look just like the Gulf War and those who understand that the Gulf War was a testing ground for advanced weapons and strategy to be used in a future, different war. Most seem awed by the "perfect" [*wanshan*] execution of the attack.[12] One writer described the new changes in information, command and control brought about by the Gulf War as a "great transformation" [*zhongda biange*],[13] and a second suggested that strategies to defend and attack computers and electronic systems could be as significant in determining the outcome of future wars as strategies to defend and attack citizens were in past wars.[14] Finally, Wang Pufeng called the Gulf War the "epitome" of information war.[15]

Since DESERT STORM, Chinese IW research has rapidly proliferated in newspapers, journals, and books. Some of the most prominent IW researchers and their billets are listed in Table 1 below.

Table 1

Important Chinese IW Theorists

Theorist	Billet/Comments
MG Wang Pufeng	Father of Chinese IW field Seminal work: *Information Warfare and the Revolution in Military Affairs*
Shen Weiguang	State Council Special Economic Zones Office (former PLA)
Wang Baocun	Academy of Military Sciences
Li Fei	*Liberation Army Daily*
Wang Xusheng	PLA Academy of Electronic Technology
Su Jinhai	PLA Academy of Electronic Technology
Zhang Hong	PLA Academy of Electronic Technology

---

[10] "Army Paper on Information Warfare," *Jiefangjun bao*, 25 June 1996, p. 6.

[11] Wang Pufeng, pp. 113–116; 123–126.

[12] Ibid., p. 203.

[13] Liu Yichang (ed.), *Gaojishu zhanzheng lun* (On High-Tech War), Beijing: Military Sciences Publishing House, 1993, p. 272.

[14] Li Zhisun and Sun Dafeng, *Gaojishu zhanzheng molü* (The Strategy of High-Tech War), Beijing: Defense University Publishing House, 1993, pp. 3–9, 184–201.

[15] Wang Pufeng, p. 144.

In addition, it has become increasingly obvious that some IW "centers of excellence" are emerging in the PLA. These centers are listed in Table 2 below.

These researchers began to congregate at a series of high-level scholarly meetings. In December 1994, the Commission of Science, Technology, and Industry for National Defense (COSTIND) sponsored a symposium entitled "Analysis of the National Defense System and the Military Technological Revolution," which was closely followed by an October 1995 meeting that dealt with "The Issue of Military Revolution." The alleged high point of Chinese IW research was a 22 December 1995 COSTIND National Directors conference, when Liu Huaqing allegedly stated:

> Information warfare and electronic warfare are of key importance, while fighting on the ground can only serve to exploit the victory. Hence, China is more convinced [than ever] that as far as the PLA is concerned, a military revolution *with information warfare as the core* has reached the stage where efforts must be made to *catch up with and overtake rivals*. (emphasis added)[16]

More recently, a group of 40 information warfare researchers met in Shenyang for a *Junshi xueshu* symposium on information warfare. The researchers, who were drawn from relevant departments of the army's general departments, military regions, armed services, scientific research institutions, academies, and units, discussed the "nature, position, role, guiding ideology, principles, modes, methods, and means" of information warfare.[17]

Table 2

Major Centers of IW Research

Center	Comments
Academy of Military Sciences Military Strategy Research Center	Main IW research center Developing IW strategies Integrating IW into overall military doctrine Dedicated to "winning information warfare in the information age" Affiliated with the Society for International Information Technologies
PLA Academy of Electronic Technology	
General Staff Department Third Sub-Department (GSD/3rd)	IW work carried out by Research Institute 61 and Information Engineering Academy
China National Research Center for Intelligence Computing Systems	
COSTIND University of Electronic Science and Technology (Chengdu)	

---

[16]"Latest Trends in China's Military Revolution," *Hsin Pao* [*Hong Kong Economic Journal*].

[17]Li Pengqing and Zhang Zhanjun, "Explore Information Warfare Theories with PLA Characteristics— *Junshi xueshu* Magazine Holds Symposium," *Jiefangjun bao*, 24 November 1998, p. 6, in FBIS-CHI-98-349, December 15, 1998.

Important Chinese IW Concepts and Terms: Definitions

When examining Chinese IW theories, the logical place to start is Wang Pufeng's seminal work, *Information Warfare and the RMA*. Wang defines information warfare as follows:

> Information war is a product of the information age which to a great extent utilizes information technology and information ordnance in battle. It constitutes a "networkization" [*wangluohua*] of the battlefield, and a new model for a complete contest of time and space. At its center is the fight to control the information battlefield, and thereby to influence or decide victory or defeat.[18]

Later, the author elaborates his definition:

> Information war is a crucial stage of high-tech war. . . . At its heart are information technologies, fusing intelligence war, strategic war, electronic war, guided missile war, a war of "motorization" [*jidong zhan*], a war of firepower [*huoli*]—a total war. It is a new type of warfare.

The author distinguishes this new type of warfare from the previous paradigm:

> Information and the capacity [to employ it] together release new energy in battle; information's "networkization" opens up a new battlefield of computers. With the "informationization" [*xinxihua*] of the army, agility and speed, mobility, and depth of attack, in a battle without a front line, all create a leap ahead of the traditional methods of warfare. The area [of the battle] grows, its speed increases, the accuracy of the attack is more acute, all of which change past conceptions of space and time.[19]

It is important to note that nothing in these definitions conflicts with American military conceptions of information warfare.

Important Chinese IW Concepts and Terms: Principles

The aim of IW in the Chinese literature is information dominance [*zhixinxiquan*], defined as the ability to defend one's own information while exploiting and assaulting an opponent's information infrastructure.[20] This information superiority has both technological and strategic components. On the one hand, it requires the ability to interfere with an enemy's ability to obtain, process, transmit, and use information to paralyze his entire operational system. This accords with U.S. military conceptions of information dominance. On the other hand, some Chinese commentators assert that information superiority is not determined by technological superiority, but by new tactics and the independent creativity of commanders in the field, placing much more emphasis on personnel and organization-related components of the conflict.

---

[18]Wang Pufeng, p. 37.

[19]Ibid., p. 2.

[20]This section draws from MAJ Mark Stokes' excellent study, *China's Strategic Modernization.*

The information battlefield itself is transformed in the PLA literature. Concepts of front and rear battlelines blur as the "multidimensional" battlefield space, integrating air, land, sea, space, and the electronic spectrum, becomes the arena of combat.[21] Within this battlefield, military units conduct "seamless operations" [*feixianxing zuozhan*], integrating sensors with weapons systems. Operational emphasis is placed on deep strike [*zongshen zuozhan*] and over-the-horizon warfare [*yuanzhan*] against command and control facilities, which are perceived to be the "vital points" [*dianxue*] of the system. The objectives of the operation are not the seizing of territory or the killing of enemy personnel, but rather the destruction of the other side's willingness to resist.

Victory on this information battlefield will shift the focus of operations. In the words of two PLA authors,

> the key to gaining the upper hand on the battlefield is no longer mainly dependent on who has the stronger firepower, but instead depends on which side discovers the enemy first, responds faster than the latter, and strikes more precisely than the latter. [The two sides] vie for the advantage in intelligence and command control, i.e. to see which side holds a larger amount of and more accurate information and is faster in transmitting and processing the information. On the other hand, they have to vie for advantage in the precision of the strike, i.e., to see which side can hit the other at a longer distance and hit the other side first at the same distance.[22]

As a consequence, detection, concealment, search and avoidance become central goals, pushing the military towards networked command and smaller, modular units.

Something Borrowed, Something Blue

One of the problems in analyzing PLA IW strategy, however, is disaggregating it from translations or outright copying of U.S. doctrinal writings, as well as Russian, German and French sources.[23] From conversations in Beijing, it is clear that the PLA has translated both FM-100-6, *Information Operations*, and JP 3-13.1, *Joint Doctrine for Command and Control Warfare*, along with a myriad of lesser documents, journal articles, and policy papers, including more abstract research on information revolution written by the Tofflers, David Ronfeldt, John Arquilla, and Martin Libicki. PLA writings selectively steal concepts and definitions from these works, though it is rare that doctrine is adopted wholesale. As a result, the terminology, definitions, and even case studies found in most Chinese writings are similar to the debate in the United States. A sample is presented in the next three paragraphs, though one could easily add hundreds of additional examples to this list.[24]

---

[21] Wang Jianghuai and Lin Dong, "Viewing Our Army's Quality Building from the Perspective of What Information Warfare Demands," *Jiefangjun bao*, March 3, 1998, p. 6, in FBIS-CHI-98-072, March 13, 1998.

[22] Ibid.

[23] Wang Pufeng, p. 141.

[24] For an article which is almost entirely derivative, see Wang Baocun and Li Fei, "An Informal Discussion of Information Warfare (Parts One, Two and Three)," *Jiefangjun bao*, June 13, 1995.

For example, at the highest level of abstraction, one PLA author describes the information age as the third important age in world history, following the agricultural age and the Industrial revolution.[25] Furthermore, he characterizes the defining feature of the latter part of the information age as the exponential increase in "data production, storage, exchange, and transmission." Both of these ideas are taken without attribution directly from the Tofflers' seminal futurist books *The Third Wave*[26] and *War and Anti-War*,[27] respectively.

In an example of direct appropriation of U.S. military operational doctrine, one PLA author defined the aim of IW as "preserving oneself and controlling the enemy," the core distillation of the U.S. military's concept of "information dominance."[28] Moreover, the same author asserted that IW included "electronic warfare, tactical deception, strategic deterrence, propaganda warfare, psychological warfare, computer warfare, and command and control warfare,"[29] which is virtually identical to the U.S. Air Force's doctrinal "Six Pillars of IW."

In conceptions of the information battlefield, the similarities continue. PLA authors discuss "integration" [*yitihua*] and seamless operations [*feixianxing zuozhan*], tying together the five dimensions of warfare—air, land, sea, space, and the electromagnetic spectrum—through the integration of sensors with mobile missiles, air, and sea-based forces. These sensors are meant to facilitate "dominant battlefield awareness," which in turn permits deep strike [*zongshen zuozhan*] against enemy command and control hubs, communication networks, and supply systems, blurring previous distinctions of a clear battleline.[30] For students of U.S. military doctrine, this conception of the battlefield is virtually identical to the core principles of Joint Vision 2010.[31]

The question, therefore, could be posed in the following manner: Is there a *Chinese IW strategy?* There are certainly important differences between the Chinese and American IW literatures. To summarize, PLA writers universally regard IW as a strictly military subject first and foremost, while Western authors largely accept the dichotomy between information warfare waged between states or militaries (i.e., cyberwar) and information warfare waged between substate actors and states (i.e., netwar).[32] Second, Chinese IW authors imbed their discussions within familiar ideological frameworks, such as Maoist guerrilla strategy and Sun Zi. In the Maoist

---

[25]Cai Renzhao, "Exploring Ways to Defeat the Enemy Through Information," *Jiefangjun bao*, March 19, 1996, p. 6.

[26]Alvin and Heidi Toffler, *The Third Wave*, New York: Bantam, 1980.

[27]Alvin and Heidi Toffler, *War and Anti-War: Survival at the Dawn of the 21st Century*, Boston: Little, Brown and Company, 1993.

[28]Su Enze, "Logical Concepts of Information Warfare," *Jiefangjun bao*, June 11, 1996, p. 6.

[29]Ibid.

[30]Cai Renzhao, "Exploring Ways to Defeat the Enemy Through Information," *Jiefangjun bao*, March 19, 1996.

[31]Office of the Joint Chiefs of Staff, *Joint Vision: 2010*, available at http://www.dtic.mil/doctrine/jv2010/jv2010.pdf.

[32]This distinction between cyberwar and netwar was coined by John Arquilla and David Ronfeldt. See *The Advent of Netwar*, Santa Monica, Calif.: RAND, MR-789-OSD, 1996.

vein, IW is referred to as the "New People's War," with particular attention paid to the idea of "overcoming the superior with the inferior." Both U.S. and Chinese authors are guilty of overusing Sun Zi, especially the notion of "winning the battle without fighting." While most of these references are nothing more than rhetorical flourishes, they do reflect two stark realities: (1) the extent to which Chinese (and U.S.) authors are struggling to find a framework for understanding IW and (2) the continuing pull of more traditional strategic frameworks. Third, Wang Pufeng and others emphasize the nontechnological aspects of information warfare to a much greater extent than U.S. military analysts, especially the need for new strategies and new organizational forms.[33] Fourth, Chinese IW theorists, by virtue of the PLA's relatively backward state, are forced by circumstance to discuss IW from the perspective of a technologically inferior military, often in opposition to a technologically advanced foe.

This latter point deserves further elaboration. One of the most interesting Chinese IW concepts is the notion of "overcoming the superior with the inferior," which draws inspiration from both Sun Zi and Maoist "People's War." A basic assumption of this line of reasoning is that the PLA will most likely face an opponent capable of achieving information dominance on the battlefield. In response, the PLA has two choices. The first is to adopt nontechnological measures to overcome technological disadvantage, such as camouflage, concealment, and deception techniques. While there is some merit in this argument, the experience of the Iraqi army in DESERT STORM does not foster much optimism that this strategy would be successful against a determined opponent.

The second choice is more interesting, and I would argue, should be much more worrisome to U.S. military planners. PLA writings generally hold that IW is an unconventional warfare weapon, not a battlefield force multiplier. Indeed, many writings suggest that IW will permit China to fight and win an information campaign, *precluding the need for military action.* When this train of thought is combined with the notions of "overcoming the superior with the inferior," one can quickly see the logical conclusion of the argument: IW as a preemption weapon.[34] According to Lu Linzhi,

> In military affairs, launching a preemptive strike has always been an effective way in which the party at a disadvantage may overpower its stronger opponent. . . . For the weaker party, waiting for the enemy to deliver the first blow will have disastrous consequences and may even put it in a passive situation from which it will never be able to get out . . .

As a concrete example, he points to the Gulf War, where Iraq's failure to launch a preemptive attack resulted in their defeat:

---

[33]This is not to say that Western authors do not emphasize the nontechnological aspects of information warfare. In fact, John Arquilla and David Ronfeldt are two prominent examples of American IW theorists who see the profound organizational and societal implications of IW and the information revolution writ large. See John Arquilla and David Ronfeldt (eds)., *In Athena's Camp.*

[34]Lu Linzhi, "Preemptive Strikes Crucial in Limited High-Tech Wars," *Jiefangjun bao,* February 14, 1996, p. 6.

In the Gulf War, Iraq suffered from passive strategic guidance and overlooked the importance of seizing the initiative and launching a preemptive attack. In doing so, it missed a good opportunity to turn the war around and change its outcome.[35]

This accords with some Western military analysts, who argue that Iraq should have attacked Allied forces in Saudi Arabia at the early stage of the deployment rather than permitting the forces of the U.S. and the other members of coalition to deploy without hindrance over a six-month period.[36]

To avert this outcome, Lu states that an effective strategy by which the weaker party can overcome its more powerful enemy is

> to take advantage of serious gaps in the deployment of forces by the enemy with a high tech edge by launching a preemptive strike during the early phase of the war or in the preparations leading to the offensive.[37]

The reason for striking is that the "enemy is most vulnerable during the early phase of the war."[38] In terms of specific targets, the author asserts that

> we should zero in on the hubs and other crucial links in the system that moves enemy troops as well as the war-making machine, such as harbors, airports, means of transportation, battlefield installations, and the communications, command and control and information systems.[39]

If these targets are not attacked or the attack fails, the "high-tech equipped enemy" will amass troops and deploy hardware swiftly to the war zone, where it will carry out "large-scale airstrikes in an attempt to weaken . . . China's combat capability."[40]

HOW COULD THE CHINESE CREDIBLY USE IW? AN UNSETTLING SCENARIO INVOLVING THE UNITED STATES AND TAIWAN

In his discussion of IW as a preemption weapon, Lu Linzhi lays out a scenario in which China employs a preemptive strike to defeat a technologically superior enemy during the latter's mobilization and deployment phase. When one reads between the lines, it becomes readily apparent that the author is describing the rough parameters of a potential confrontation between China and the United States. This becomes even more clear in the following revealing passage, where he frankly discusses the technological imbalances between China and its thinly disguised "high-tech enemy":

> Reconnaissance positioning satellites, AWACs, stealth bombers, aircraft carriers, long-range precision guided weapons . . . the enemy has all that; we don't. As for tactical guided missiles, electronic resistance equipment, communications,

---

[35]Ibid.

[36]A discussion of the coalition forces' early vulnerabilities can be found in Michael Gordon and General Bernard Trainor, *The General's War*, Boston: Little, Brown, and Company, 1995, pp. 57–64.

[37]Lu Linzhi, "Preemptive Strikes."

[38]Ibid.

[39]Ibid.

[40]Ibid.

command and control information systems, main battlefield aircraft, main battlefield tanks, and submarines, what we have is inferior to the enemy's.[41]

When one imagines scenarios in which the PLA would be concerned with preemptively striking U.S. forces during the deployment phase for early strategic victory, it is difficult to avoid the obvious conclusion that the author is discussing a Taiwan conflict. For the PLA, using IW against U.S. information systems to degrade or even delay a deployment of forces to Taiwan offers an attractive asymmetric strategy.[42] American forces *are* highly information-dependent, and rely heavily on precisely coordinated logistics networks, such as those operated by TRANSCOM. If PLA information operators using PCs were able to hack or crash these systems, thereby delaying the arrival of a U.S. carrier battle group to the theater, while simultaneously carrying out a coordinated campaign of short-range ballistic missile attacks, "fifth column," and IW attacks against Taiwanese critical infrastructure, then Taipei might be quickly brought to its knees and forced to capitulate to Beijing. The advantages to this strategy are numerous: (1) it is available to the PLA in the near term; (2) it does not require the PLA to be able to attack/invade Taiwan with air/sea assets, which most analysts doubt the PLA is capable of achieving for the next ten years or more; and (3) it has a reasonable level of plausible deniability, provided that the attack is sophisticated enough to prevent tracing.[43]

CONCLUSION

To sum up, the available evidence suggests that the PLA does not currently have a coherent IW doctrine, certainly nothing compared to U.S. doctrinal writings on the subject. While PLA IW capabilities are growing, they do not match even the primitive sophistication of their underlying strategies, which call for stealth weapons, joint operations, battlefield transparency, long-range precision strike, and real-time intelligence. Yes, the PLA is acquiring advanced telecommunications equipment through its commercial operations, even BC4I gear, but it is not clear that this equipment or subcomponents are being incorporated into PLA units, much less integrated into the military's system as a whole. Therefore, IW may currently offer the PLA some attractive asymmetric options, some of which may be decisive in narrowly circumscribed situations, but the Chinese military cannot reasonably expect anything approaching "information dominance" for the foreseeable future.

---

[41]Ibid.

[42]Two PLA authors explicitly endorse what they call "asymmetric information offensives." See Wang Jianghuai and Lin Dong, "Viewing Our Army's Quality Building from the Perspective of What Information Warfare Demands," *Jiefangjun bao*, March 3, 1998, p. 6, in FBIS-CHI-98-072, March 13, 1998.

[43]The plausible deniability of a PLA IW attack will increase markedly by the end of 1998, when a Trans-Eurasian landline cable will be completed. Currently, all international Internet gateways out of China connect to the North American backbone. When the Trans-Eurasian connection is open, however, Chinese hackers will be able to "wipe" their IP headers in Europe, making it extremely difficult for U.S. information operators to trace their origins.

CHINESE INFORMATION WARFARE TERMINOLOGY

xinxi zhanzheng—information warfare

junshi geming—revolution in military affairs (RMA)

zhixinxiquan—information dominance

yitihua—integration

feixianxing zuozhan—seamless operations

zongshen zuozhan—deep strike

turanxing yu kuaisuxing zuozhan—sudden and quick strikes

dianxue—vital points

yuanzhan—over-the-horizon warfare

bingdu—viruses

wangluohua—networkization

xinxihua—informationization

feixianxing zuozhan—"a war without a front line"

zhiming daji—mortal strikes

xinxi gaosu gonglu—"information superhighway"

ruan shashang—soft destruction

kuayue—leapfrogging

10. CHINA'S DEFENSE INDUSTRIES: A NEW COURSE?[1]

John Frankenstein[2]

"We must deepen the structural reform and make continuous efforts to improve the mechanisms of management and operation. We must strengthen unity and cooperation, fully bring into play the role of provincial and municipal offices of science, technology and industry for national defense, energetically promote the "64-character spirit of pioneering an enterprise" advocated by Central Military Commission Chairman Jiang Zemin, especially the spirit of patriotism, the spirit of seeking truth from facts and blazing new trails, the spirit of working hard and devoting one's services, and the spirit of unity and cooperation, in close conjunction with the reality of our scientific and technological industries for national defense, take over and carry forward the fine and age-old tradition and workstyle of "working hard and relying on one's own efforts, being scientific and matter-of-fact in one's approach, going all-out in work, and making selfless sacrifices" of China's defense industries, and strive to make the building of spiritual civilization in scientific and technological industries for national defense outpace that in all other trades and professions and to reap a double bumper harvest in the building of material and spiritual civilization."

—General Cao Gangchuan

Director, Commission on Science, Technology & Industry for National Defense, 1997[3]

"Grim."

—Yu Zonglin,

Director, National Defense Department, State Planning Commission on the condition of Chinese defense industries. 1996.[4]

---

[1]Acknowledgments: Thanks to Ms. Susan Aagaard Petersen, Asia Research Center, Copenhagen Business School, and Ms. Teresa Dong and Ms. Tang Man Wai of City University of Hong Kong for invaluable research assistance; and to Dr. Ann Markusen, Dr. Bates Gill, Mr. Peter Almquist, Dr. Heather Hazard and Dr. Harlan Jencks for helpful comments and critical reading of an earlier version of this paper or portions thereof. Support for this paper was provided by the Council on Foreign Relations, New York. An early version of this paper was presented to the Council's Study Group on Military Industrial Restructuring, the Arms Trade and the Globalization of the Defense Industry, New York, December 1997. Besides the specific citations below, the paper draws broadly on John Frankenstein and Bates Gill, "Current and Future Challenges Facing Chinese Defense Industries," *China Quarterly*, No. 146, June 1996; and Jörn Brömmelhörster and John Frankenstein (eds.), *Mixed Motives, Uncertain Outcomes: Defense Industry Conversion in China*, Boulder, CO: Bonn International Center for Conversion/Lynne Rienner, 1997. The author is, however, solely responsible for any errors of fact, analysis or judgment.

[2]John Frankenstein writes and consults on Asian issues. His most recent book is the edited volume *Mixed Motives, Uncertain Outcomes: Defense Industry Conversion in China*, Boulder, CO.: Lynne Rienner, 1997. A former U.S. Foreign Service Officer, he has held academic positions in the United States, Hong Kong, and Europe.

[3]"Enhance Sense of Strategy, Elevate Level of Science and Technology for National Defense," *Jiefangjun bao*, February 24, 1997, in FBIS-CHI-97-048, March 28, 1997.

[4]From China Defense Industry, May 24, 1996, cited in Defense Science Technology & Industry (DSTI), Hong Kong, May 16–31, 1996.

CONTRADICTIONS

A first glance at the Chinese defense industrial complex (CDIC) suggests that it suffers from the same afflictions and has attempted the same remedies as the defense industries of other countries. The CDIC faces major declines in military procurement, problems with second and third tier suppliers, and falling employment. Attempted fixes include rationalization, consolidation and "conversion."[5] But a closer examination shows that both the problems and solutions reached in the People's Republic have "special Chinese characteristics."

Those characteristics arise from the economic and military contexts in which the CDIC operates. CDIC is caught up in a contradictory relationship between the policies of economic reform and the goal of "revitalizing the economy" on the one hand, and the People Liberation Army's (PLA) attempts to modernize on the other. The developing synthesis is far from neat and contains numerous contradictions, even though the aim of both economic reform and military modernization—wealth and power—is the dream shared by Chinese modernizers from the Self-Strengtheners of the 19th century down to the current leadership.

FAILURES AND REFORM

The CDIC, part of the larger state-owned enterprise (SOE) sector, has more than its share of SOE problems. For most of the 1980s and 1990s it was told to sink or swim in the sea of commerce ("convert") *and* to continue resource-consuming, low-profit military production. This dual-track strategy appears to have been unsuccessful. One outcome of the CDIC's failures: the PLA, the CDIC's main customer, has turned to the outside world for the technologies on which it bases its future, as CDIC can not develop, much less make, those advanced weapons systems. Another result: "conversion" as part of the CDIC development strategy will be cut back if not abandoned. "Converted" enterprises will continue as commercial enterprises, but not as part of the CDIC.

Yet for almost two decades Beijing sought solutions to its problems in "conversion" and organizational shifts. Briefly, the strategy of conversion was to promote "spin-ons"—that is, to utilize gains achieved through commercially viable production for civilian purposes to modernize defense production and defense technology. The outcome of this strategy was mixed. "Conversion" as a business proposition appears to have been successful only in a few cases. Much of the effort came to be justified in terms of maintaining employment and "social stability."

Over time it became apparent that what the Chinese would call a "strategic shift" in the CDIC would be necessary, and at the March 1998 Ninth National People's

---

[5]For a global view, see "Linking Arms: A Survey of the Global Defense Industry," *The Economist,* June 14, 1997; "A Farewell to Arms Makers," *The Economist*, November 22, 1997, p. 79; GAO Report "Defense Trade: European Initiatives to Integrate the Defense Market," GAO/NSIAD-98-6, October 1997. For an insider's perspective, see Norman A. Augustine, "Reshaping an Industry: Lockheed Martin's Survival Story," *Harvard Business Review*, May-June 1997, pp. 83–94.

Congress (9NPC), a program of major, long-term restructuring was announced. That long-term program, if realized, will change the face of the CDIC.

The key elements of this reform include:[6]

The "civilianization" of the Commission on Science Industry and Technology for National Defense (COSTIND) and CDIC consolidation. COSTIND, formerly a military body which reported to the Party's Central Military Commission, was given ministerial status under the State Council. General Cao Gangchuan, who had been its chief, was replaced by Liu Jibin of the Ministry of Finance, a civilian with defense industry experience. COSTIND will be in charge of military production, supervise the corporations of the CDIC, and indeed take over the governmental and ministerial responsibilities of those corporations. The corporations themselves will be reformed as enterprise groups, reconcentrating on defense production under COSTIND. At the same time, their commercial ties with their parent bureaucracies will be reduced. We can surmise that defense procurement in the future will be conducted through contracts rather than command economy allocation mechanisms. "Converted" enterprises and other components of the CDIC that produce only civilian products will be divested or otherwise separated from their CDIC parents and turned over to local authorities and the market. We would add that this last step echoes the reforms slated for the non-defense SOE sector: the state will retain control of the top 1000 or so SOEs while the remainder will be "set free" in the marketplace.

A new General Department in the PLA high command. General Cao (ex-COSTIND) was appointed head of a new PLA general department, the General Armaments Department (GAD). The GAD retains COSTIND's direct military responsibilities, including weapon systems management and research and development (R&D). The GAD will also absorb sections concerned with military equipment from other General Departments. A PLA source volunteered that the move of the Equipment Department (ED) from the General Staff Department (GSD) to the GAD came about because the GSD/ED had jurisdiction over only equipment for the ground forces, while the new GAD will have authority for weapons systems management in all the service branches of the PLA.[7]

PLA Inc. closed down. In a related development, in July 1998 the PLA was ordered to withdraw from its commercial activities. This move is discussed below.

But first we shall take a brief look at CDIC's past organization, its economic environment, and trends within the People's Liberation Army. We then shall turn to an examination of how the CDIC is responding to these challenges.

---

[6]For some of the details, see "Luo Gan Explains Restructuring Plan," *Ta kung pao*, March 7, 1998 in FBIS-CHI-98-068, March 9, 1998. The text is a speech given by Luo, Secretary General of the State Council, at the First Session of the Ninth NPC on March 6, 1998.

[7]See "China Defense White Paper," *Xinhua*, July 27, 1998, in FBIS-CHI 98-208, July 27, 1998.

STRUCTURE

The CDIC is made up of ministries and enterprises under the Chinese state (the State Council and its central ministries, plus provincial and municipal governments). PLA enterprises served more to generate funds to cover shortfalls in the military budget and are—or were—not central to the CDIC. The CDIC has evolved from a collection of secret, numbered "machine-building" industrial ministries headed by military officers, to an array of civilian-run, profit-seeking corporations within horizontally integrated enterprise groups (*jituan*) which have increasingly sought to diversify their activities beyond their core military businesses. Currently the major CDIC organizations are:

* China Ship Construction Corporation (CSCC)—Commercial/naval vessels

* Aviation Industries of China (AVIC)—Aircraft

* China North Industries Group (NORINCO)—Armor, artillery, small arms

* China Aerospace Corporation (CASC)—Missiles, satellites

* China National Nuclear Corporation (CNNC)—Nuclear technology

The reforms were set underway in March 1998, but the shapes—and perhaps even the names!—of these organizations apparently have yet to be determined.

SOME HISTORY

Historically, the CDIC has occupied an important part of the state sector of the economy. Of the 156 "key industrial projects" that formed the centerpiece of Soviet assistance to the PRC in the 1950s, 41 were for weapons production.[8] Between 1966–1976, a "Third Front" strategic relocation moved or constructed a major proportion of the defense sector (55%, according to Chinese officials) in remote areas of southern and western China.

The rationale of the Third Front was strategic relocation of key defense plants away from vulnerable coastal areas. It was a huge expense, absorbing perhaps 50% of Chinese national investment during the period 1966–1976. It left a legacy of isolation and technological backwardness which still burdens the CDIC. In private conversations, Chinese defense industry analysts acknowledge that the Third Front project—like Mao's other campaigns—led to great waste. Indeed, the construction of the Third Front effort coincided with the Cultural Revolution and must have been affected by "the ten years of chaos." In retrospect, the Third Front was a mistake

---

[8]The historical background is covered in John Lewis and Xue Litai, *China's Strategic Seapower*, Stanford: Stanford University Press, 1994, p. 76; and John Frankenstein, "The People's Republic of China: Arms Production, Industrial Strategy and Problems of History," in Herbert Wulf (ed.), *Arms Industry Limited*, Stockholm: Stockholm International Peace Research Institute/Oxford University Press, 1993. See also Liu Xiaohua, "Zhu Rongji Discusses Matters of Vital Importance With Military—Inside Story of Reorganization of China's Five Major Military Industry Departments," *Kuang chiao ching*, No. 305, February 16, 1998, in FBIS-CHI-98-065, March 6, 1998, for a discussion of the CDIC's historical bureaucratic alignments.

MBI	Ministry/Corp. (1982)	Ministry/Corp. under State Council(1988)-	Ministry/Corp. under State Council(1993)	Corporation/Group under COSTIND* (1998 - 9th NPC)
2	Nuclear Energy Ministry	Ministry of Atomic Industry	China National Nuclear Corp. (CNNC)	CNNC
3	Aviation Ministry	Ministry of Aerospace (MAS) (combines Aviation & Space Industry Ministries)	Aviation Industries of China (AVIC) (MAS separated into AVIC and CASC)	AVIC
4	Electronics Ministry	Machine Building & Electronics Industry (MMBEI)	Ministry of Electronics Industry (MEI) (MMBEI divided into MEI and [civilian] Ministry of Machine Industry)	Merged with Ministry of Posts and Telegraphs to form Ministry of Information Industry
5	Ordnance Ministry	MMBEI	North Industries Group (NORINCO) (Ex-Ordnance, MMBEI)	NORINCO
6	Ship Construction Corp.(CSSC)	CSSC	CSSC	CSSC
7	Space Industry	MAS	China Aerospace Corporation (CASC)	CASC
8	Missiles (Merged with #7 in 1981)	—	—	—

NOTE: #1 MBI dealt with civilian production.

*COSTIND shifts from military chain of command to come under State Council.

CDIC Evolution

based on an inappropriate reading of the Soviet experience. More than the centrally controlled industries were involved: provinces and localities were urged to build their own "small" Third Fronts for self-sufficiency in small arms and ammunition manufacture. The *Encyclopedia of the Chinese Economy* gives figures which indicate the scale of the Third Front burden: total investment of about RMB200 billion, employing 16 million people to build some 29,000 factories (with RMB120 billion in fixed assets), of which about 1300 were medium-sized and about 600 were large-scale, key "backbone" (*gugan*) enterprises.[9]

SIZE

Official numbers are vague as to the size and scope of the defense industry.[10] According to the *Encyclopedia of the Chinese Economy*, the State Council–run defense sector encompasses about 1000 enterprises (each consisting of multiple factories, marketing organizations and research units), and 200-plus major research institutes

---

[9]For the Chinese figures, see Chen Daisun (ed.), *Zhongguo jingji baike quanshu* (The *Encyclopedia of the Chinese Economy*), Beijing: *Zhongguo jingji chubanshe*, 1993, p. 135. See also Bates Gill, "Defensive Industry," *Far Eastern Economic Review*, November 30, 1995, p. 62; and Barry Naughton, "The Third Front: Defence Industrialization in the Chinese Interior," *China Quarterly*, September 1988, pp. 351–386. See Qiang Zhai, "Beijing and the Vietnam Conflict, 1964–1965: New Chinese Evidence," Cold War International History Project Bulletin special issues on "The Cold War in Asia" (draft version), Winter 1995/1996, pp. 233–250, for additional Chinese documentation on the Third Front.

[10]Since the 1998 reforms are so recent and as yet incomplete, we caution that most of our discussion of the CDIC refers to the sector as it was *before* the reforms.

PLA/CDIC LINES OF CONTROL
A SIMPLIFIED CHART

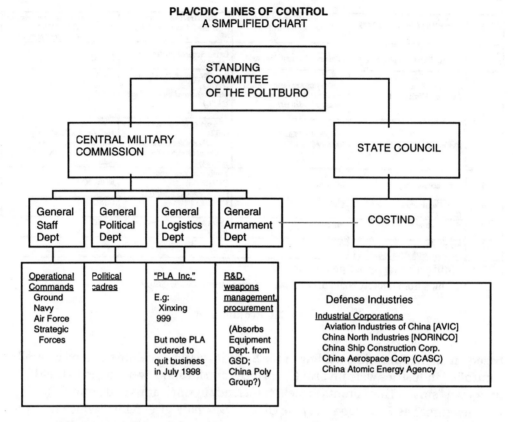

The PLA is deployed in 7 military regions, which supply the infrastructure to the various commands. MRs and operational commands also were or are involved in "PLA Inc."

(RIs), employing 300,000-plus engineers and technicians as well as some 3 million workers.[11] Given the proliferation of operations encouraged by the Chinese reforms, shifting ownership and control patterns, and the deliberate vagueness of Chinese statistics, particularly concerning this "forbidden area," it is difficult to be more precise than these general figures.

Individual sectors of the defense industry are quite diverse. According to accounts published in the mid-1990s, the aviation industry includes six universities and colleges, over 30 research institutes and more than 200 trading companies and enterprises employing more than 500,000 workers. Enterprises under such an organization are not small; rather, they are vertically integrated conglomerates. For example, the Chengdu Aircraft Engine Corporation, a "backbone" enterprise, employs 20,000 engineers and workers in 16 factories, 4 research institutes, and 11 branch companies, with more than 40 joint ventures and 6 "window" enterprises set

---

[11] *Zhongguo jingji baike quanshu,* p. 1754.

up in the open coastal cities.[12] One suspects, but cannot confirm, that the reduction of this kind of organizational complexity is one of the goals of the structural changes introduced at the 9th NPC.

In sum, China built a substantial defense industrial complex based on the Soviet technologies of the 1950s. But the complex was at once embedded in, yet isolated from, the larger command economy. That, plus the cutoff of Soviet aid in 1960, effectively left China with a defense industrial system stuck in the past, unable to innovate and dominated by politics. Reverse engineering and copy production was the norm. To be sure, and to its credit, China did develop some advanced technologies on its own, especially nuclear weapons, satellites, missiles (including SLBMs) and nuclear-powered submarines. But these were developed as special projects under the patronage of top officials—almost as one-off demonstrators—at a time when China perceived itself to be under nuclear threat from the U.S. and the USSR. They were isolated from the larger military and industrial system and, thanks to their patrons, from much, but not all, of China's tumultuous politics. The projects required enormous investments, but perhaps with the exception of land-based missiles, which are making a transition to solid-fuel boosters, but were never deployed effectively in significant numbers, they have never been modernized to any significant degree. Technological and operational shortcomings, problems with maintenance, lack of expertise (including the lack of a critical mass of technicians) and questions of expense all have contributed to China's difficulties with these systems.[13]

THE PLA

Although "PLA Inc." does not properly fall within the category of "defense industries," a brief comment is appropriate. PLA economic organizations started as units devoted to internal, self-supporting supply, distribution and military logistics—the guerrilla legacy—but expanded their operations to the civilian sector. These highly varied commercial ventures, perhaps as many as 20,000, are run by top command organizations, the seven military regions, the 24 group armies and subordinate local units. Some of these companies are large, truly commercial businesses (services and manufacturing, ranging from property and food processing to clothing, pharmaceuticals, steel pipe fabrication and telecoms), while many others are engaged in more local operations (transport, recreational and entertainment facilities [some reportedly of a dubious nature!]). Many provide employment for army dependents and income to support troop living conditions.[14] The

---

[12]China Aviation News, June 2, 1994, cited in *Defense Science, Technology & Industry Digest (DSTI)*, Hong Kong, June 1994, p. 5.

[13]See Lewis and Xue, *China's Strategic Seapower*, for an account of the struggles China went through to develop China's nuclear submarines, which still spend most of their time tied up dockside. A glance through *China Defense: Research and Development*, China Defense Science & Technology Information Center, Hong Kong and Beijing, 1988, shows that most of China's leading military scientists and technicians were born before 1920. In fact, many of those responsible for the development of China's strategic weapons had studied and worked in the U.S. and returned to China during the McCarthy era.

[14]Again, this is an echo of what is happening elsewhere in Chinese society. Many SOEs have set up "collectives" to provide employment for SOE worker family members.

conventional wisdom is that cash generated by these activities mostly stays with the originating unit for reinvestment, troop support and other local purposes, although a certain percentage of profits—in some cases as high as 70%—is supposed to be remitted to the Center.[15]

The impetus for the expansion of the PLA into commerce came from the military budget shortfalls of the 1980s, when military modernization was downgraded to the last of the "Four Modernizations." For instance, one of the goals given the PLA's *Sanjiu*, or 999, pharmaceutical conglomerate is to "play a positive role in developing the army's production and in making up for the inadequacies in military spending."[16] Other military analysts have written of "contradictions" between army requirements and budget allocations.[17] But the economic freedoms granted the PLA also led to massive corruption, and in July 1998 the PLA was ordered to divest itself of its commercial operations. Given the number and diversity of PLA businesses, it is not clear how this order will be carried out.

We should not be surprised by the PLA's commercial involvement. It is an example of "bureaucratic entrepreneuralism" that one finds throughout the Chinese economy. Indeed, if the Academy of Sciences can run a construction company, why shouldn't the PLA run hotels, sell medicine, provide transport services, and peddle surplus weapons? One must remember that the PLA, virtually a separate society within China, has a long tradition of self-reliance and self-support. It simply diversified and commercialized its internal economy as it expanded into the unregulated "white space" of the economy. Furthermore, the very structure of the Chinese economy unavoidably drew the PLA into business, for there are areas in which the PLA has a monopoly (e.g., air traffic control or radio frequency allocations in cellular telephone bands).

In any event, the evidence is that the PLA enthusiastically embraced this new battle front. Indeed, there was a Chinese joke to the effect that China only has a navy, because the air force and the army have all jumped into the sea. (The Chinese expression used to describe going into business, "*xia hai*" translates as "going down to the sea [of commerce]".)[18]

---

[15]See Arthur Ding, "China's Defence Finance: Content, Process and Administration," *China Quarterly*, No. 146, June 1996, pp. 428–442. There are, of course, many ways to calculate profits, and one suspects that PLA units have figured out how to maximize what they can retain. See also Tai Ming Cheung, "Can PLA Inc. be tamed?" *Institutional Investor*, July 1996.

[16]See "Military Organizes Military Transnational Group," *Zhongguo xinwen she*, December 9, 1992, p. 33.

[17]Gu Jianyi, "*Junshi jingji rougan xianshi wenti fenxi*" (On Current Problems in Military Economy), *Junshi jingji xue (Military Economic Studies)*, June 1995, pp. 32–35.

[18]Recounted in Min Chen, "Market Competition and the Management Systems of PLA Companies," in Brömmelhörster and Frankenstein, *Mixed Motives, Uncertain Outcomes*, p. 205.

PLA INCOME

But how much did the PLA "make" from these activities? This is one of the more intriguing guessing games in the PLA-watching trade.[19] In the mid-1990s it was estimated that the PLA "made" about US$5 billion from its commercial activities, though it was never clear if this represented sales, income or profit.[20] If sales, this would have put "PLA Inc." into the bottom half of the Fortune 500, close to the sales of the Quaker Oats Company. Other estimates range from a low of US$720 million to over US$7 billion.[21]

Reports in the Hong Kong press commenting on the business ban suggest that the PLA take was somewhat lower: around RMB15 billion a year (US$1.8B).[22] The Hong Kong Chinese language press reports that the central government will reimburse the PLA to the tune of RMB10–15 billion (US$1.2B–US$1.8B) per year to make up for lost revenues from curtailed PLA businesses, though one commentator has put the figure at RMB30 billion (US$3.6B).[23] RMB15 billion represents about 18 percent of the *official* 1997 defense budget of RMB81.3 billion. Since Beijing claims that about one-third of the defense budget goes for personnel, and that PLA Inc. revenues went mainly for personnel expenses, we can see that PLA Inc. revenues represent a substantial subsidy—roughly another 50 percent—to the personnel budget. But we don't know precisely how PLA Inc. revenues were allocated, and, as noted above,

---

[19]Not only is there some controversy about the contribution "PLA, Inc." makes to the Chinese military budget, there are disputes among analysts about how to calculate the budget itself. The U.S. Arms Control and Disarmament Agency (ACDA) calculates that Chinese defense spending (overall) between 1985–1995 followed a rising curve between US$53.5 billion and US$63.5 billion (in 1995 constant dollars). Official Chinese numbers are, needless to say, considerably below those amounts. Even though the official military expenditure figure has increased each year by 13%, the 1995 figure was US$7.7 billion and only US$9.8 billion for 1997. In any event, while analysts differ about the budget itself and the *amount* of "PLA, Inc.'s" contribution, the consensus is that it is supplemental to larger central government expenditures. Units generating cash get to keep most of it. As Gill remarks, "a significant share of extra-budgetary earnings is allocated toward the improvement of living standards . . . and other socioeconomic responsibilities of the unit. See ACDA, *World Military Expenditures and Arms Transfers 1996*, Washington, D.C., Table I, p. 65. For discussion on the perils of Chinese military expenditures analysis, see Bates Gill, "Chinese Defense Procurement Spending: Determining Chinese Military Intentions and Capabilities," Wye Conference on the PLA, 1997; Wang Shaoguang, "Estimating China's Defence Expenditure: Some Evidence from Chinese Sources," *China Quarterly*, September 1996; Arthur Ding, "China's Defence Finance: Content, Process and Administration," *China Quarterly*, June 1996; Gill, "Chinese Defense Procurement Spending," p. 8. We would note in passing that the U.S. military also generates significant "non-allocated funds" from on-base activities (PXs, recreational facilities) which are plowed back into troop benefits.

[20]See, for instance, David Shambaugh, "The Cash and Caches of China's Brass," *Asian Wall Street Journal*, September 22, 1994.

[21]Cited by Bates Gill, "Chinese Defense Procurement Spending."

[22]Kuang Tung-chou, "Premier Promises To Increase Military Funding To Make Up for 'Losses' After Armed Forces Close Down All Its Businesses," *Sing tao jih pao*, July 24, 1998 in FBIS-CHI-98-205, July 24, 1998. "Enterprises Belonging to Seven Military Regions Must Cut Ties with the Military Within Three Years, Analysts Say Prospects for Some Hong Kong Companies Should Be Reassessed," *Xin bao*, July 27, 1998 in FBIS-CHI-98-208, July 27, 1998.

[23]See "Government Reportedly Planning To Compensate PLA Firms," *Ming Bao*, July 25, 1998 in FBIS-CHI-98-207, July 26, 1998. See also Willy Lam, "PLA Gets Payoff for Business Loss," *South China Morning Post*, August 3, 1998.

most analysts believe that the "real" Chinese defense budget is considerably more than the official figure, perhaps as much as at least double.[24]

Still, reports from Hong Kong insiders suggest that returns from most PLA industries and services have been flat or in decline for the past two or three years. Tai Ming Cheung suggests PLA-run services—hotels, transportation, trading—were the most profitable, while factories run mostly in the red.[25] But the off-budget expense (covering losses from failing factories, housing, other social costs) may be considerable, perhaps as much as RMB100 billion (US$12B).[26] Removing this burden from the army's shoulders makes economic sense. But there is another, political reason for getting the PLA out of business: corruption.

The military's economic activities gave rise to a considerable amount of corruption, ranging from the sale of army vehicle license plates to smuggling. The army, Chinese critics say, has been diverted from its military mission and professionalism, even by its more legitimate business dealings. By the mid-1990s, amid much criticism from top army leaders of the PLA's enthusiasm for commerce, there were moves to prohibit combat units from business and to otherwise limit PLA commercial activities. The July 1998 order proscribing PLA business came, in fact, as part of a larger assault on official corruption by the Party, and should be seen as the outcome of the confluence of several currents affecting both China as a whole and the PLA: anti-corruption drives, pressures for professionalism (in both the civilian and military worlds), and, in the background, the larger economic reform movement to cut official ties with the commercial world.[27]

Still, what will happen to some of the high-fliers in PLA Inc.? There are numerous smaller firms run by the sons and daughters of high military officials—"princelings" involved in what one might term "*nomenklatura* capitalism"—which benefit from their connections. Companies with access to the highest levels—the arms-dealing Poly Group (which reported to the General Staff Department and whose leadership has included relatives of Deng Xiaoping), the General Logistics Department's

---

[24]The China Defense White Paper (op. cit.) says that 35% of the budget—RMB28.5B—goes for personnel. If we add a hypothetical RMB15B to this, and assume PLA manpower at 3,000,000, we find that the PLA "spends" RMB1450 (US$175) per year per soldier. We leave it to the reader's imagination to find this amount credible.

[25]See Tai Ming Cheung, "The Chinese Army's New Marching Orders: Winning on the Economic Battlefield," in Jörn Brömmelhörster and John Frankenstein (eds.), *Mixed Motives, Uncertain Outcomes: Defense Conversion in China*, Boulder, CO: Lynne Rienner Publishers, 1997.

[26]Willy Lam writes in the *South China Morning Post*, August 3, 1998: "Provincial and municipal administrations will also be footing bills totaling more than 100 billion yuan when they take over PLA businesses and assume more responsibility for providing social welfare and employment for soldiers."

[27]See Zhang Ruosang & Luo Yuwen, "Jiang Orders PLA-Owned Firms To Close," *Jiefangjun bao*, July 22, 1998 in FBIS-CHI-98-204, July 23, 1998. This article explicitly links the closing of PLA, Inc. to corruption. See also the wrap-up on corruption in the PRC, Susan V. Lawrence "Excising the Cancer," and related stories, *Far Eastern Economic Review*, August 20, 1998, pp. 10–14. For examples of other official Chinese commentary linking the ban on PLA business with corruption, see "Major PLA Units Vow To Cease Businesses," *Xinhua*, August 3, 1998 in FBIS-CHI-98-217, August 5, 1998; "China: PLA Logistics Director Calls for Commercial Disengagement," *Xinhua*, July 29, 1998, in FBIS-CHI-98-210, July 29, 1998. For insightful Hong Kong commentary, see Lung Hua, "China: PLA Business Ban To Hit Central Government—Zhongnanhai Trends," *Xin bao*, July 29, 1998, in FBIS-CHI-98-216, August 4, 1998; and Ren Huiwen, "'Inside Story' of Antismuggling Drive—Beijing Political Situation," *Xin bao*, August 7, 1998, in FBIS-CHI-98-219, August 7, 1998.

Xinxing, and Sanjiu, for example—have increased the scope and depth of their activities to the point where they have become indistinguishable from other highly diversified Chinese holding companies. If the Center continues to push for economic results over all other criteria and manages to push through these reforms, it is highly likely that PLA firms will lose their military coloration all together. Still, one of the more interesting questions to come out of the 9th NPC CDIC reorganization and the PLA business ban is whether PLA firms like Poly, Xinxing and 999, plus PLA service branch operations, such as the PLAAF's United Airlines, will come under COSTIND's supervision or otherwise be removed from their PLA parent, mirroring steps taken elsewhere in the CDIC.

THE DILEMMA OF THE STATE SECTOR

The CDIC is part of China's crisis-ridden state-owned enterprise sector. We do not want to present a detailed account of the SOE situation here, and we would acknowledge that there some SOEs which are in fact doing well. But on the whole, even though the Chinese leadership has in the past called the SOE sector the "core of the economy," that core is rotten.

In the mid-1990s, some Chinese economists were saying that 70 percent of SOEs were "in the red."[28] Industrial debt (that is, money owed by factories to each other within the sector) had reached RMB631 billion—over 14% of GDP and 34% of industry valued added—and unsold inventory had reached a value of RMB401 billion—9% of GDP and 22% of industry value added. At a news conference during the 1996 National People's Congress, Wang Zhongyu, minister of the State Economic & Trade Commission, said that SOE losses had increased by over one-third.[29]

A 1997 World Bank report noted that overall SOE industrial overcapacity was more than 40%. Chinese government surveys taken in mid-1997 show overcapacities ranging from 70% for excavators to 50% for TVs, engines and bulldozers and 40% for chemicals—significantly, these are industries in which the CDIC plays a part. About 50% of SOE industries suffer net losses. Preliminary 1997 figures suggest that unsold inventory levels may have reached RMB730 billion or about 10% of GDP.[30] In fact, official 1996 figures suggest that the SOE sector as a whole ran at a net loss, with total losses of RMB79 billion exceeding profits of RMB42 billion by 88%.[31]

---

[28]"State Culprit for Surge in Company Losses," *South China Morning Post (Business)*, November 15, 1995, p. 4; *South China Morning Post*, March 7, 1995; *South China Morning Post*, March 10, 1995; *China Daily*, March 5, 1995; *China Daily*, March 6, 1995; *China Daily*, March 12, 1995; and *Jingji ribao*, March 1, 1995. For a useful discussion of SOEs from a managerial perspective, see T. MacMurray and J. Woetzel, "The Challenge Facing China's State-Owned Enterprises," *The McKinsey Quarterly*, No. 2, 1994, pp. 61–74.

[29]See "Losses Grow for Chinese Industry," *International Herald Tribune*, March 15, 1996, p. 15.

[30]World Bank, *China's Management of Enterprise Assets: The State as Shareholder*, Washington, D.C.: World Bank, 1997, p. xi. The specific industry figures are from Directorate of Intelligence, "China's Economy in 1995–1997," Washington, D.C.: CIA, December 1997, p. 6–7. The 1997 inventory figures are extrapolated from data in Pamela Yatsko and Matt Forney, "Demand Crunch," *Far Eastern Economic Review*, January 15, 1998, p. 45.

[31]*China Statistical Yearbook 1997*, Table 12-17, p. 439.

Inefficient concentration simply adds to the slow-growth misery of the SOEs: the World Bank notes that the SOEs account for only about one-third of industrial production but provide two-thirds of urban employment and absorb about three-quarters of investment.[32] Not surprisingly, the profitable portion of the SOE sector hogs resources: the major SOEs—some 512 key enterprises, much less than 1% of the sector—control 55% of the total assets, 60% of sales and 85% percent of SOE pre-tax profits.[33]

Furthermore, the Chinese financial system is shaky. China's stock markets are more of a crap shoot than anything else, and Chinese banking procedures—accounting, credit analysis—are weak and are further enfeebled by politics and cronyism. Although China's foreign reserves and domestic savings rates are high, nonperforming loans are currently estimated to be equal to 20–30% of GDP, and, according to the World Bank, the entire banking system has negative net worth.[34] Even though the further reforms promised by the Ninth NPC would appear to move the Chinese economy forward, the 1997–1998 financial turmoil in Asia has sparked gloomy prognoses for the Chinese economy and undoubtedly has given pause to the Chinese leadership.[35] Overall, not a healthy picture.

CDIC FINANCIAL PROBLEMS

Chinese reports concerning the defense sector throughout the 1990s are similarly downbeat, with many commentators noting losses and overcapacity. Indeed, in the early 1990s economic boss Zhu Rongji was reported to have claimed that the majority of the SOEs in trouble were from the defense sector.[36] In 1995, *People's Daily* noted that "over the past decade and more, the country's munitions factories have seen a constant drop in the production orders of military goods; as a result, there are more hands than work in these munitions enterprises, the proportion of loss-making enterprises are expanding. . . . "[37] In mid-1996 the director of the national defense department of the State Planning Commission called the situation facing the defense sector "grim" and "very arduous." Weapons production would continue at only about one-third of the sector's capacity. A "considerable number" of defense

---

[32]World Bank, *China's Management of Enterprise Assets*, p. xi.

[33]"Yearender: State Enterprise Reform Enters Crucial Stage," *Xinhua*, December 24, 1997, in FBIS-CHI-97-358, December 24, 1997.

[34]Steven Mufson, "Time Runs Short for China Banks, the Economy's 'Soft Underbelly'," *International Herald Tribune*, November 24, 1997, p. 13; and "When China and India Go Down Together," *The Economist*, November 22, 1997, p. 73.

[35]See, for instance, the range of opinion expressed in Robert Dernberger, "China's Transition to a Market Economy: Back to the Future, Mired in the Present, or Through the Looking Glass to the Market Economy," and Wing Thye Woo, "Crises and Institutional Evolution in China's Industrial Sector," both in Joint Economic Committee, *China's Economic Future: Challenges to U.S. Policy*, Armonk, N.Y.: M. E. Sharpe, 1997; Mark Clifford *et al.*, "Can China Reform Its Economy?," *Business Week*, September 29, 1997, pp. 39ff; and George Wehrfritz, "Will China Be Next?," *Newsweek* (European Edition), December 1, 1997, pp. 32–33.

[36]Tai Ming Cheung, "On Civvy Street: China's Lumbering Arms Makers Face Market Rigours," *Far Eastern Economic Review*, February 6, 1992, pp. 40–43.

[37]Huang Shaohui *et al.*, "A Successful Practice of Converting Military-Oriented Industries to Civilian Production," *Remin ribao*, December 14, 1995, in FBIS-CHI-96-004, December 14, 1995.

enterprises were losing money and appeared unable to solve their financial problems.[38]

THE MILITARY DILEMMA

As if these structural and economic issues were not enough, the military context also poses problems for the CDIC. An industry must serve its customers—in the case of the CDIC, the PLA. But the PLA has requirements that the CDIC cannot meet.

To understand any army, one has to go beyond counting the number of troops and weapons systems. One has to look at doctrine—the definition of the enemy and how force would be used—as well as the state of readiness and logistical capability. Yet a quick "bean counting" glance at manpower and inventory levels is the place to start. And when we look at the PLA, the first two things that strike us are the army's size—about 3 million (although perhaps as much as a third are essentially civilians in uniform providing services such as medical treatment)—and the huge number of antiquated systems it has deployed: 24 group armies deploying 12,000 main battle tanks and armored personnel carriers based on 1950s and 1960s Soviet models (but only two fully mechanized divisions); and an air force based on Chinese adaptations of MiG 19s (3000 highly cannibalized airframes) and MiG-21s.[39] Not withstanding China's recent but small purchases of Su-27s and other modern military equipment, some military observers privately term the PLA a "military museum" and a "junkyard army." These huge numbers reveal not strength but weakness—the cost of full-scale force modernization would be overwhelming. At the same time, the announcement that up to half a million troops will be cut by the year 2001 is a step in dealing with this issue (though, we would add, about 14 divisions—approximately 150,000 troops—will simply swap their cap insignia in a transfer from the PLA to the People's Armed Police).[40]

But China's strategic focus has shifted to claims in the littoral and the South China Sea, where Chinese claims clash with those of many ASEAN states—and toward Taiwan. The importance of the South China Sea is not so much that it *may* contain oil and gas fields, but rather that it is a key choke point in the flow of oil from the

---

[38]Yu Zonglin, in *China Defense News*, May 24, 1996, cited in DSTI 16-31, May 1996. We would add that authoritative Chinese sources are silent on actual weapons production figures, so we must fall back on estimates. These indicate a significant decline in major systems production: annual warplane production has fallen from a high point of about 200 aircraft in 1982 and 1983 to about 80 in the 1990s; no bombers have been produced since 1990. Warship production is by its nature slow, and, in China, at low volume—6 destroyers and 13 frigates, and nothing heavier, have joined the fleet since 1990. Main battle tank production fell from about 650 in 1984 (half were, however, exported) to 100 in 1994. See Frankenstein and Gill, op. cit., for more detail. Michael Swaine cites a variety of sources which give somewhat lower production estimates for aircraft in his contribution "China" in Zalmay Khalilzad (ed.), *Strategic Appraisal*, Santa Monica: RAND, MR-826-AF, 1997, pp. 185–221.

[39]See the order of battle presented in Michael Swaine, "Chinese Military Modernization: Motives, Objectives and Requirements," U.S. Joint Economic Committee, *China's Economic Future: Challenges to U.S. Policy*, Armonk, N.Y.: M. E. Sharpe, 1997.

[40]See "Chi Haotian on Defense Policy," *Zhongguo xinwen she*, February 4, 1998 in FBIS-CHI-98-037, February 6, 1998.

Middle East to Japan, Korea, and China itself. Taiwan, of course, remains unfinished business from the Chinese civil war. But the "threat" has changed.

Today, the short-term threat is more ambiguous: not so much outright, head-to-head conflict, but rather the ability of outside forces—particularly the United States—to constrain or thwart China's freedom of military action in pursuit of her strategic goals, especially vis-à-vis Taiwan. And in the longer term, China's security problems may become more complex. China will have to deal with the emergence of a Japan with as much political and military clout as it has economic importance. And China's attention must once again turn toward her land borders as energy, vital to China's continued development, begins to flow from Central Asia and Siberia (Chinese oil interests have major investments in Kazakhstan). Furthermore, there has been a major shift in thinking about the way modern warfare is conducted, away from massive conflict to increased maneuver, highly targeted attack using precision weapons, and information warfare. The Gulf War was an early and already dated example of this "revolution in military affairs" (RMA).

Thus China faces the need to modernize the force, to deal with the widespread obsolescence of its inventory, and to overcome the PLA's "short arms and slow legs" in the face of mobile potential enemies—in the Chinese phrase, "fight a modern war under hi-tech conditions." It is easy enough to list China's weaknesses: lack of combat experience in the officer corps; rudimentary logistics and "C4I" (command, control, communications, computers and intelligence); no airborne platforms for early warning and mid-air refueling; limited blue water and amphibious capabilities; questionable abilities in anti-submarine warfare; low ground force mobility; and poor joint force coordination (including virtually no close air support capabilities), to name just a few.

However, it is also clear that the Chinese leadership is aware of these problems—or at least they should be. The library of COSTIND's China Defense Science and Technology Information Center subscribes to virtually every important foreign military journal in print, from *Jane's Defence Weekly, Aviation Week, US Naval Institute Proceedings, Parameters*, and other English-language materials to French, German, and Russian publications. Ever since the Gulf War the Chinese military press has been full of articles about the "revolution in military affairs," "five-dimensional warfare," "information war," "building weapons of soft destruction," and the like; Chinese leaders constantly exhort the army to move toward hi-tech.[41]

---

[41]Just a few citations here: Zhao Chengmou, "A Prediction of Future Battlefields and Weaponry in the Early 21st Century," China Defense Science & Technology Information Center Papers #4 & 5, 1990; *Jiefangjun bao*, March 20, 1991; "High-Tech War and Military Strategy," *Jiefangjun bao*, May 29, 1991; "Watch Closely the Revolution of Military Technology in the New Era," *Jiefangjun bao*, October 23, 1995 in FBIS-CHI-95-231, October 23, 1995; Wang Baocun "Military Transformation in an Information Era," *Jiefangjun bao*, April 21, 1998 in FBIS-CHI-98-126, May 6, 1998; Wang Jianghuai and Lin Dong, "Viewing Our Army's Quality Building From the Perspective of What Information Warfare Demands," *Jiefangjun bao*, March 3, 1998 in FBIS-CHI-98-072, March 13, 1998. These articles are more exhortations to achieve high-tech capabilities than celebrations of PLA achievements. One gets the impression that at least some of the Chinese leadership regards the RMA as a kind of "Sunzi-esque" magic bullet that will allow the PLA to achieve a low-cost, asymmetrical advantage over a high-tech enemy. In this one can see the continuing influence of Mao's "People's War" approach—indeed, for the historically minded, much of the current discussion reminds one of the Self-Strengtheners' fixation on "superb and secret weapons."

Area	Item
Aircraft	Air combat simulator under development by AVIC's "Blue Sky Aviation Simulator Technology Development Center; Aviation Industries of China Research Institute no. 613 of Luoyang conducting R&D in photo-electronic detection and tracking systems, head-up displays, and helmet aiming and display systems.
Aircraft, ship-borne	PLA Navy Aviation Technology Academy studies flight safety, air stream dynamics at the stern of a moving vessel through modeling and actual flight operations of a helicopter flying from a missile destroyer; ship-borne aircraft flight dynamics studied at symposium sponsored by Beijing Aerospace University; participants were from PLA Navy, Electronics Ministry, AVIC, other universities.
Computing	CAS national research center works on the "Shuguang 3000" massive parallel super computer; Aviation RI completes airborne high speed computers and airborne computer networking project; Space Industry Number 771 RI develops 5.8 kg "supercomputer" using 32-bit RISC CPU developed by the RI; Second Artillery engineering design RI develops "4-dimensional engineering construction computer management and simulation system"; VSLI research.
C4I	R&D on advanced transmitters operating in the 900 MHz range; telemetry advances; General Staff Dept. research institute develops army-wide on-line communications network and advanced command automation; Zhuhai Kexing Development Co. develops command and control central and remote display systems using GPS; Academy of Military Sciences operations institute develops an expert artificial intelligence system to aid commanders; research institutes under the General Staff Dept, COSTIND, 2nd Artillery and aerospace jointly develop a ruggedized microcomputer for field use; reports of successful test of real-time remote-sensing image processor; Ministry of Machine Building Industry Research Institute no. 55 announces completion of China's first liquid crystal display panel; Beijing MR group army achieves transportable field combat command automation from army down to regiment; PLA's first "wire communications automated duty management system" operational in Guangzhou MR [1996]; Electronics #7 RI works on multiple mobile/wireless communications systems; establishment of fiber-optic systems.
Cruise missiles	No. 8359 Research Institute completes a cruise missile assembly and testing facility.
Fire control automation	Guangzhou MR field artillery unit claims complete automation of fire control for all systems from field guns to anti-aircraft weapons; Southwest computer industry company develops automated artillery fire communications and command system.
Logistics	Guangzhou MR engineering institute develops field combat railroad platform vehicle, which serves as a temporary loading platform; it can also be used as a bridge.
Long-range navigation	"Changhe-2 Long-Range Radio Navigation and Positioning system" in operation; PLA Airforce develops airborne GPS system
Naval systems	Seminar on ship artillery and missile systems held by China Shipbuilding Engineering RI; delivery announcement of newly-designed supply ships for PLA navy; Shipyard builds first 10,000 ton "national defense mobilization vessel"
RPVs	R&D on RPVs conducted by PLA; at Northwest China Industrial University (termed China's largest r&d and production base for pilotless aircraft); National Defense S&T University and Guangzhou MR Logistics Dept. develops a drone with real-time photographic capabilities; Xian drone center develops a new "B-7" drone with a 40 km/60 minute range; Xian Aisheng Technology Corp develops unmanned reconnaissance aircraft equipped with cameras and infrared scanners.
Radars	An electronics research institute under the Chinese Academy of Sciences, working on synthetic aperture radar technology since the mid-1970s claims progress on airborne and satellite SAR applications; China National Aerospace Industry Corp develops warning radar; Chinese Academy of Engineering Sciences develops inverse synthetic aperture radar technique; anti-stealth wide-band radar

Table 1—Selected Chinese Defense R&D, 1993–96

Military R&D concentrates on many of these issues. See the table above for a selection of Chinese military R&D efforts. The open sources available to this writer suggest that the PLA is extremely far, even at the R&D level, from "post-modern" battlefield command systems that feature multiple sensor links and automatically configurie multi-channel/frequency hopping secure communication networks. Still, a wide range of R&D is going on within academic institutions and research institutes.

The question is, however, whether this R&D can be translated into production and deployment. The record of the "hardware" producing part of the CDIC suggests that it will be difficult to move into wide production of this type of "software." Rather, as was the case with strategic weapons, we could hypothesize that China's version of the RMA will be developed in special organizations.

In other words, China will follow a "pockets of excellence" approach, as seen in the creation of a small number of rapid reaction units (RRUs—army units, smaller than divisions, with more modern equipment, logistics and mobility). This reflects the Chinese penchant for experimentation. It also echoes China's "trickle west" coastal development strategy: focused investment that may have a demonstration effect. Thus, in a larger sense, the "dual economy/dual society" phenomenon that we see in China is now reflected in China's military development. We are likely to see continuing expansion of these small-scale modernization efforts, which have, in fact, been underway for much of the last decade.[42]

Still, the PLA has a long road to go down to reach a military status commensurate with the PRC's political and economic importance. The PLA's capabilities have improved and that modernization in the long term is firmly in the Army's sights. But modernization is, in fact, a target moving faster than most players can keep up with.[43] Chinese military writers certainly are up-to-date on Western thinking. Whether the PLA can make the adjustments necessary to incorporate large-scale modernization even to Western levels of the 1980s is, however, highly uncertain.

THE DILEMMA OF TECHNOLOGY TRANSFER

Thus, in the short term, at least, China has turned outward for its modern military technology, a violation of the Chinese imperative of self-reliance. The PRC's shopping list is long and reveals what the CDIC has been unable to develop and produce: advanced jet fighters (both purchase and production licenses for the Su-27K), airborne early warning and other avionics, modern air defense systems, precision-guided munitions, armor technology, larger surface combatants equipped with hypersonic sea-skimming anti-ship missiles, submarines, and dual-use communications technology. The sources are worldwide. Much comes from Russia, but Israeli, West European, and even U.S. firms have been suppliers.[44]

But Chinese officials, from the Self-Strengtheners to top PLA generals, have long emphasized that China must avoid dependence on foreign sources, or otherwise be stuck in a developmental trap, unable to adapt or proceed beyond the level of expensive imported technology.[45] In other words, there is a severe strain between the Chinese drive for self-reliance and the reality of foreign dependence. For

---

[42]We will not discuss issues surrounding China's nuclear forces here, since that would take us into realms of theology beyond our immediate interest. Suffice it to say that China has a limited "counter-value" deterrent capability. Or to put it another way, China's ICBM guidance systems are good enough to target "soft" cities, not good enough to target "hard" missile silos. However, a study of international military critical technologies by the U.S. Department of Defense grades China as having essentially world-class nuclear capabilities. Still, that some Chinese political leaders would casually ask, during the 1996 missile exercises around Taiwan, whether the U.S. would trade Los Angeles for Taipei suggests that some in Beijing may not fully understand the implications of nuclear war.

[43]Even the Chairman of the NATO Military Committee, German General Klaus Naumann, has expressed concerns that European armies may not be able to keep up with the innovations deployed by U.S. forces. In particular, he worries that NATO members "can no longer cope with the speed of the revolution" in reconnaissance, command and control, information systems and precision weapons. See *Aviation Week & Space Technology*, October 6, 1997, p. 23.

[44]Joseph C. Anselmo, "China's Military Seeks Great Leap Forward," and Nickolay Novichkov, "Russian Arms Technology Pouring Into China," both in *Aviation Week & Space Technology*, May 12, 1997, p. 69–73. See also "China: Military Imports From the United States and the European Union Since the 1989 Embargoes," Washington, D.C.: General Accounting Office, GAO/NSIAD-98-176, June 1998.

[45]Old Marshal Zhang Aiping raised the issue in the March 1983 *Hong Qi [Red Flag]* in words virtually identical to those of Feng Gueifen, the 19th century Self-Strengthener. It's an old issue, and old issues, like streetcars, tend come around.

instance, even as China signs up to co-produce Su-27s, other military factories, no doubt encouraged and perhaps protected by their local political authorities, continue ahead with at least two other fighter programs. These tensions between foreign procurement and competing domestic programs can only lead to an inefficient allocation of resources.

Furthermore, success in the technology transfer process is not guaranteed. Western technology executives at a recent defense conversion meeting in China noted privately that they wouldn't take some of the factories they were shown even if they were given away. A Beijing-based American executive with long experience in dealing with the CDIC considers most Chinese defense plants "dismal . . . right out of the Middle Ages," burdened with "enormous overheads," and unable to reach economies of scale, in part because of a lack of managerial and technical expertise. They have "a false faith in their abilities," possessing the skills of "hammer and chisel engineering in the age of CAD/CAM." In a general critique of the process, noting poor results and incomplete assimilation, officials at the State Science and Technology Commission complain:

> Our sense of technological innovation is underdeveloped. People do not fully realize the import of technology is just the beginning, that what really matters is absorption, assimilation, and independent innovation. . . . As a result, the relationship between the import of technology and self-development is a tenuous one.[46]

The reality of the situation constrains Chinese military capabilities. As Paul Godwin of the National Defense University has written,

> China's military planners face an increasingly difficult dilemma. A national strategy focused on limited, local war along its borders and its maritime claims, accompanied by the requirement to sustain its nuclear forces, has created requirements for technologies which the military technology base cannot develop and the industrial base cannot yet produce. . . . For the foreseeable future. . . . China's armed forces will have to continue to plan on the basis of the assumption of obsolescence.[47]

THE CDIC RESPONSE: "CONVERSION"

One CDIC response to this poor combination of circumstances was *"jun zhuan min"*—"defense conversion." The term, at least as used in the West, encompasses many concepts: the use of military production assets to produce for the civilian market; the production of dual-use items; spin-off, or the adaptation of military technologies for civilian uses; diversification; demobilization of personnel and facilities; surplus weapons management; reengineering an entire sector through

---

[46]Shi Dinghuan, Yang Tiancheng, and Mu Huaping, all from the Department of Industrial Science and Technology, State Science and Technology Commission, "Proposals to Improve Import of Technology and Accelerate Technological Innovation," *Zhongguo keji luntan* (Forum On Science And Technology In China), No. 4, July 97, pp. 8–12, in FBIS-CHI-97-283, October 10, 1997.

[47]Paul Godwin, "Military Technology and Doctrine in Chinese Military Planning: Compensating for Obsolescence," in Eric Arnett (ed.)., *Military Capacity and the Risk of War*, Stockholm, Sweden: Stockholm International Peace Research Institute/Oxford University Press, 1997.

consolidation ("shrinking smart")[48]; or, more broadly, a process of industrial disarmament. It is also clear that for many in the Western defense industry, "defense conversion" is a four-letter word. Norman Augustine, president of defense giant Lockheed-Martin, has called the process a strategy "unblemished by success."[49]

But for the CDIC, "conversion" appeared to offer a solution to many of the sector's problems. It received attention at the highest levels, was included in the mechanisms of the central Five-Year Plans and came under the direction of a Three Commission Liaison Group for Defense Conversion, which brought together officials from the State Planning Commission, the Science and Technology Commission, and COSTIND (when it came under the military), plus a State Council office that deals with problems of the Third Front industries.[50] The fate of this Commission, given Premier Zhu Rongji's bureaucracy busting, is not known. But even as Beijing was attentive to conversion, the responsibility for much conversion work has been shifted down to the province level, particularly in the Third Front areas of Guizhou and Yunnan. If a result of the 1998 reforms is that "converted" enterprises will be cut off from their CDIC parents, it will effectively signal an end to the attempt to combine military and civilian production and perhaps also be another indication that the post-Deng leadership of Jiang Zemin and Zhu Rongji are distancing themselves from Deng's policies. And it will mean that much of the writing about "conversion" in China will shift from the policy inbox to the historical bookshelf. But given China's claims of considerable success in its conversion efforts—Beijing has asserted that 70–90% of CDIC output goes for civilian purposes—a closer look is still warranted.

A Xinhua news agency release from September 1997 is typical of official Chinese media reporting on conversion:

> For every 10 such taxis, seven are manufactured by former weapons factories. These factories also turn out one fifth of the total national output of cameras and almost two thirds of motorcycles. Even the rice on the dinner table may be one of the 325 kinds of grains that have been treated by nuclear radiation technology, formerly the preserve of the military.[51]

Other Chinese statements give more detail of a sort.[52] For instance, in 1994, the ordnance industry claimed that 90 percent of its industrial output in southwest

---

[48]"Shrinking smart," a useful short-hand term, comes from J. T. Lundquist, "Shrinking Fast and Smart," *Harvard Business Review*, November–December 1992. The article looks at possible remedies for the ills of the U.S. defense sector.

[49]K. L. Adelman and N. R. Augustine, "Defense Conversion," *Foreign Affairs*, Spring 1992.

[50]See Paul H. Folta, *From Swords to Plowshares? Defense Industry Reform in the PRC*, Boulder, CO: Westview Press, 1992; and Jörn Brömmelhörster and John Frankenstein (eds.), *Mixed Motives, Uncertain Outcomes*.

[51]"PRC Defense Industry Turning Swords Into Ploughshares," *Xinhua*, September 29, 1997, in FBIS-CHI-97-272, September 29, 1997.

[52]Transparency is not a notable feature of the Chinese economic landscape. CDIC officials can be downright cagey and misleading when asked to go beyond general statements. For instance, the manager of an aircraft plant I visited claimed not to know the selling price of his production. Several young Chinese analysts I have spoken with do not appear to understand basic analytical techniques or why, when percentages are cited, one would like to know the numerator and denominator of the calculation. Even when company records are available they are difficult to interpret. An American executive told me that most of these records are worthless in any event: "unreadable." In one case, not atypical, it took

China was for the civilian market—motorcycles, minicars, heavy-duty trucks, cameras, refrigerators and other white goods, optical and electrical instruments, and machinery for oil production.[53] By 1996, sales had reached US$1.5 billion, 95% of total sales.[54] In 1995 the official *China Daily* reported that 1994 sales of civilian products made by the ordnance sector increased 31 percent over 1993, to reach 18.5 billion yuan (about US$2.2 billion), with a forecast for another 30 percent increase during the year.[55] In a paper delivered at a conference on conversion held in Beijing in mid-1995, Zhang Weimin, a North China Industries Group [NORINCO (G)] vice-president, gave an upbeat account of the group's development. It had grown to encompass 157 large and medium-sized factories, more than 30 research and development institutes, 200 sales companies, and 60 subsidiaries trading in 100 countries (including the United States, Dubai and Russia). Overall, he said, the group's joint ventures produced 40 percent of the motorcycles sold on the Chinese domestic market, and aimed to produce 450,000 mini-cars and 20,000 heavy trucks per year by the end of the century.[56] The diversity of production is great—for instance, the aviation and space industries produce everything from satellites and aircraft to motor vehicles to typewriters, pots and pans, and even religious objects, such as the cast bronze Buddha on Hong Kong's Lantau island.

Following practices found on the nondefense side of the economy, the CDIC formed new enterprises through consolidations and mergers. COSTIND sponsored a holding company in Hainan which drew on investments from across the various CDIC ministries and corporations (we understand that the company failed). Both industry-based and regionally based groups, such as the Huanghe electronics group and Guizhou's defense conglomerates, were established. NORINCO has formed joint ventures in motorcycle production and communications. Third Front firms have been urged to make a "triple jump"—"jump out of the backwater, skip to coastal areas, pole vault overseas"—and by the early 1990s it was claimed that they had established 800 enterprises along the coast and in cities and special economic zones. These operations are strictly commercial, designed to cash in on the boom in the "China Gold Coast"; the Shenzhen special economic zone just north of Hong Kong is full of such "window firms" (e.g., a NORINCO plastics factory making manufacturing molds and toys and, in an interesting example of exploiting machining competencies, an aviation factory workshop that makes watch cases for export). The Guangzhou HuaMei (i.e., China and America) Communications Ltd. joint venture, established to set up a high-speed broadband telecommunications network, reflects some of the new horizontal flexibility in organization we see throughout the Chinese economy. The U.S. side is SCM Brooks Telecommunications, while the Chinese side, Galaxy New Technology, is a cross-ministerial enterprise between the Ministry of

---

accountants over 6000 man-hours to validate a potential joint venture (JV) partner. Thus we are thrown back to anecdotal evidence and press reports for our data.

[53]See "'War Industry' Produces More Civilian Products," *Xinhua*, June 9, 1994, p. 36.

[54]"PRC Defense Industry Turning Swords Into Ploughshares," *Xinhua*, September 29, 1997, in FBIS-CHI-97-272, September 29, 1997.

[55]See "Arms Maker Produces More Civilian Goods," *Xinhua*, January 19, 1995, p. 36.

[56]Untitled paper distributed at the International Conference on the Conversion of China's Military Industries, sponsored by the OECD and CAPUMIT, Beijing, June 1995.

Electronics, the Ministry of Posts and Telecommunications and COSTIND. The joint venture is of interest because the technology transferred has dual-use potential.[57] And in 1997, just before the 15th Party Congress, Chinese officials were promoting direct foreign investment in defense electronics enterprises; the pitch for foreign investment in the defense sector continues.[58]

SUCCESS?

In other words, there has been, as elsewhere in the Chinese economy, a good deal of experimentation going on. But how successful has it really been? Jin Zhude, the former head of the COSTIND affiliate China Association for the Peaceful Use of Military Industry Technology (CAPUMIT), has written that success in defense conversion would include management reform, adapting to domestic and international markets, attracting foreign investment, moving from capital-intensive to technology- and information-intensive industry, and developing "mainstay" or "pillar" industries that can drive development such as automobile manufacturing.[59]

Have these goals been met? It doesn't seem so. At a defense conversion conference in China, a NORINCO official confided to a visitor that "90" out of "100" defense firms were having problems meeting their payrolls. The regime, he said, "was not telling the whole truth." A Chinese newspaper report on Sichuan-based defense industries found that "most factories are on the verge of bankruptcy," and concluded that in their turn to the market, "prospects for success were dubious."[60] More understated, another Chinese military researcher noted "the policy of stimulating civilian industry through military R&D and arms production, if not a complete failure, has many limitations itself."[61] A 1996 report from the Hong Kong–based PRC-sponsored *Wen Wei Bao* noted that "nearly half" of the 80% of the CDIC that have attempted conversion "have failed to reap reasonable profits or have suffered losses."[62]

Officials involved in defense conversion have criticized the CDIC's record. One noted that the defense sector, habituated to allocated capital and handicapped by a corporate culture that promoted "self-reliance for all things," carried excess

---

[57] The case raised a number of issues about U.S. export licensing procedures. See "Export Controls: Sale of Telecommunications Equipment to China," Washington, D.C.: General Accounting Office, GAO/NSIAD-97-5, November 1996. This JV also illustrates some of the complications of the recent reforms: how will this JV be governed and who holds what shares, now that the MPT and the Ministry of Electronics have been merged into the Ministry of Information Industry and COSTIND is no longer a military organ?

[58] See Tom Korski, "China: PLA Sets Sights on New Type of Recruit," *South China Morning Post*, January 19, 1998; and G. Arslanov, "Foreign Investments Help China to Convert Defence Industry," *Tass*, January 18, 1998.

[59] Jin Zhude, OECD/CAPUMIT conference paper, 1995.

[60] Pei Jiansheng, "Market Solution Eludes Remote Military-Industrial Complex," *China Daily Business Weekly*, November 6–12, 1994, p. 7. The remarkably frank story ran in *China Daily*, an official English-language paper aimed at foreigners.

[61] Fan Wei, "Arms Procurement and National Economic Development in China," unpublished manuscript, March 1995.

[62] Yin Tan, "China Quickens Pace of Defense Production Conversion Into Civilian Production," *Wen wei pao*, April 16, 1996, in FBIS-TAC-96-006, April 16, 1996.

inventory, found cooperating and networking with other organizations extremely difficult, had no market knowledge, produced low-quality goods, and lacked cost- and customer-consciousness.[63] Defense sector scientists told a visitor to a 1995 defense industry conference in Beijing that defense conversion faced major problems: lack of capital, high indebtedness, and poor market research, which led to the production of low-quality but overpriced, noncompetitive goods and a belief that defense conversion was a technical production problem requiring a hardware fix when in fact it demanded a software fix—changes in management and strategy.

The State Planning Commission (SPC) also is not happy with defense conversion progress. According to a report in the *Chinese Defense Industry News* quoting a SPC official, very few—about 12%—of the CDIC had the ability to develop "pillar products" by themselves, while 40% did not have "pillar products" at all. CDIC civilian products, the official said, were of medium to low technological levels and quality; product mix was low; and because of limited investment, production runs lacked economies of scale, leading to low, if any profits.[64] It would also appear that Chinese banks are not overly impressed with defense conversion efforts. In late 1995, it was reported that Beijing was considering setting up a RMB50 billion fund (US$6.1 billion)—on top of RMB10 billion already spent—to underwrite defense conversion efforts. The reason, according to Wu Zhao, the president of CAPUMIT, was clear: "Banks have become more independent. Although lending quotas are available, sometimes local banks refuse to lend the money to local military industrial enterprises."[65] If Chinese banks, bolstered by guarantees, won't bet on conversion, who should?

Authoritative statements suggest that there is some anxiety about the process, and, with 20-20 hindsight, suggest that by 1996 the conversion process was under the kind of reexamination that would lead to the 1998 reforms. Certainly the economic environment in China in the 1990s posed problems for the inefficient CDIC. Jin Zhude pointed out that economic austerity moves (primarily tightened credit), designed to cut inflation, had weakened the conversion program: "Since 1993, as China strengthened macroeconomic controls . . . the development momentum of conversion has clearly weakened, and it is necessary to make new restructuring and plans." There is more than a hint of crisis in his urgings that the defense sector become more efficient. Otherwise, it "again could become a major burden on or hindrance to [*baofu*, a very strong word with negative implications] China's economic development, as it was in the 1970s."[66] A State Planning Commission official noted that if the CDIC cannot continue their failing high-cost, low-profit conversion operations, their products will be "squeezed out" or eliminated through competition. The SPC's solution would be to continue structural reform, "centering

[63]He Jianhua, unpublished 1995 defense conversion conference paper.

[64]Yu Zonglin, *China Defense Industry News*, May 24, 1996.

[65]Vivien Pik-Wan Chan, "State Considers Fund To Aid Defense Conversion," *South China Morning Post*, November 7, 1995. Local CDIC managers told a visitor to Chongqing in 1995 that even COSTIND support was inadequate, thus the need to go to banks.

[66]Jin, in OECD/CAPUMIT conference paper (1995). In the Chinese text this is *youyou keneng chengwei yinxiang guomin jingji fazhan de baofu, zhongjian 70 niandai de lishi.*

on raising modern weaponry research and development (R&D) and production capabilities, while solving the problem of excess capacity for old products and insufficient capacity for new products. *Only in this way will [the CDIC] be able to meet the requirements of defense modernization.*[67] In other words, concentrate on weapons.

SWORDS INTO . . . ?

Indeed, "conversion" had another purpose beyond improving the efficiency of the CDIC. It was, as Paul Folta has put it, "Swords into plowshares . . . and better swords."[68] Defense conversion in the Chinese context was not only intended to deal with the CDIC's economic situation, it was also part of China's military modernization program by aiding Chinese military production capabilities. This has not exactly been a secret. Defense plant managers interviewed by Paul Folta said they expected profits from civilian production to aid defense production modernization. Conversion activities would also allow military production costs to be spread over a wider base.[69] The basic strategy is spelled out in the so-called 16-character slogan, attributed to Deng Xiaoping:

Jun-min jiehe,	Combine the military and civil,
Ping-zhan jiehe,	Combine peace and war,
Jun-pin youxian,	Give priority to military products
Yi min yang jun.	Let the civil support the military.

These intentions have been expressed many times throughout the 1990s. For instance, speaking at a defense conversion conference held in Beijing in 1991, a senior engineer in the aerospace industry said: "Among the dual tasks of serving national defense construction and serving economic construction, the national defense construction goes first and the economic construction second. . . . [the aim of] both military and civil use is a long term stable military system." A Hubei provincial official said that the "guiding principle" for conversion was to "treat [military and civilian] equally without discrimination, [but to] put military production first, and give it appropriate preferential treatment" (by which he meant tax breaks and first call on infrastructure).[70] This dialectical view of the relationship between military and civilian industrial capacity was one of the major themes of a long 1996 article in a major ideological journal by Chinese defense minister General Chi Haotian, who called for converting enterprises to keep military mobilization and production needs in view: "On condition that production of military supplies is not

---

[67] Reported in China Defense Industry News, May 24, 1996. Emphasis added.

[68] Folta, p. 1.

[69] See Folta, p. 168.

[70] Unpublished papers by Zhang Hongbiao of the Aerospace Ministry and E' Wanyou, Vice Director, Hubei Economic Commission, circulated at the International Cooperation in Peaceful Use of Military Industrial Technology Conference, Beijing, 1991.

affected, facilities and equipment of [defense] scientific research should enthusiastically service the country's economic construction."[71]

DUAL USE—SPIN-OFF *AND* SPIN-ON

Others promoted dual-use technologies in the conversion effort. Indeed, some 1995 commentary suggests that for some, conversion means "spin-on" as much as it does "spin-off." No less than former Central Military Commission (CMC) Vice-Chairman General Liu Huaqing was reported by *Liberation Army Daily* to argue at a January 1995 national conference on cooperation and coordination work in the military industries that "we must seize the opportune time of the end of the Eighth Five-Year Plan [1995] and the Ninth Five-Year Plan to push our national defense science and technology and weaponry onto a new state." According to the paper, he said that contributions from civilian industry, the Chinese Academy of Science and the university system are "component parts" and the "foundation for developing science and technology industries for national defense." China, he said, "should pay attention to turning advanced technology for civilian use into technology for military use...."[72] A Xinhua commentary, also dating from January 1995, noted a number of "spin-on" developments: civilian industry "solved a large number of sophisticated technology problems crucial to the production of nuclear weapons, nuclear submarines, guided missiles and satellites [and] new materials...."[73] And Huai Guomo has pointed out: "The trend toward the interchangeability of military and civilian technology is increasing, and this provides a solid technological basis for the rapid modernization of national defense and the constant upgrading of weaponry."[74]

Arguing that dual-use makes a contribution to the civilian economy while keeping military production lines warm, an official of the Chinese Defense Science and Technology Information Center noted that in times of crisis this capability can strengthen deterrence. Remarking on foreign experiences with dual-use, he said "the development of dual-use technologies provides the opportunity of developing military technologies in disguised form . . . [and] can save the costs in weapons

---

[71]Chi Haotian, "Taking the Road of National Defense Modernization Which Conforms to China's National Conditions and Reflects the Characteristics of the Times—My Understanding Acquired From the Study of Comrade Jiang Zemin's Expositions on the Relationship Between the Building of National Defense and Economic Development,"Qiushi [Seeking Truth], April 1996.

[72]"Liu Huaqing Urges Development of Defense Technology," Xinhua, January 30, 1995, p. 30ff.

[73]Xinhua, January 25, 1995, p. 24.

[74]Cited in Xiang Wang, "Development of Modern Technology and Defense Conversion: Interview with Huai Guomo, vice minister of the Commission of Science, Technology and Industry for National Defense," Conmilit (Xiandai Junshi), No. 296, May 1993, p. 4. This perception must have been reinforced when Huai was a visiting fellow at the Stanford University Center for International Security and Arms Control, located in the heart of Silicon Valley, in 1993. In any case, some of COSTIND's joint ventures with U.S. firms, such as Hua Mei Telecommunications, may have been set up with "spin-on"—as well as minimization of potential problems with export controls—in mind. See Bruce Gilley, "Peace Dividend," Far Eastern Economic Review, January 11, 1996, pp. 14–16.

systems and [military] modernization."[75] A Chinese scientist, who formerly designed ICBM engine systems, told this writer privately that money from civilian satellite launches goes toward the salaries of scientists and engineers working on the military side of the Chinese missile program.

In other words, one *intention* of conversion appeared to be not only to utilize redundant facilities, but also to apply the benefits of conversion—funds and improved technology—to the maintenance of the defense industrial base and to further military modernization. The aim, according to an article in *Military Economic Research*, is to put the defense industries "on the development road of the socialist market economy, and to guarantee the unbroken improvement of military production capabilities."[76] If industrial demilitarization (or industrial disarmament) is one definition of conversion, we do not find it in the Chinese case.

A PARADOX

Thus the Chinese conversion effort on the whole appears to be a muddle—not much success in moving toward success on the civilian market, not much success in modernizing defense production. Indeed, it would seem that in those cases where defense conversion has been economically successful, there is an unwillingness to remain a military supplier. Indeed, the fear that the conversion effort would crowd out military production has been expressed by Chinese officials. One acute Chinese observer formerly involved with defense conversion efforts has written:

> Profits from military production tend to be lower, requirements higher and quantities less than in civilian production. Motivating enterprises and maintaining an adequate industrial base for military production under a tight defense budget are therefore questions that have become important. . . . [77]

His concerns appear to have been realized. There are the management problems of "one factory, two systems," in which the cash-generating civilian side of the operation has taken precedence over the military production side.[78] CDIC managers, who have the bottom line responsibility, appear to be rather forthright about this issue. A defense industry manager complained to *China Defense Conversion News* that he was caught in a bind, trapped between the new environment brought about by the economic reform program and the requirement that he keep military production going: his factory had to "expend 50% of our efforts

---

[75]Unpublished paper by Li Hechun, China Defense Science and Technology Information Center, circulated at the International Cooperation in Peaceful Use of Military Industrial Technology Conference, Beijing, 1991.

[76]"On the Development of Conversion in Large- and Medium-Sized Defense Enterprises," Junshi jingji yanjiu, February 1997, p. 14.

[77]Chai Benliang, "Conversion and Restructuring of China's Defense Industry," in Brömmelhörster and Frankenstein, p. 80.

[78]See Du Renhuai, *"Jungongye shixing 'Yi Qi Liang Zhi' zhanzhi zhidu shentao"* (Military Enterprises Carry Out 'One Enterprise, Two Systems'), in *Junshi jingjixue*, July 1997.

to produce barely 5% of our output [i.e., military products]."[79] Similar comments have been offered to Western visitors—a NORINCO official volunteered that the organization really didn't like to meet military production quotas because it was not as profitable as more lucrative commercial opportunities. Another CDIC official noted that he could obtain profitable market prices for his civilian goods, but had to accept marginal payments for his military production.[80]

The paradoxical outcome of the Chinese defense conversion strategy—that is, promoting conversion to increase defense capabilities but ending up with successful conversion enterprises not being willing to supply defense needs—is mirrored elsewhere in the Chinese case. The implicit strategy behind Beijing's effort to attract foreign direct investment (FDI)—"the open door"—was not to enmesh China's economy with the rest of the world, bur rather to develop import-substitution industries. But the outcome was that China's growth has became heavily dependent on exports and FDI. Of course, China is hardly alone in finding that grand strategies might have unintended consequences.

CONVERSION AND "SOCIAL STABILITY"

But there may be yet another more compelling rationale behind the defense conversion effort: the maintenance of employment. Indeed, a criterion of "success" outlined by Jin Zhude was "dealing with problems of social security" in the defense sector. When a newspaper reporter asked him about the results of defense conversion, he said that while a little profit was being made overall, "the remarkable thing" was that "unemployment was avoided."[81] Jin had earlier written that "timely conversion" had many benefits: it "avoided unemployment, close downs of factories, and social instability caused by sharp reduction of military orders."[82] His remarks were echoed by CAPUMIT official Wu Zhao, who noted that CDIC employment would fall as the PLA cut its forces: "We have to speed up the pace of defense conversion, or mass unemployment will threaten social stability." [83]

Indeed, concern over "social stability" runs throughout commentary, pro and con, about Chinese economic reform in general.[84] In 1993, the Chinese Academy of Social Sciences issued a 300-page report warning of social and ethnic clashes as the

---

[79]*Zhongguo junzhuanmin bao (China Defense Conversion News)*, April 2, 1993. My thanks to Evan Feigenbaum of Stanford University for drawing this news clip to my attention. Interestingly enough, this "one factory-two systems" problem has also been noted by foreign joint venture managers—but it is the JV side of the operation that tends to get exploited for its technology and skills.

[80]Bates Gill, "Defensive Industry"; CBNC Asia program on Chinese defense conversion by Robert Stern, 1995.

[81]T. Poole, "China's Army Storms Hong Kong With Goods for Sale," *The Independent,* July 1993.

[82]Jin Zhude and Chai Benliang, "Strategic Thinking of Chinese Conversion in the 1990s," International Cooperation in Peaceful Use of Military Industrial Technology Conference, Beijing, 1991.

[83] Vivien Pik-Wan Chan, op. cit.

[84]This view has been expressed to the author personally by both Chinese and foreign observers. See the useful summary in Steven Mufson, "Chinese Reforms: 'Neoconservatives' Cast a Cold Eye—Challenges to Deng Flow Into the Mainstream," *International Herald Tribune*, November 14, 1995, p. 2, and Chen Yizi, "A Realistic Alternative for China's Development and Reform Strategy: Formalized Decentralization," *Journal of Contemporary China*, No. 10, Fall 1995, pp. 81–92.

economic reforms brought change. Jiang Zemin highlighted the issue in his "12 major relationships" issued in 1995. This concern also runs across the Chinese political spectrum, from liberals concerned about social pressures to conservatives, who see a major source of instability in the masses—perhaps 100 million—of displaced rural folk who have flooded into China's burgeoning coastal cities in search of menial work. We would add that this concern seems to underscore managerial decisionmaking in China. A study of decisionmaking in Beijing-area SOEs by Yuan Lu shows concern for stability a constant undertone.[85] The same issues show up in interviews with joint venture managers, who have mentioned sabotage of machines as a possible consequence of worker unrest. A factory manager in central China once told a potential Hong Kong investor that at least half of his employees were redundant. But he didn't let them go. Why? Because, he said, he cared for his workers. And for his life.

To be sure, China is a "high friction" society. Peasant uprisings and urban disturbances regularly mark Chinese history. Today, rural disturbances, strikes, and ethnic clashes in the West and in Tibet continue. We would not argue that China is on the verge of massive unrest or break-up. But as Chairman Mao once said, a single spark can start a prairie fire. For a regime that operates in the shadow of June 1989 and has, in effect, lost cultural hegemony, the issue of social stability is particularly sensitive.[86]

LESSONS?

These concerns also point up the importance of considering the reform of the Chinese defense industries in the full context of the current Chinese situation. The CDIC is embedded in a system that is undergoing severe transition and has encountered a host of difficulties. Attempts to deal with these problems have generated unintended consequences.[87]

Are there lessons here? Maybe. Certainly the Chinese case demonstrates the difficulties of defense industry reform, not to mention the problems of transition from a planned to a market-oriented economy. We would not deny that there have been market successes in the program (many the outcome of foreign investment or

---

[85]Yuan Lu, *Management Decision-Making in Chinese Enterprises*, London: Macmillan, 1996. Yuan Lu, formerly the head of research at the EU-sponsored management training institute in Beijing, is now at the Chinese University of Hong Kong.

[86]On this last point, see, for instance, An Yunqi, "Surveys on Changing Concepts of Value of Chinese Workers," *Dangdai sichao [Contemporary Thinking]*, April 1997, in FBIS-CHI-97-108, April 20, 1997. The article reports on attitude surveys carried out among urban workers between 1986–1996. The surveys reveal decreasing political awareness and increasing concern for personal values; increasing urban income disparities; and large concealed underemployment, not to mention unemployment. All of these factors contribute to social apathy and even social unrest. Even Party members are not immune: "What merits our attention is the fact that the number of party and youth league members who cling to social and political ideals is on the decline each passing day."

[87]All of these problems with the CDIC may have contributed to shakeups in the CDIC bureaucracy: the 1996 retirement of long-time COSTIND head Ding Henggao and his replacement by LtGen Cao Gangchuan; the 1998 reorganization of a civilianized COSTIND under the State Council; and the subsequent creation of the General Armaments Department, headed by General Cao.

foreign cooperation)—shipbuilding (major exports here, but also increased competition from Korea), consumer electronics, aircraft subassemblies (work for Boeing) and motorcycles.[88] But successes have come when the enterprises involved have not strayed far from their core competencies. As Norman Augustine has written,

> Defense conversion must be approached cautiously. The strategy can work, as indeed it has in some instances when companies have moved in adjacent commercial markets: markets that share technologies and customers with a company's established business. But conditions must be right, and expectations must be realistic.[89]

Might there be other solutions to the CDIC dilemma? Might increased arms exports at least keep the weapons lines open? After all, moving into exports is a conventional way of extending the product life cycle and keeping factories open. But in fact, the value of Chinese arms exports have been declining since their peak during the Iran-Iraq war, when China supplied both sides. According to the Stockholm International Peace Research Institute, Chinese deliveries of major weapons systems dropped from US$3.1 billion in 1987 and US$2.2 billion in 1988 to about US$1.1–1.2 billion per year in the 1990s.[90] Arms Control and Disarmament Agency (ACDA) figures show the same trends, though the valuations are slightly different. The high point came in 1988, when China transferred US$3 billion worth of arms. By 1995 those deliveries had dropped to US$600 million. China's major customers were and remain in the Middle East and South Asia, and the categories of weapons system delivered were concentrated on armor, jet fighters, artillery and missiles.[91] But the PRC's current clients—e.g., Myanmar—are not a big market, and in any case, Chinese arms, while cheap, are not that militarily competitive when stacked up against those of Russia. Whether China has a niche market in nuclear and missile technologies with countries like Pakistan and Iran is another question, but if so it is likely that the dollars stay with the marketing organization. In any event, Chinese non-proliferation pledges at the 1997 U.S.-China summit in 1997 will further erode this market.

---

[88]For instance, the Changhong Electronics Plant in Sichuan, formerly a defense electronics operation, has switched over to color TV manufacture and apparently dominates the market. See "Chengdu Aircraft Uses High Technology to Grab New Vitality," *Science & Technology Daily*, May 28, 1998 in FBIS-CHI-98-167, June 16, 1998, for a listing of advanced imported technologies acquired for aircraft subassembly work. The article does not mention that Chengdu Aircraft also makes J-7 fighters. NORINCO joint ventures in the motorcycle industry have led to market dominance in that sector and are usually a stop on any industrial tour, but according to an interview with the Japanese manager of one of those model joint ventures (in 1995), profits were still a way off, and in 1998, the Thai backer of one of these plants withdrew from the venture. See Karen Cooper, "CP Pokphand losses mount to US$109m," *South China Morning Post*, May 22, 1998. In August 1998, Motorcycle Center of America, a distributor, announced the cancellation of a deal with another related Sino-Thai joint venture because promised financial statements were not forthcoming. See "Motorcycle Center of America Inc. Terminates Acquisition of Hai-Nan Jia-Thai Motorcycle Company," *PRNewswire*, August 14, 1998.

[89]Norman Augustine, "Reshaping an Industry," p. 87.

[90]In constant 1990 U.S. dollars. See SIPRI arms trade database, 1995.

[91]ACDA, *World Military Expenditures and Arms Transfers 1996*, Tables IV and V.

TAKING THE LONG VIEW

Thus we must take a somewhat long perspective on CDIC issues. The reforms kicked off at and after the 9th Party Congress will take years to sort themselves out. Access to capital is one of the key issues facing the CDIC, though this problem is not limited to the Chinese defense sector. Some of the more aggressive central corporations in the CDIC—NORINCO, for example—have sought to raise capital in both domestic and international capital markets, and no doubt other firms have tried the same. Joint venturing is another way to attract capital. But for most of the CDIC, bank loans appear to be the favored method of finding working capital. Since it is a habit in SOEs to enter bank loans in the account books as income (which never has to be paid back), it is no wonder, as we have seen, that Chinese banks are not crazy about CDIC "opportunities." And given the recent ups-and-downs of Asian stock markets and a general decline in foreign direct investment in China, finding capital by other means is not going to be easy.

The March 1998 reforms suggest that Chinese policymakers have decided to further rationalize the CDIC structure, to "shrink smart," by keeping core defense capabilities within the CDIC and getting rid of nondefense business. How the reforming CDIC will handle the tight vertical integration that marks the socialist industrial structure and how the CDIC supply chain will be managed is not clear. CDIC managers talk about the need to modernize plant and equipment, or set up new lines (if not factories) because the defense plants themselves are obsolete ("the one factory, two systems" issue). To make refrigerators, build a refrigerator plant. And to make Su-27s, take a green field approach: the planes China will build as part of its deals with Russia will not be built in old factories.

Policies that allow structural rationalization and more efficiency, better human resource management practices (particularly workforce training), and flexible managerial leadership are essential to make any headway. (See the Harbin Aviation case below.) Indeed, these are precisely the same remedies that the modernizing civilian sector is adopting. And, of course, these are the recipes recommended by "the market." But even so, the prospect for much of the CDIC is not secure. Will the reforms planned for the CDIC result in further isolation from the more dynamic civilian sector? And while we've seen that parts of the CDIC have benefited from participation in the international division of labor (particularly aviation), a more self-contained CDIC may find it difficult to access those benefits. Finally, it is not enough simply to have grand plans for restructuring. In business, as in war, the human factor is crucial. New structures will demand new managers and more technical talent—and most observers find China facing challenges in developing human capital.

It will be a difficult road ahead. Perhaps the future can be found in the views of Wang Zhongyu, minister of the State Economic & Trade Commission. Speaking at the 1996 National People's Congress, he pointed out: "The relationship between reform, development and stability must be properly handled." A number of consolidations and bankruptcies were being carried out as an experiment. The aim, he said, was to develop a system of "the survival of the fittest."[92] Indeed, the road to wealth and power is not an easy one.

CASE: HARBIN AVIATION

The Chinese press tends to report good news, especially about defense matters. Still, a long-time observer of Chinese reporting about defense industry issues cannot help but note shifts in contemporary coverage. To be sure, the general tone remains policy-oriented and upbeat. But now it is also politically correct to talk about shortcomings and even describe failures. One lengthy example can be found in reports on Harbin Aviation, one of the key defense industries in China's northeastern Heilongjiang province, an area heavily dependent on SOEs and the defense sector.

A major producer of H-6 (Tu-16) mid-range jet bombers, Harbin Aviation in 1971 produced 200 planes with a claimed output value of RMB1.5 billion. By 1982, total output value had fallen to RMB70 million, about 5% of the output value 11 years previous. Military output accounted for only 5% of this figure.

Harbin's initial response was typical of Chinese state-owned enterprises: opportunistic and somewhat primitive diversification. Make anything that might generate cash, including washing machines, noodle machines, pots and pans, even horseshoe nails. But the market for these items was saturated and production ended up unsold—in warehouses, not households. Needless to say, this outcome was hardly satisfactory.

The enterprise leadership then took a long—and from our experience, an unusually strategic—rethink. They essentially moved to establish themselves in niche markets: minivans, small civilian aircraft and mid-tech pharmaceutical packaging machines, even as they kept up military production. They reformed the enterprise personnel system by instituting a labor contract system, increasing pay differentials and instituting ways of handling surplus labor. By putting excess workers into training and opening up a diversified range of sideline services, supplier and production subsidiaries could absorb surplus labor according to market demands for their products (a common enough practice in nondefense SOEs). They also took steps to start charging for welfare services the enterprise formerly provided for free.

Significantly, Harbin Aviation also became involved in foreign investment and foreign contract work. The enterprise undertook to manufacture parts and subassemblies for foreign aircraft manufacturers, including British Aerospace. They licensed technology from Aerospatiale and Turbomeca for the French Dauphine II helicopter in several configurations. Foreign sales were in the offing.

Heilongjiang papers reported that total output value for 1995 was projected to be RMB30 billion, with profits and taxes paid above RMB5 billion. Not only was this a considerable nominal increase, but more important, compared to the numbers noted above, a four-fold increase in the ratio of profits and taxes to overall output

---

[92]See "Losses Grow for Chinese Industry," *International Herald Tribune*, March 15, 1996, p. 15.

value, from the 4–5% range to over 16%. The enterprise was also reported to be one of the nation's top 500 companies in terms of sales.

None of this transformation, of course, happened overnight. The process took most of the 1980s, required substantial support from central, provincial and municipal authorities as well as outside investment.

The lessons the Chinese press draws from this experience could have come from any basic MBA text, especially the importance of good leadership, initiative, training for personnel at all levels, market research leading to high-value-added products, quality control, flexible organization, and a developed sales and service force. And starting up separate, new lines for civilian production. Harbin Aviation thus is a model for "conversion." Other defense sector operations in the province need to learn from the company. The reports say that while Heilongjiang's aviation industries are "in the fast lane," other defense enterprises are "bogged down in a ditch."[93]

For analysts interested in the aims of conversion, however, the implications of the Harbin experience go further than illustrating a successful example of economic reform. The "Dauphine" helicopters come in artillery spotting, armed, shipboard and high plateau models for "both domestic and foreign market requirements." A Hong Kong magazine highlighted a gunship model of the aircraft in an article which discussed operations that might be mounted against Taiwan.[94] Subcontracting spare parts for foreign manufacturers "enables [the enterprise] to learn advanced technology free of change," leading to increased export sales and an increase in the quality of military aircraft produced: "This shows that Harbin Aviation has not relinquished its old aircraft manufacturing business, and it also demonstrated how a sideline business can provide reverse nurture to the main industry."[95]

---

[93]Xu Feng and Sun Yongxiang, "Report on Heilongjiang Defense Enterprise Troubles," *Heilongjiang Ribao* (*Heilongjiang Daily*), December 1995–January 1996, in FBIS-CHI-96-032, 15 February 1996. Additional details are drawn from Yitzhak Shichor, "Converting the Military-Aviation Industry to Civilian Use" in Brömmelhörster and Frankenstein.

[94] "Helicopter Gunships of China's Ground Forces," *Kuang chiao ching*, 16 April 1998 in FBIS-CHI-98-128, 8 May 1998.

[95]"Defense Conversion Survey at Harbin Company," *Fendou* [*Struggle*], Harbin, April 1995, in FBIS-CHI-95-108, 6 June 1995.

11. SYSTEMS INTEGRATION IN CHINA'S PEOPLE'S LIBERATION ARMY

Rear Admiral Eric A. McVadon, U.S. Navy (Retired)[1]

In many circles, the People's Liberation Army (PLA) is best known for its obsolescence, shortcomings, failures, and ineptitude. However, among others, especially some circles in the West, the PLA's strengths and successes are highly touted. There is a measure of truth in the exaggerated convictions of both these categories of observers of PLA prowess, or the absence thereof. It is not enough to suggest, as some might, that the PLA does well in areas of low technology and poorly in high-technology areas, although there is certainly a kernel of truth in that oversimplified explanation. The PLA has, of course, excelled in some areas where at least a modicum of modern technology is involved. Success for the PLA, however, has been extremely elusive in areas where integration of systems and technologies is required. This is not to suggest that the whole story of PLA technological successes and failures can be linked to systems integration. The story is far more complex than that. Nevertheless, an examination of the issue based on systems integration capabilities is a useful and instructive one to pursue. This is undertaken in three parts: (1) an introductory look at two prominent examples of PLA modernization programs from the perspective of systems integration, (2) a definition of systems integration and the various levels into which military systems integration may be divided, and (3) an examination of PLA aspirations for systems integration, the status of the effort including problem areas, and conclusions about what the future might hold.

I. TWO METHODS OF COPING WITH SYSTEMS INTEGRATION IN THE PLA

What the PLA Cannot Do Well: The Strategic Rocket Force Story

The PLA has been unable to undertake modernization of all, or even most, of its numerous and diverse forces and units, either simultaneously or seriatim. It simply lacks the wherewithal to do so. This inability to carry out sweeping or steady modernization stems from inadequacies in its research and development facilities,

---

[1] Rear Admiral Eric A. McVadon, U.S. Navy (Retired), is a consultant on East Asian security and military issues and Director of Asia-Pacific Studies for National Security Planning Associates, a subsidiary of the Institute for Foreign Policy Analysis. He was defense and naval attaché at the American embassy in Beijing, 1990–1992. His navy career included extensive experience in politico-military affairs and air antisubmarine warfare. He now conducts research and analysis, writes, and speaks widely in North America and East Asia on Asia-Pacific regional security and defense issues.

sparse funds for equipment acquisition, limited ability to acquire and assimilate technology, and lack of availability of educated personnel, training methods and devices, and more. One preferred solution or work-around has been to modernize selected PLA forces and units. Some of these selective modernization efforts have been referred to as "pockets of excellence." The pockets vary greatly in magnitude and scope. One of the largest and deepest of the pockets (double-entendre intended) is the Strategic Rocket Force (SRF) or Second Artillery, as it is best known in the PLA.

The example of the SRF is particularly interesting because it is a conspicuous pocket of excellence—arguably the most conspicuous—and an area where systems integration is minimal. In fact, it can be said that the Chinese designed their Strategic Rocket Force in such a way that the force has great utility without reliance on the integration of systems beyond the level of the individual missiles. Unlike the American intercontinental ballistic missile (ICBM) force with its complex structure of warning systems, decisionmaking arrangements, command and control networks, coordination of a triad of platforms, etc., China's system for its nuclear missiles is extremely simple and unintegrated. The PLA's SRF is not linked to systems in space for detection of an enemy's incoming missiles. The execution of a prompt, complex, coordinated nuclear strike is not contemplated by China. Instead a rather leisurely retaliation is envisioned, using a handful of missiles, possibly days or weeks after surviving an attack. The situation is similar for the short-range ballistic missiles (SRBMs) of the SRF, for example the Dongfeng-15s (CSS-6 in American terminology or M-9s, as they are more widely known) used against Taiwan in 1995 and 1996. This is not to suggest that the Dongfeng-15s and China's other ballistic missiles are not complex systems, requiring many properly operating components to achieve success. Nonetheless, these SRBMs and their TELs (transporter-erector-launchers) can reach a launch point and be fired at a target location without reliance on an integrated system. The point is that China's most conspicuous military technological success story has a conspicuous characteristic: essential absence of requirements for complex, external systems integration. This suggests that the Chinese have recognized their weakness in systems integration and avoided this burgeoning but complex technological discipline. The structure of the SRF, with respect to both its nuclear and conventional arsenals, takes into account China's lack of skills in systems integration.

The PLA's shortcomings in systems integration are no secret, either inside the PLA or among foreign observers. Dr. Paul Godwin of the U.S. National War College has referred to articles, essays, and seminars (taken from the *Liberation Army Daily*, military professional journals, and seminars sponsored by an organ of COSTIND— the Chinese Commission on Science, Technology, and Industry for National Defense) on this subject over roughly five years of this decade. He noted that these essays and seminars reflect that the PLA is unable to aspire to early integration of advanced technologies into their operational systems—and recognizes that inability.[2] The SRF example seems to take this conclusion one step further. The PLA, out of necessity, can be quite good at making do, at achieving imaginative and

---

[2]Paul H. B. Godwin, "Military Technology and Doctrine in Chinese Military Planning," p. 43.

effective work-arounds. However, in doing so, the PLA knowingly comes up well short of world-class capability, in large measure because of inherent inability to undertake, much less achieve, the sort of systems integration that has become a hallmark of truly modern armed forces, especially nuclear forces.[3]

What the PLA Can Do: The Tale of Three Ships

It would, of course, be absurd to suggest that the PLA has neither attempted nor achieved any significant form of system integration. Nevertheless, examples are not easy to discover. The logical explanation for the lack of visibility is neither that none exist nor that secrecy has obscured the existing examples, but it is true that the PLA generally makes a conscious effort to hide examples of system integration. There is a full realization by the PLA that disclosure of its efforts and achievements in this area reveals important secrets about its capability—secrets that go far beyond simpler revelations of the identity of all or most of the weapon systems and other obvious equipment on a combatant ship. For example, it is one thing to reveal (essentially unavoidably) that a ship has an air search radar, a fire-control radar, a surface-to-air missile battery, and antiaircraft guns. On the other hand, how those components can operate together, and if that operation is effective, is a much higher magnitude secret. This secrecy is reinforced by the xenophobic Chinese view of control of military information. (It should be remembered as well that PLA secretiveness stems not only from a desire to hide systems capabilities but also from an effort to keep the curious from learning just how embarrassingly rudimentary some PLA systems are compared to those of modern armed forces.)

The PLA Navy, consequently, has been considerably less than forthcoming with respect to the degree of integration of weapon systems that exists on its three most advanced missile-equipped warships: the *Jiangwei*-class frigates, the later versions of the *Luda*-class destroyer (including the single *Luda* III), and the most recently commissioned ships, the *Luhu*-class destroyers. Even after separate visits to the three ships by the author and colleagues over a period of several years and the review of available reference materials, we cannot be confident of what is precisely the case concerning systems integration. However, the correlation of these direct observations on board the ships with reported information on sales and installations leads to considerable confidence in the conclusion that all three classes of ship have combat direction systems installed and operating systems to integrate the sensors and weapons of these ships, at least within the ship.

The fact that relatively new surface combatant ships have combat direction systems may not seem a remarkable conclusion, but the class of frigate built just before the Jiangwei program began, the *Jianghu*, does not seem to have had such a system. The author asked during a 1991 tour of a *Jianghu*-class frigate to visit the space where

[3]The Chinese nuclear missile arsenal serves well under this somewhat strange arrangement: no warning system, no alert status, warheads stored apart from missiles, a minimal force, etc. Its shortcomings, including those brought about by China's lack of skills in systems integration, are acceptable—possibly even desirable—because the utility of that arsenal is directly dependent on its not being used. Arguably, China has not sought an integrated system, or suffered from the lack thereof, because it does not seek to achieve an integrated nuclear capability that fosters reliance on a ready ability to retaliate. Another way of putting this is that China has made a virtue of necessity.

such equipment was located and was told that no such space existed. (Such answers were not unusual at that time and were not always factual, so this could not be considered as conclusive information.) A query by the author in the early 1990s to a Thai Navy officer concerning the *Jianghus* built by China for his navy resulted in essential corroboration that the ship lacked a facility comparable to the combat information center, as that officer understood the concept. Norman Friedman obtained similar information concerning a *Jianghu*-class frigate delivered in 1990 to Thailand.[4] These three examples of "negative information" are not conclusive evidence that *Jianghu* do not, or did not, have some sort of combat information center, but they strongly suggest that remarkable possibility.

Possibly the most that can be concluded from this limited information concerning installation practices and timing is that at some point, not later than early in the 1990s, China began as a matter of course to incorporate what, by that time, was called a *combat direction system* in its warships. (Other inconclusive evidence suggests that this practice may have begun earlier in the *Luda*-class destroyers.) In any event, the system doubtlessly now being installed in Chinese combatants, significantly, is not of Chinese design. It was developed by Thomson-CSF of France. The prototype was assembled in the mid-1970s and is now called TAVITAC (*Traitement Automatique et Visulation TACtique*). According to *Jane's Naval Weapon Systems*, two of the early systems were sold to China.[5] The Chinese version of the system is called ECIC-1, the existence of which reflects an apparent ability at least to replicate a system for integrating tactical information, displaying the data, and permitting the designation of weapons systems to targets. (Incidentally, China's audacity and ingenuity in unauthorized reverse-engineering may be indicated by the fact that, according to *Jane's*, China is not licensed by Thomson-CSF to produce the system.)

Jane's describes the early TAVITAC (through the early 1980s) as a mainframe system for tactical data handling. The upgraded version, TAVITAC 2000, is said to have a "star" architecture but as still relying on a basic mainframe design structure, short of the superior technology and versatility of the fully distributed architecture of the TAVITAC NT which Thomson has sold to Kuwait. The PLA Navy's newest guided missile destroyers are equipped with either the ECIC-1 or the TAVITAC 2000. This system compiles a picture of the tactical situation using inputs from radars and other sensors both on the ship and from remote sources (another ship, for example). To connect with off-board sensors, it uses a data link, which the U.S. Naval Institute reference book *Combat Fleets of the World* terms Link-W[6]—said to be similar to the Link 11 of Western navies. Several hundred targets can be tracked. The system designates these targets to weapon systems and purportedly provides some measure of assistance in reaching judgments concerning the tactical situation, i.e., assessing

---

[4]Norman Friedman, "Chinese Military Capacity: Industrial and Operational Weaknesses," in Arnett, pp. 69–70.

[5]E. R. Hooton, *Jane's Naval Weapon Systems*, Surrey, U.K.: Jane's Information Group Ltd., 1998.

[6]A. D. Baker III, *Combat Fleets of the World 1998–99*, Annapolis, Maryland: Naval Institute Press, 1998, p. 120.

urgency related to developing situations and assigning priorities to engaging targets with various systems (missiles and guns).

The system typically uses five or six vertical or horizontal consoles consisting of keyboards and large displays. The following description compiled from *Jane's Naval Weapons Systems*[7] may offer the technically inclined some idea of the relative sophistication of the system:

> The TAVITAC 2000 system uses the Thomson-CSF MLX-32 computer which is built around the Motorola 68030 and 68040 32-bit microprocessors. Each TAVITAC 2000 system has two mutually redundant computers with six Mbytes of memory, one acting as master and the other as a hot standby, and capable of 3 Mips. The system uses Ada software language, the UNIX System V operating system, and features a duplicated VME bus 10 Mbytes Ethernet-standard local area network. A rugged disk storage offers a database management capability for map displays, ship resources, and management.

Another interesting, if less technical, indication of the level of sophistication of the combat direction system installation in the *Luhu* is that France has installed the TAVITAC 2000 in its impressive, stealthy new *Lafayette*-class frigates. Based on information from several sources,[8] it appears that this combat data system, incongruously, is also present on the version of the *Lafayette* frigates recently delivered to Taiwan. (As with the PLA Navy, the ROC Navy is not inclined to allow visitors on these new frigates, especially in the combat direction center.)

What We Learn from the SRF and Ship Sagas

The story of the PLA's Strategic Rocket Force and the tale of the PLA's three most modern warships are instructive with respect to the state of systems integration in the PLA. The Chinese know both that systems integration is important in the building of modern armed forces and that the PLA is very weak and inexperienced in this field. This recognition has led them, in the case of the SRF, to avoid their weakness, even if it means having a nuclear deterrent "with Chinese characteristics," implying in this case a need to work around severe limitations. In the case of the destroyers and frigates, recognition of the weakness in systems integration technologies forced China to go elsewhere, to France in this case. It was not feasible to design a workable guided missile destroyer's combat direction system if it had to be done with Chinese characteristics. At a minimum, a modicum of advanced system integration technology was necessary to have a moderately combat-capable warship. It is noteworthy that the evidence derived from these cases and the impression one derives from a broader look at the systems integration picture for the PLA leads to the conclusion that China has gone no further with respect to systems integration than to identify the problem and PLA shortcomings. No significant aptitude has been displayed for successfully attacking the problem indigenously.

---

[7]Hooton.

[8]Friedman, p. 69; A. D. Baker III, *Combat Fleets of the World 1995*, Annapolis, Maryland: Naval Institute Press, p. 730; Hooton, *Jane's Naval Weapon Systems*, table at end of TAVITAC section.

Several aspects of the issue remain quite murky, however. The importation of systems like TAVITAC and even the Chinese copying and production of that system do not necessarily imply that the systems work well on board the PLA Navy ships. For example, the diverse radar and sonar systems and other sensors, some obtained from various countries and others produced in China, undoubtedly present formidable challenges in data interfaces, input coordination, and systems compatibility. Computer hardware and software differences among the various systems and the translation and comprehension of manuals and computer programs offer further challenges. It is hard to imagine that the PLA Navy has succeeded in putting all this together and formed a seamless combat direction system.

II. DEFINING SYSTEMS INTEGRATION AND ITS LEVELS OF APPLICATION

Comprehending the Concept

Like the currently popular terms *asymmetry* and *RMA* (*revolution in military affairs*), *systems integration* has become something of a buzzword in defense affairs circles. Further, the term is used or misused widely, often to reinforce whatever point the writer or speaker wants to make, with little concern for consistency or accuracy. Consequently, prior to delving more deeply into the matter of systems integration in the PLA, it is useful to examine informed efforts to define and understand this term. The term is relatively new in its current ubiquitous military usage, especially in the PLA. Few, if any, defense specialists have spent more time and effort on defining the term systems integration and focusing its application than those in the U.S. Department of Defense who are dedicated to the acquisition of systems, development of technologies, and the formulation of the U.S. policies related thereto. The introduction to a definitive document produced by that office, the U.S. Department of Defense publication entitled *Militarily Critical Technologies List* (*MCTL*),[9] states:

> *Systems integration* enables the harmonious and productive working of disparate components and the interfaces that connect them. Each weapon system requires the use of specific hardware and software and the integration of new technologies or advances in existing technology subsets to increase overall system performance, improve manufacturing or reduce costs.

> *Systems integration* is an ongoing process. Good integration includes traceable assurances that the components and functions will fit together and operate in concert. In the past, weapons systems designers have successfully improved both the hardware and software in an interactive process, and then integrated both to effect simultaneous improvements. Excessive integration adds cost and time without yielding a significant improvement in the product or system. Too little integration results in products or systems that do not function as advertised.

---

[9]Office of the [U.S.] Under Secretary of Defense for Acquisition & Technology, *Militarily Critical Technologies List*, Washington: National Technical Information Service, Springfield, VA 22161, June 1996, p. 2.

Technology integration can be treated as a subset of systems integration. High technology weapons systems are fundamentally driven by availability and integration of technologies . . . . *The tools and techniques for preparing, mixing and matching the various components are also critical technologies because they are key to achieving the desired qualities* [final emphasis added].

This careful, if lengthy, description of systems integration and explanation of its importance outlines the purpose and nature of the process of achieving it, points to the pitfalls, and, possibly most important for this examination of the subject, states that the very methods employed in the integration are *critical technologies in themselves.* As has long been recognized, China, and especially the PLA, suffers from technological disadvantages in areas such as electronics, computers, and software. That is difficult enough, but, more tellingly, China, it seems, has yet to begin to grapple seriously with the next very complex step: mastering the critical technologies of systems integration, referred to in the extract from the MCTL.

Paul Dibb, formerly Director of the Australian Joint Intelligence Organisation and Deputy Secretary of the Australian Department of Defense, wrote recently: "Not only is implementation or planning for systems integration almost totally deficient in the [Asia-Pacific] region, there is also a very limited capacity to modify and adapt current combat systems that are vital to operational effectiveness." Noting that systems-integration technology has eluded even Japan, Dibb asserts that the failures "are even more pronounced in China and India."[10]

The Scope of Systems Integration: Five Levels Applicable to the PLA

For China's military leaders, contemplating the largely unexplored sweep of systems integration as it applies to China's armed forces, from the broadest context down to individual units or troops in combat is, undoubtedly, a daunting task—as it is for the outsider trying to grasp the scope of this problem for China. There is, of course, no fixed set of categories or levels for the application of systems integration. However, the following attempt to divide the sweeping problem into five comprehensible levels of applicability may be useful as a device to try to understand both the scope of the problem and its many facets.[11]

Military systems integration at the regional or global level. The highest plane of the systems-integration challenge that faces China could be termed the *big picture* level. This level of the integration problem is perhaps illustrated well by examining the saga of the U.S. aircraft carrier battle groups deployed to the region in March 1996 as a response to the second round of SRF M-9 missile "tests" and leading up to the first popular election of a president in Taiwan. Put starkly, the Central Military Commission (CMC) in Beijing had to depend on American announcements from Washington and Honolulu and reports from CNN to learn that the carrier battle groups had been deployed and where they might be operating. Then the PLA had no

---

[10]Paul Dibb, "The Revolution in Military Affairs and Asian Security," *Survival,* Vol. 39, No. 4, Winter 1997–98, pp. 102–103.

[11]The following scheme of five levels of system integration is an artificial device conceived by the author to facilitate treatment of a complex subject and should not in any way be viewed as reflecting concepts employed by the PLA or the U.S. Department of Defense.

means to verify the presence of the battle groups or to determine their positions. Unknown were even the general locations, much less the latitude and longitude or course and speed of individual ships. It is likely that to this day no one in China knows when the second carrier battle group arrived, where it operated, how its ships and aircraft were disposed, and precisely when it departed the area. From the U.S. perspective, this was a very comfortable situation. Its forces, because they could not be located by China without U.S. complicity or "cooperation," were operating essentially in a form of sanctuary. Further, there was the luxury, if desired, to announce or leak something about the location and disposition of forces and have that serve almost as well as if it were wholly factual. This example illustrates well China's predicament should it wish to react to such events more than a few score miles off its coast, lacking an integrated system that can present a dynamic tactical picture over the area of concern.

The character of this problem of location and identification possibly can be even better appreciated by a quick look at China's existing capabilities to determine the tactical situation in the ocean areas off China and Taiwan. A PLA Navy maritime reconnaissance aircraft, on an extended mission, might, after arriving on station, search an area of 20,000–30,000 square miles using radar and electronic intercept equipment—assuming that the U.S. ships (or other naval units of interest) were not evading detection. Yet the area that might warrant searching (where the carriers could be in positions to close the target area and launch strikes or conduct other missions, for example) is 400,000–600,000 square miles—ten times the optimistic search area of a single aircraft mission. Moreover, the searching aircraft, typically using radar, can be denied the ability to detect the ships of interest by being turned away by intercepting aircraft and through various electronic and spoofing means.

China looks to a future when the PLA can operate its ships and aircraft several hundred miles from its coastlines and, in doing so, protect its maritime interests. This blue-water endeavor may take rudimentary form, ensuring that ocean commerce is not disrupted, or it may attempt something more sophisticated, such as an effort to achieve sea denial or control. Whatever form China chooses, the admittedly rather rudimentary example provided here illustrates that the PLA cannot begin to determine the tactical situation it faces in its surrounding ocean areas, absent an integrated surveillance system. That system should be able to detect threats or other contacts of interest (and discriminate between them) day and night and in all weather, and even under circumstances where the targets are attempting, through sophisticated or simple means, to avoid detection. This system would then have to be integrated with means to identify and evaluate the detected contacts, eliminate false contacts, and correlate the many detections in such a way as to compile target tracks and to be able to forecast future positions—assuming some action with respect to the targets was contemplated. This is but one example of a level of systems integration that would require interfaces and correlation on a grand scale among sensors and other systems widely separated in distance, design, and character.

This level of the "big picture" system integration problem also would apply to aspirations by China to compile a picture of satellites in space that might threaten

China or that China might hope to threaten. Furthermore, as alluded to at the outset, China is severely constrained, whether by design or necessity, in its nuclear deterrent policy by its lack of capability to obtain warning of a nuclear missile attack. All of these examples of the highest level of system integration are almost certainly out of China's reach for the foreseeable future. Beijing may continue to conclude that it will simply have to tolerate this shortcoming in big-picture system integration. Whichever direction China chooses with respect to this very complex technological problem and, especially the systems-integration aspects thereof, will be a critical determinant of the character of China as a major military actor in East Asia. As physicist and security expert Norman Friedman put it after extensive research, "While China's recent history features a number of socio-political set-backs that have crippled the military technology base, the rest of the world has been racing forward at a remarkable rate. As a result China is not only far behind the state of the art in electronics and command and control; Chinese planners may be unable even to conceive of appropriate solutions to the problem of closing the gap."[12]

Integration among platforms in a warfare area. The next level of systems integration involves meshing various platforms and components within a specific warfare area. These areas might include, for example, antisubmarine warfare (ASW) or air defense.

An effective ASW capability optimally would include:

— **aircraft** (fixed-wing and helicopters) dropping sonobuoys and monitoring the radio signals from those buoys for submarine-generated noise or for echoes from explosive charges dropped in conjunction with the buoys, or, in the case of helicopters, using passive and active dipping sonars. Aircraft able to proceed at high speed to investigate suspected contacts generated by other means and deliver attacks with homing torpedoes;

— **surface ships** (probably destroyers and frigates) employing hull-mounted sonars and trailing variable-depth sonar equipment. Ships able to coordinate at least local area antisubmarine operations and to deliver attacks with homing torpedoes;

— **submarines** (preferably quiet nuclear-powered vessels which have faster submerged speeds for unlimited periods) able to occupy the same acoustic water layers as the target submarine and with the ability to communicate contact information to the local ASW coordinator and, critically, to antisubmarine aircraft that are able to close contacts rapidly and deliver attacks, as described above;

— **sea-bottom acoustic arrays** (sets of hydrophones on the ocean floor) positioned in strategic areas of concern and linked to monitoring stations ashore with the capability to detect, identify, and track targets that could then be prosecuted by aircraft or possibly by ships or submarines in some circumstances.

Air defense systems are more intuitively obvious to most readers, but they might include as a minimum:

---

[12]Friedman, p. 67.

— **land-based, sea-based, and/or airborne radars** that can provide requisite coverage against low-flying aircraft, are resistant to jamming or other electronic deception, can be defended against anti-radiation missiles, and can detect, identify, track, and facilitate intercept and engagement of targets;

— **interceptor aircraft** with appropriate speed, altitude capability (service ceiling), maneuverability, and equipped with engagement radars or other sensors and air-to-air missiles sufficiently capable against the intruding aircraft so that they may be fired effectively before the intercepting aircraft can be evaded or neutralized; and

— **land-based or sea-based surface-to-air missiles**, some of which can engage targets at considerable distances and others capable of short-range engagements.

It must be said forthrightly that the PLA Navy and PLA Air Force have not achieved significant proficiency in any of the component technology areas described for ASW or air defense systems. Moreover, they certainly have not made substantial progress in any form of ASW or air defense system integration, save rudimentary direct communications (e.g., voice radio) between ASW aircraft and surface ships or interceptor aircraft and controlling radar sites as they carry out rather old-fashioned ground-controlled intercepts, for example. At the warfare-areas level of systems integration, there has been essentially none of the "harmonious and productive working of disparate components and the interfaces connecting them," as described in the U.S. Department of Defense MCTL.

China's prospects for changing the situation at this level of systems integration were addressed by Erik Baark in his recent examination of science and technology policy and technological innovation in Asia. He wrote, "When the interaction of military and civil sectors in China . . . is evaluated in terms of technological capabilities, it appears that there are still some serious bottlenecks, in particular the lack of innovation networks which, in practice, could serve to link . . . R&D to manufacturing. The networks which formally exist . . . suffer from a fragmentation which leaves little in terms of critical mass for the development of integrated weapon systems."[13]

Integration of various components to constitute a weapons platform. The next level in the hierarchy of systems integration is that of combining components to constitute a combat aircraft, a combatant ship, a battle tank, or similar platform—not the individual weapon systems on the platform but the overall platform that mounts the various sensors and weapons. Considerable skill and experience are required to combine successfully a fuselage, hull, or vehicle body; a propulsion unit; the electrical and other auxiliary systems required; and the suite of sensors and weapon systems, generally including very sophisticated electronic components and computers and all the linkages needed. The problem is multiplied many times over if the various assemblies are from several countries with varying origins and conforming to different technical standards. (Further, Chinese military aircraft, ship,

---

[13]Erik Baark, "Military Technology and Absorptive Capacity in China and India: Implications for Modernization," in Arnett (ed.), p. 109.

and vehicle manufacturers have not generally been able to produce identical versions of units of the same model, even when in serial production. Consequently, in many cases, "black boxes" cannot readily be exchanged from spare-parts bins or among units to troubleshoot, effect repairs, or carry out preventive maintenance, for example.)

The general nature of the problem can be appreciated by considering the PLA Air Force's (PLAAF's) proposed future fighter aircraft, something intended roughly to approximate a modernized F-16. Paul Dibb, using this developmental F-10 as an example, refers to these as *hybrid systems*, "combining platforms, radars, avionics, and missiles from different suppliers." He goes on to write that "China's next-generation fighter, the troubled J-10 [often referred to as the F-10 in the West], has a Chinese airframe, Israeli avionics, and Russian engines."[14]

The *Luhu*-class destroyer serves possibly as an even more dramatic example. The first ship of that class, *Harbin*, while under construction at the Jiangnan Shipyard in Shanghai, was described reliably (by Chinese sources) as featuring 189 of China's "achievements in the development of naval equipment," incorporating *more than 40 advanced foreign technologies*, and fitted with over 50,000 sets of equipment. The ship was described as having "equipment developed or produced by 19 provinces [of China], 11 ministries, commissions and corporations, and 100 manufacturers, research institutes, and entities." The anti-air missile launcher and its Mach 2.4 missiles (an actual *Crotale* launcher on the first ship and a Chinese version of that system on the second, further complicating things) were developed in France by Thomson and Matra Missiles. An air search radar (TSR3004) and the combat data system (the TAVITAC, described previously) were also developed by Thomson, but the long-range air search radar is Chinese. The A244S homing torpedoes are of Italian origin.[15] The C-801 anti-ship cruise missiles were designed and manufactured in China. The origins of many other advanced components large and small (e.g., electronic countermeasures equipment), and notably including the sonar systems, are unknown to the author.

If these statistics and lists, many of which were cited publicly by PLA Navy representatives with the intent to impress the Chinese public and foreigners as well, were not stunning enough, it should be noted that only two ships of this class were built. There was little if any time and opportunity to profit from lessons learned in this extremely complex area of systems integration. Certainly much of the effort that went into achieving some level of compatibility and devising interfaces for all these disparate components can be applied in varying degrees to the upcoming new class of destroyers, (*Luhai*, previously known as the *Dalian-C*) but those new ships, the first of which is already on the way, will incorporate new technologies that China has been able to develop and acquire.[16]

---

[14]Dibb, p. 99.

[15]Tseng Hai-tao, "Commander Jiang Wants to Accelerate Naval Construction, China's Newest Warship Emerges," *Kuang chiao ching*, August 16, 1996, in FBIS-CHI-96-209, October 29, 1996, pp. 2, 5; information on the identity and origin of the long-range air search radar is from Baker, p. 120.

[16]Conversations between the author and knowledgeable PLA Navy officers in 1997.

This illustrates yet another aspect of the problem. The PLA has great difficulty forecasting what systems and technologies can be obtained, when they will become available, and which of those available it wishes to use. There are several reasons: (1) China's own research and development effort is spotty and inadequate, (2) many countries restrict what they sell to the PLA (and are prone to change their minds), (3) foreign equipment is expensive and suppliers often demand payment in hard currency, (4) China is reluctant to repeat the experience of becoming overly dependent on foreign suppliers (as it was on the Soviet Union in the 1950s and on several Western countries in the 1980s before Tiananmen interrupted that episode of technology transfer), and (5) the PLA has difficulty assimilating and incorporating new technologies. Once more, the *Luhu* class and its follow-on are illustrative. Looking at the critical matter of ship propulsion, the new follow-on destroyer, although similar to the *Luhu*, has a wider hull (broader beam) to accommodate a different, somewhat larger and bulkier marine gas-turbine engine. That is because post-Tiananmen sanctions imposed by the U.S. have precluded acquisition of additional GE LM 2500 engines like those in *Harbin* and her sister ship.[17] The follow-on ships will have Ukrainian G525000 gas turbines, said to be selected using "a combination of technical and political factors."[18]

Dr. Paul Godwin, an experienced and recognized specialist on the PLA at the U.S. National War College, summed up a part of China's plight at this level of systems integration as follows: "The simple fact that all the PLA's advanced weapon platforms depend on imported technologies for their power plants, weapons, and electronics is a clear indicator that China's research centers have yet to produce weapon platforms based on indigenous technology that match those the advanced industrial states were manufacturing by the 1970s. For such a military technology and industrial base, advancing into the technologies required for the 21st century is a daunting task. This task is made even more intimidating by China's continuing quest for military self-sufficiency."[19]

Furthermore, the path to integrating these technologies is almost never direct and efficient. The sagas involved in obtaining the avionics and engines for the J-10 chronicle more than a decade of frustration, dashed expectations, disappointments, engineering changes, schedule and cost overruns, and failures. All these have done far more than delay the progress of this seemingly plagued fighter aircraft program. They have also severely taxed Chinese aeronautical designers and engineers in an area that is already their near nemesis: the challenge of integration into a single tactical airframe of all these diverse systems from international sources. A similar situation exists with respect to the *Luhu* and many other areas where China has undertaken the daunting task of combining components from diverse suppliers in an effort to deploy a system far more advanced than that which China is able to design and build indigenously.

---

[17] *Ibid.*

[18] Tseng, p. 3.

[19] Godwin, "Military Technology and Doctrine in Chinese Military Planning," p. 59.

Unless one cynically attributes it all to a quality of stubborn persistence in those who guide these PLA programs, it must be concluded, from the frequent and repeated resort to these methods, that no better method appears feasible to them. There is no question that the encumbrances of this erratic and uncertain method are not trivial. They are a critical factor in limiting China's successes in producing ships, aircraft, and other platforms that even approach or approximate the level of modern weapon systems. The *Luhu*-class destroyer is a ship that Western navies and the Japanese Maritime Self-Defense Force would have been proud to put to sea 20 or more years ago. The J-10, if it gets past the prototype stage, will likely hold its own against the F-16, an aircraft that first flew in the U.S. Air Force about a quarter of a century ago. But the designers and builders of tactical aircraft and combatant warships for the PLA are stymied by the problems of integrating systems to constitute a modern fighting platform.

Integrating components from diverse sources that form a weapon system. This level of systems integration involves the use of components to comprise a specific functional weapon system—an individual weapon system on a platform, not the overall platform (ship or aircraft). The classic case of this level of integration is the combining of a detection system (a radar, for example) with a system to aim the weapon (a fire-control system, for example) and one or more devices that deliver lethal fire on a target (missiles and the associated launcher and/or guns and the projectiles they fire). The air and missile defense system of the *Luhu*-class destroyer is a good example. The components of that system were described above.

As mentioned, the long-range air-search radar is of Chinese origin (although it probably is derived from earlier radars developed in other countries). The *Luhu* has no means of engaging air and missile targets at the ranges (up to a hundred miles or more) that this *Sea Eagle* radar could gain contact. It would be desirable to have an automatic interface between the Chinese radar and the French systems that can provide a tactical display, track the targets, and assign missile batteries or guns to the target (TAVITAC 2000). The Chinese have probably at least attempted to install such a linkage, but, for the PLA, trying and succeeding to the degree that combat reliability is attained may be two quite different things.

The Luhu's missiles and launcher are French (the *Crotale* system or the Chinese version thereof, as mentioned previously), probably facilitating the interface between the combat direction system and the anti-air missile system, both designed by Thomson. However, the long-range 100mm and short-range 37mm guns are Chinese, as are the Type 347G (*Rice Lamp*) fire-control systems[20] for the guns. (The unmanned, wholly automatic 100mm gun, with a firing rate of 30 rounds per minute per barrel,[21] closely resembles an Italian main-battery gun. The 37mm guns are rapid-fire weapons—760 rounds per barrel per minute,[22] although the weapon resembles another Italian gun, the PLA Navy has expressed great pride in its

---

[20]A. D. Baker III, *Combat Fleets of the World 1997-98*, p. 120.

[21]Conversation between the author and a PLA Navy captain on board the *Luhu*-class destroyer *Harbin* (but not normally assigned to the ship) while on a port visit to San Diego.

[22]Baker, p. 115.

development.) The establishment of an extremely reliable and effective interface between the combat direction system and the rapid-fire 37mm guns, two systems of widely differing origins, is especially crucial. These guns and their fire control system, something like the *Phalanx* close-in weapon system (the Gatling-gun-like CIWS) developed and used extensively in the U.S. Navy,[23] are the last resort to down anti-ship cruise missiles that leak through other air defenses. Failure of this complex sequence of target designation, precise fire control, and then faultless fire of a fusillade of projectiles from the guns would be devastating, likely resulting in heavy damage or the disabling of the ship. So, the PLA faces the need to resolve this extraordinarily difficult integration problem in a situation where there is no second chance or backup for that crucial system in combat.

Other crucial examples of systems integration problems at the level of a specific weapon system include melding the radar in the Russian Su-27 interceptor aircraft and the Chinese air-to-air missiles (AAM). The PLAAF would seek the capability to employ its PL-series missiles in addition to the Russian missiles purchased as part of the Su-27 sales arrangements. PLAAF leaders must ponder the likelihood that Russian-made missiles might be subject to easy defeat by Russian pilots, if there were once more frictions over the northern border—or that the Russian designers might have revealed missile-countering secrets to Americans or others. In the obverse of this example, the PLAAF may wish to make its Russian-made heat-seeking and radar-guided AAMs compatible with the J-10 or other Chinese-built fighter aircraft. To illustrate how far the PLAAF may have to go to be able to achieve these complex meldings, one need only recall the observations by some knowledgeable observers that the PLAAF apparently has not yet achieved a consistent capability with the Su-27s, acquired early in this decade, to employ air-to-air missiles in all-weather conditions and beyond visual range. The PLAAF does not have a good record or reputation for success in endeavors such as this, or even in much less complex tactical integration.

However, despite problems and gaps, there has been some progress in this level of systems integration by the PLA, even if spotty, as amply illustrated by the presence of the TAVITAC systems in the three latest PLA Navy warships. This example of limited success brings to the fore other critical questions: Will the PLA expend the effort and consistently provide the resources needed to keep these (largely foreign) systems operative? Will the advanced skills required for operation and maintenance be taught on a continuing basis? Will preventive maintenance and repairs of casualties be undertaken promptly and correctly or will these things be ignored, as has so often been the case in the PLA? Is there a full understanding by PLA operators and technicians of the detailed workings of the systems and the complexities of the

---

[23]The U.S. CIWS, incidentally, has been sold to Taiwan for use on frigates. The fact that this system is available as last-resort protection for Taiwan's frigates makes it all the more imperative that the PLA Navy strive to feel confident in its own systems. In any sort of confrontation at sea that may come about between the PLA Navy and ROC Navy, the relative effectiveness of the anti-ship cruise missiles of the two sides is almost certain to be the crucial factor. These two systems are likely to be major determinants in the matter of defense against cruise missiles.

systems integration? Most Western observers believe that the answers range from no to maybe, at best, for questions such as these.

The existing systems integration efforts at this level are important for reasons beyond the successes with individual weapon systems. These efforts are likely to reveal the pace and trend of PLA systems integration—whether it will be slow or rapid, and whether it will tend toward failure and epidemic frustration or success and contagious enthusiasm.

Integration of high-tech systems with obsolescent and low-tech equipment. The final of the five levels of systems integration is of special applicability to China, and a situation where success of the integration may not be nearly so dependent on crucial interfaces, automaticity, and the like. For decades to come, the PLA will continue to have a very large inventory of old equipment and a much smaller inventory of advanced systems. The effort to employ selected high technology as a force-multiplier or enhancer for obsolescent equipment will surely receive a great deal of attention and may even become a very important and widespread aspect of PLA systems integration, even if the PLA (prudently) accelerates retirement of obsolete equipment. Lieutenant Colonel Lonnie Henley wrote in April 1996 that China is likely to pursue "the integration of high-tech conventional forces with guerrilla, militia, and paramilitary forces. There is already considerable discussion of this approach, generally under the rubric of 'people's war under high-tech conditions'."[24]

PLA leaders and writers on military affairs have indeed suggested this concept in an even broader sense than that described by Henley, albeit somewhat obliquely. They look to certain advanced or innovative means as a way to leap-frog over their technological backwardness and to find chinks in the digital armament that the U.S. and other modern armed forces tout as the way battles of the future will be fought. A PLA major general wrote:

> We must use a practical combination of information warfare and Marxist and Maoist military thought to guide information warfare and issues in military construction [building of the force] . . . . [T]he military [PLA] must emphasize the study of ways to use inferior equipment to achieve victory over enemies with superior equipment.[25]

> We must use all types, forms, and methods of force, and especially make use of nonlinear warfare and many types of information warfare methods which combine native and Western elements to use our strengths.[26]

A PLA senior colonel offered:

---

[24]Lonnie Henley, "China's Capacity for Achieving a Revolution in Military Affairs," an unpublished paper prepared for the Annual Strategy Conference at the U.S. Army War College concerning the status of China in the 21st century held 23–25 April 1996. Lieutenant Colonel Henley, U.S. Army, has considerable experience related to the PLA.

[25]Wang Pufeng, "The Challenge of Information Warfare," in Michael Pillsbury (ed.), *Chinese Views of Future Warfare*, p. 319. Major General Wang is a former Director of the Strategy Department, Academy of Military Science, Beijing. His paper was excerpted from *Zhongguo junshi kexue*, Spring 1995.

[26]*Ibid.*, p. 325.

> The basic way to defeat a powerful opponent with a weak force in a high-tech war is to bring the overall function of [the weaker force's] operational system into full play . . . and, through the integration of the above two aspects, attain the goal of turning the inferior into the superior and finally defeat the enemy.[27]

Another major general provided this perspective:

> Large quantities of high-tech weapons on battlefields pose serious challenges to traditional methods of operation . . . On the other hand, traditional methods will be reinvigorated and adapted to new operational conditions. With technical development of precision all-weather targeting, stealth weapons, precision guidance, and night fighting, traditional warfare can also be enhanced.[28]

Open to debate is how seriously these concepts should be taken. Is there substance behind the words? Are these largely just examples of PLA writers who feel the need to tackle concepts they have read about in Western military journals and apply them to the PLA? Henley's experienced analysis of various writings on this issue includes the thought that there is a "bit of a 'me too' tone" to it all. Certainly that is an aspect of this concept that should not be ignored. Peripheral (unmodernized) PLA units naturally want to believe that they are, or can become, part of an effective fighting force, and their leaders desperately need something hopeful to tell their troops to keep morale from collapsing under the realization that their ill-equipped units would be little more than cannon fodder on the modern battlefield—or its naval or air warfare equivalent. As things stand now, we simply do not have solid evidence that the PLA is, as a matter of doctrine, pursuing this method to enhance the effectiveness of less-capable forces. Available evidence (other than the type of rhetoric cited above) points in the opposite direction, that the PLA is concentrating new systems and the integration (however rudimentary) of those systems in its "pockets of excellence": the elite rapid-reaction, or "fist" (*quantou*) units, as well as the naval and air equivalents thereof.

Michael Pillsbury has shed light on another shadow lurking in this concept, at least as it is described in Chinese military writings. In commenting on an article concerning the potential enhancement of Chinese airpower along these lines, Pillsbury wrote, "This peculiar misperception seems to mean that obsolete 30-year-old fighter aircraft (the majority of China's air force) can be made effective by adding a few AWACS aircraft and electronic jamming aircraft, which China is in the process of acquiring." (It is not possible to determine with confidence whether the writings on which Pillsbury comments are in fact an example of a PLA misperception. They may instead be wishful thinking, or an attempt to mislead potential adversaries— possibly Taiwan more than the U.S.—to believe that the PLAAF will soon be much more potent.) The point here is that there is a tendency among developing armed forces, to which the PLA is not immune, to believe, all too optimistically, that the

---

[27]Shen Kuiguan, "Dialectics of Defeating the Superior with the Inferior," in Pillsbury (ed.), *Chinese Views of Future Warfare*, p. 217. In 1997, Senior Colonel Shen was a professor at the [PLA] Air Force Command Institute in Beijing. His paper originally appeared in *Zhongguo junshi kexue*, Winter 1994.

[28]Wu Guoqing, "Future Trends of Modern Operations," in Pillsbury (ed.), *Chinese Views of Future Warfare*, p. 349. In 1997, Major General Wu was Director, Department of Operations and Tactics, Academy of Military Science, Beijing. His paper was originally published in *Zhongguo junshi kexue*, Summer 1994.

addition of this or that advanced system will catapult it into modernity. Although the record is full of examples where that sort of effort has failed, hope springs eternal.

Of course, there will be specific situations where the integration of advanced systems with far less sophisticated equipment will produce enhancements, or where some very capable modern systems can provide cover and protection for older, more numerous, less capable platforms. There will be situations where obsolete and modern equipment will be present during the same engagement, with benefit accruing in some instances and confusion the likely result in others. However, it is much more likely in most cases that systems integrated in such a manner, whether by design, necessity, or inadvertence, will be no stronger than their weakest links.

Indeed, in many cases the attempt may prove counterproductive because valuable high-technology systems will be adulterated, wasted, or even sorely jeopardized. Various levels of technology can, of course, be integrated, but assurance of an outcome that is likely to succeed under combat conditions is quite another question. The success of such an integration effort is much more probable with components that are balanced and compatible. This level of systems integration, the concept of combining advanced systems to give new life to the PLA's "military museum," would seem to hold very limited promise for the PLA. Given the plight of the PLA, these dim prospects for success are unlikely to discourage isolated attempts to bring about miracles. However, widespread efforts along these lines are highly improbable.

III. DREAMS, REALITY, AND PROSPECTS

PLA Aspirations for Systems Integration: Hopes and Dreams

The concept of systems integration and its complexity are, no doubt, understood and appreciated by many of the more perceptive and better-educated officers of the PLA. Their insights are being passed along to many others through articles in *Jiefangjun bao* (*Liberation Army Daily*) and various Chinese military journals. A 1996 article is illustrative.[29] The writer describes the trend toward integration of military intelligence, tactical decision-making, and attack against opposing forces, recognizing the contradiction between the modern demand in combat for near-real-time actions and manual operations. (His reference to crude manual operations is, of course, one of the devices used by such writers to point out the situation in the PLA without having to do so directly.) He observes that such unautomated systems that exist independently, even if operated manually at the upper limit of human capability, cannot satisfy the requirements of modern warfare. Quoting from the article: "The resolution of these contradictions is certain to be the setting up, with computers as the nucleus, of an integrated system of intelligence, decision-making, and attack; forming an organic whole of automated information processing, computer-automated judgments or auxiliary decision-making and automated weapon operation; reducing information links; and having machines replace manual

---

[29]Wei P'ing, "Trend Toward Integration of Intelligence, Decision-making, Attack," *Hsien-tai chun-shih*, No. 229, February 11, 1996, pp. 49–51.

operations." There is further explanation in some detail of the characteristics of intelligence reconnaissance technology, the development of the science of decision-making, and the features of modern precision weapons. The author asserts that the Aegis system on U.S. and Japanese warships is representative of one type of integrated system and the German Cheetah motorized antiaircraft artillery system typifies another. The author concludes that such systems lower the requirements for weapons and equipment and offer better prospects for "big victories."

Many Chinese writings on subjects similar to this one are largely regurgitation of articles in U.S. or other Western military journals—a point often made by Westerners examining the state of PLA progress in this area. The use of the Aegis and Cheetah examples surely raises that specter in this case. Whether that is the case here is interesting to speculate about, but does not detract from the issue at hand. The point here is that the PLA, and certainly the PLA elite, is fully acquainted with the jargon and concepts of systems integration, at least at the level of detail of such newspaper and journal articles. There can be no remaining question as to whether the PLA recognizes that victory in future conflicts against modern forces will be virtually unachievable unless it can attain a goodly measure of the system integration goals to which it aspires.

The PLA Navy (PLAN) has not lagged behind in *conceptualizing* integrated systems within individual ships, between ships, and even more broadly. The expression frequently employed by PLA Navy writers is *digitized naval warfare*. They define this term to include digitized communications and information systems, computer data processing systems and terminals, and links to combat platforms. The vision is of a digitized chain of command and control to sharply increase the combat performance potential of all naval warfare platforms and weapons. As a 1996 article by two PLA Navy officers in the journal *China Military Science* stated:

> Single-ship and unit offensive-defensive capability, single-unit coordination capability, and inter-unit joint operations capability are all growing sharply [meaning in modern naval operations, not necessarily in the PLA]. And as to coordinated naval operations, with submarine, aircraft, and surface ship communication systems being linked up, mutual information transmission problems being solved, and the current joint operational difficulties and problems within and among units being overcome, all combat platforms can be effectively linked into one operating entity, to strike enemies with maximum combined force [again, hardly a description of today's PLAN] . . . . Such comprehensive operating systems can link the whole establishment together. So we can predict that the naval C^3I system grounded in satellite and computer technology will become the priority of soft systems development in the tide of the new military revolution, as well as being linked up and made compatible with air force and army C^3I systems.[30]

PLA writers often reveal more in their writings than the literal meaning of the words. They apparently are constrained in candidly addressing the shortcomings of the PLA

---

[30]Shen Zhongchang, Zhang Haiying, Zhou Xinsheng and Shi Yukun, "The Impact of the New Military Revolution on Naval Warfare and the Naval Establishment," *Zhongguo junshi kexue*, No. 1, February 20, 1996, pp. 57–60. Senior Captain Shen is Director of the Science and Technology Department, Naval Military Academic Studies Institute; Lieutenant Commander Zhang and Lieutenant Zhou are affiliated with the Institute. Mr. Shi edited the article.

Navy. One method of getting around these constraints and raising the issue of PLAN inadequacies is to write about what is happening to navies generally as they modernize. That technique has been employed liberally in the article cited above. Put another way, the developments described in this article will be recognized by PLA Navy officers, the principal readers of this journal, as capabilities that the PLA does not have or concepts that the PLA is just beginning to come to terms with. Consequently, the authors do not have to state directly that their navy is lacking in the types of equipment described and the integration thereof. When they write about the conduct of future naval warfare, their purpose is clear: They are sending the message that the PLAN is behind and needs to catch up. The final sentence of the excerpt is illustrative; it should not be seen primarily as a prediction of the future of naval warfare but should be interpreted as a call to action. Stripping away the artifices of PLA writing, its meaning might be as follows: *The PLA Navy must turn its attention to the acquisition and integration of C^3I based on satellite and computer technology. This should be a priority in PLAN soft systems development and should be compatible with air force and army C^3I systems.*

The following excerpt from the same article uses this technique once more and pointedly employs a description of U.S. Navy practices to suggest what he considers the proper direction for the PLAN:

> All soft systems [sic] have become a key indicator of ship combat performance. So in the modern warship development process, all soft systems, particularly communications equipment, target detection equipment, and electronic warfare systems, are growing not only ever more numerous but also increasingly complex, becoming the key components of weapon systems. The U.S. Navy, when designing and building navy vessels, gives priority consideration to electronic equipment, equipping many of its ships with electronic jamming units to increase their defensive capability. Tactical intelligence data systems are comprehensive operating systems that the U.S. Navy has developed to a high degree of perfection on most of its surface ships. They not only can direct all weapon operations of a ship but also can use data links with other ships in the fleet to coordinate and command the weapon control systems of friendly ships and planes.[31]

Indeed, we can trace these types of rhetorical devices to the highest levels of the PLA. The Director of the Commission on Science, Technology, and Industry for National Defense (before COSTIND's reorganization in 1998), Cao Guangchuan, wrote in 1997 of the Central Military Commission's stress on scientific and technological development for national defense:

> [I]t is necessary to effect the change from aiming to win limited wars fought under ordinary conditions to winning limited wars fought under conditions of modern technologies, especially high technologies. Second, in terms of army building, it is necessary to effect the change from quantitative expansion to qualitative improvement, from building labor-intensive forces to building technology-intensive forces . . . . We must persist in putting scientific research before actual development, keeping track of the development of high technologies in the world, mainly relying on

---

[31] *Ibid.*

our own efforts, and attach importance to the digestion and assimilation of imported technologies and on innovation.[32]

It seems safe to assume that Cao's words reflect real CMC and COSTIND priorities and the state of technology in the PLA, and that they were not words written to influence or mislead a Western audience. This conclusion seems all the more plausible because many other officers and officials have written along similar lines in various journals and other publications, including some to which foreigners do not normally have access. Several conclusions might be drawn from his revealing words, but the fact that he and others, in 1997, were offering such elementary advice and guidance on such things as the sequencing of research and development may be the most interesting features. The tone of his pronouncements makes it clear that the PLA is still at a very early stage in assimilating advanced technologies, much less integrating these technologies into sophisticated systems. This is, of course, no surprise to those who follow and analyze the PLA's modernization efforts. However, it does provide confirmation and a richer context for understanding where the PLA stands today with respect to systems integration and how far it has to go. Reliance on foreign technology is evident, as is the conflicting (and understandable) desire for self-reliance. The final sentence makes clear not just the need to acquire "imported technologies" but also, by attaching importance to digesting that technology, the writer reveals (not surprisingly) that this matter of usefully assimilating such imports (and integrating them into combat systems) is an abiding concern.

The views are also shared by representatives of the defense industries that are charged with producing the systems. The President of China Ordnance Corporation wrote in 1997 in the *People's Liberation Army Daily* about integration efforts:

> Over the next few years, the ordnance industry will persist in integrating the development of new equipment with the revamping of existing equipment and in integrating development of our own efforts with the import of advanced technologies.[33]

In the same group of 1997 articles in the *People's Liberation Army Daily* as the two pieces cited above, the Supervisor of War Military Projects under the Ministry of Electronics Industry wrote:

> The year 1997 is a crucial year in the Ninth Five-Year Plan for the development of military electronics. The tasks are very arduous . . . In this year we will continue to conscientiously implement the spirit of the instructions of the Party Central Committee and the Central Military Commission by . . . strengthening basic research and anticipatory research, *improving the ability of systems integration . . . and striving to achieve a giant leap in the development of military electronics . . . We have come to soberly realize that making a success of this key project will involve many difficulties*

---

[32]Cao Gangchuan, "Enhance Sense of Strategy, Elevate Level of Science and Technology for National Defense," *Jiefangjun bao*, February 24, 1997.

[33]Zhang Junjiu, "Take Military Production as the Foundation, Make New Breakthroughs in Ordnance Industry," *Jiefangjun bao*, February 24, 1997.

because this gigantic system is unprecedented in terms of scale and matters involved [emphasis added].[34]

These two officials, as they wrote these apparently obligatory articles for a PLA newspaper feature entitled "Accelerate Development in Scientific and Technological Industries for National Defense, Meet Challenge of World Military Development," were forced by the lack of previous PLA achievements in this area to look solely to the promise of future progress. They did not have the option of, instead, describing previous signal successes. One suspects that, were these officials or their successors to be called upon in 1998 to write similar articles, similar words of hope for future progress would still fit best. Beyond the judgment that substantial progress is, in many regards, a hope rather than a fact, there are the inescapable conclusions drawn from these words that tasks of this nature, certainly including systems integration, are truly formidable obstacles for the PLA. The writers' description of the scope of the task is, indeed, not couched in optimistic terms but rather in a way that appears more to offer an excuse for why success will not be achieved in the foreseeable future.

Where the PLA Stands on Systems Integration and Technological Innovation: A Reality Check

Although China, and especially the PLA, would prefer to be self-reliant in both the acquisition of technology and systems integration, the inadequacy of China's research and development infrastructure has presented a major obstacle to realization of that goal, despite efforts to reform the scientific and technological sectors. China has tried to foster this process by treating technological know-how as a commodity and having its exchange and diffusion controlled by market forces. The unhappy result, however, has been essentially a failure (based on several complex factors) to create a demand for domestic technology by industry. Among the reasons for the failure was the lack of useful technological flow from Chinese domestic sources. The counterproductive result has been heavy demand for foreign technology and excessive reliance on that source.[35]

Chinese research and development successes are rare. As described at the beginning of this paper, China has achieved success in its nuclear weapons program and in the development of ballistic missiles with both nuclear and conventional warheads. This success, it must be said, has gone well beyond simple expansion of the technologies obtained from the Soviets in nuclear warhead technology and missilery. However, in virtually all other areas, China's achievements in the employment of advanced military technologies and systems integration have been sharply limited by the absence of indigenous technology and by the inability to acquire and incorporate

---

[34]Wang Jingcheng, "Concentrate a Superior Force to Fulfill Key Tasks of Scientific Research," *Jiefangjun bao,* February 24, 1997.

[35]Erik Baark, "Military Technology and Absorptive Capacity in China and India," pp. 94–97.

advanced technologies from other countries,[36] primarily European nations and Russia. Dual-use technologies and spin-offs therefrom have produced surprisingly little benefit to the Chinese military technology base,[37] especially in light of the mixing in recent years of military and commercial industry through defense conversion efforts and other methods to strive for efficiencies and profits.

There is little reasonable prospect for short-term rectification of these problems with respect to the low technology base for Chinese military industry. As one careful observer put it, "China's ability to develop military technology indigenously is limited by the poor organization of the military industry, which can be improved only by revamping its organization—a step which may be possible only in the context of political reform."[38] Others are even less optimistic concerning Chinese attainment of advanced technological and systems integration skill, feeling that somehow there are cultural or societal barriers that have produced the current situation and that will perpetuate it indefinitely. Obviously, such assertions are essentially impossible to prove or disprove. Some observers point to examples of success among other similar societies, but those who believe that China has irreconcilable problems remain unconvinced and have probably persuaded some Chinese that China is far better off simply to make do as best it can with foreign systems, particularly in those areas where systems integration and other esoteric skills are most prominent.

The scope of the problem, if not the possible underlying factors, seems fully appreciated by the PLA leadership. CMC Chairman Jiang Zemin issued directions (apparently in early 1997) that the armed forces must undertake a sweeping program to improve the knowledge of cadres at all levels with respect to science and high technology. To implement the direction, the General Staff Headquarters distributed a document entitled "Three-Year Plan for Cadres of the Whole Army to Study High-Tech Knowledge."[39] The official Xinhua News Agency reported an announcement by the General Political Department of the PLA about six weeks later that the army would increase the recruiting of college graduates. The change in officer procurement practices was described as "one of the important measures . . . to implement the strategy of relying on science and technology to build up the army and to accelerate the modernization process."[40] These initiatives suggest the depth of the frustration that the PLA is experiencing in dealing with the technological revolution in warfare.

Many Westerners question the efficacy of policy pronouncements such as those described above. They argue that Chinese leaders fail to recognize that truly innovative science and technology are highly unlikely to thrive in today's China—in

---

[36]Eric Arnett, "Beyond Threat Perception: Assessing Military Capacity and Reducing the Risk of War in Southern Asia," in Arnett (ed.), p. 9.

[37]*Ibid.*

[38]*Ibid.*, p. 13.

[39]Gu Boliang, "General Staff Headquarters Distributes Three-Year Plan for the Whole-Army Cadres to Study High-Tech Knowledge," *Jiefangjun bao*, February 6, 1997, p. 1.

[40]Xiao Pu, *Xinhua Domestic Service*, March 19, 1997, in FBIS-CHI-97-082, March 23, 1997.

other words, that these policies are empty rhetoric. In any case, it is clear from Chinese statements made at very senior levels that the Chinese technology base, especially that applicable to military systems, and the ability to integrate such technology into the systems of modern warfare are woefully lacking. These inescapable conclusions are prompted by the words of Chinese officials who, somewhat inadvertently, reveal the depth of the problem by the very sweep of the solutions they propose.

American specialists have arrived at similar conclusions. The *Militarily Critical Technologies List* (*MTCR*) published by the U.S. Department of Defense (referred to earlier in this paper) contains assessments of foreign capabilities in various technology areas. These assessments represent the consensus of a technical working group composed of members from U.S. industry, government, and academia, including selected members of the U.S. intelligence community.[41] In the field of electronics technology, China is evaluated by this group as possessing only limited or "some" capability (the two grades at the low end of the *MCTR* assessment scale) in each of six areas of evaluation. (For example, China is credited with limited capability in the technology associated with electronic components and with some capability in microelectronics.) The *MCTR* evaluation states: "China has been slowly developing capabilities during the past five to ten years and will probably accelerate the rate of development during the next five to ten years in an attempt to catch up militarily and commercially with others."[42] In the subsection on microelectronics, including integrated circuit design and the electronic packaging technologies required to achieve the needed high speed, high power, and ability to function in severe environments for basic building-block microcircuits, China is said to trail Russia and to be generally on a par with the East European countries.[43]

In the area of information systems technology, China is assessed in the *MTCR* to have only limited or some capability in ten sub-areas and no capability in the modeling and simulation sub-area. The sub-areas in which China is comparatively deficient include high-performance computing, intelligent systems, networks and switching, signal processing, software, and transmission systems.[44] In the area of information warfare technology, a field with which the PLA is infatuated, the U.S. specialists' evaluation is particularly damning. China is assessed as having a limited capability in the sub-area of electronic attack and no capability in the other three militarily critical sub-areas of IW.[45] The picture is only slightly better with respect to space systems technology.[46] Perhaps the most revealing aspect of these evaluations is that in all the areas that are directly pertinent to integration of military systems to produce effective combat systems, China consistently receives assessments in the

---

[41] Office of the [U.S.] Under Secretary of Defense for Acquisition & Technology, pp. iv., 2-3.

[42] *Ibid.*, p. 5-2.

[43] *Ibid.*, p. 5-12.

[44] *Ibid.*, p. 8-2.

[45] *Ibid.*, p. 9-2.

[46] *Ibid.*, p. 17-2.

240

lowest categories of capability, failing to receive a ranking in the top two categories of capability in even one of the many sub-areas.

An experienced analyst of PLA modernization summed up China's position this way: "The Chinese lag even further behind [than in other military areas] in circuit design, system integration, networking, operating systems, and development of software applications . . . [T]he American armed forces are moving rapidly along a path that China is not prepared to follow . . . It is not just a matter of available technology, or even of creativity in the application of technology. The greatest impediment to China achieving an information-based revolution is its authoritarian political system."[47] This analyst points out that in Chinese publications there is "virtually no discussion of intelligence processing and fusion systems such as the U.S. All-Source Analysis System (ASAS), or of dedicated communications links for intelligence dissemination . . . This requires high-capacity, robust communications links, standardization of data formats and transmission protocols, interoperability of intelligence communications among different systems and services, powerful information processing systems at the lowest command levels, and a commitment to the free flow of intelligence information to tactical commanders . . . [A]vailable sources do not indicate any effort by the Chinese to implement such an elaborate and open intelligence environment. So the overall prognosis is that the PLA may achieve the kind of capabilities demonstrated by U.S. forces in the Gulf War [almost a decade ago], though it is likely to take at least ten and probably twenty years for it to do so." Dr. Paul Godwin of the U.S. National Defense University seconded this when he wrote: "The PLA also lacks both the logistical support systems and command, control, communications, and intelligence (C^3I) infrastructure necessary to sustain combined-arms operations."[48]

The American consensus as described above is corroborated by a Russian analyst, Viacheslav A. Frolov. Given the intimate connections with China with respect to technology transfer, the Russian perspective is particularly worthwhile. Frolov wrote in 1998:

> The vulnerability of the PLA's C^3I system is its obsolete command and communications links and lack of any measures for anti-electronic warfare. For the former, the strategic C^3I system has effective coverage of the PLA ground forces only up to divisional level. The system is heavily reliant on radio and security telephones. Only recently have satellite communication channels been created at the Group Army level, and computerized links at the divisional level. The tactical C^3I is carried mainly by semiconductor [UHF] radios, providing only limited communication capability, usually within a range of 2.5 to 10 kilometers. Space-based communications systems and global positioning systems are seen as a crucial step to enhance the PLA's C^3I system. Currently China's six communication satellites have allocated very limited channels to the PLA. To rectify this situation, a proposal has been tabled to create a network of defense satellite communications.[49]

---

[47]Henley, p. 11.

[48]Paul H.B. Godwin, "Military Technology and Doctrine in Chinese Military Planning," p. 60.

[49]Viacheslav A. Frolov, "China's Armed Forces Prepare for High-Tech Warfare," *Defense & Foreign Affairs Strategic Policy,* January 1998, p. 7. Frolov appears particularly well suited to offer this evaluation. He is

Systems integration is, of course, not restricted to computers and weapon systems. As Paul Godwin mentioned (cited just above), the PLA has severe shortcomings with respect to logistical support systems. Paul Dibb notes that the integration of complex information systems in real time depends critically on a new approach to maintenance and the support in a combat environment of systems capable of remaining operational full time and in all weather conditions. He goes on to remark that very few Asian countries seem to acknowledge the vital nature of integrated logistic support (ILS).[50] As with other areas of technology assimilation and systems integration, there is acknowledgment (but little more) in PLA writings of sweeping new requirements for integrated logistic support. A mid-1996 article in the *Liberation Army Daily* states:

> On a digitized battlefield, a combat unit, combat support unit, combat duty support unit, and other combat systems have become an integrated whole with functions like battlefield intelligence, command, control, telecommunications, attack, damage and casualty evaluation, and so on, and this has promoted logistical support integration. On the one hand, an integrated logistical support system is capable of breaking through boundaries between logistical support systems of different services; comprehensively optimizing the disposal and utilization of logistical resources; raising logistical resources utilization efficiency; and preventing duplicate disposal and waste of logistical resources, thus comprehensively enhancing logistic support capability and efficiency and making logistical support conform with integrated combat operations.[51]

As has been seen in many other such writings, the author is obviously describing something the PLA does not have. The article seems to advocate PLA adoption of a sophisticated integrated logistic system. There are, however, several reasons to question whether the PLA will or should undertake such sweeping logistical reform at this early stage of force modernization. Not only is the task daunting and enormously expensive (as well as costly in other resources), but there is also the question of whether such a system, modeled along Western lines, is truly appropriate to the PLA's likely missions and circumstances. Not to be ignored is the realistic consideration of whether such a system in the PLA, even if instituted, would simply collapse of its own weight in a short time—an example of too much, too soon.

Certainly logistic enhancements are needed in the PLA. The question is whether the grander ILS schemes envisioned by Chinese writers, who are largely paraphrasing the logistics literature of the U.S. and other Western armed forces, are appropriate to the circumstances. The PLA, according to most outside observers, might first come to grips with the less grandiose (yet still complex and critical) concepts of logistic support, preventive maintenance, timely and efficient repair and rework, etc. This does not apply solely to support of indigenously produced equipment. The PLA does not provide adequate logistic support and maintenance of imported military equipment and systems, generally opting not to procure sufficient and appropriate

---

described as a Sinologist by background and as Project Manager at MAPO Military Industrial Group, the manufacturer of MiG fighter aircraft, in Moscow.

[50]Paul Dibb, p. 94.

[51]Gong Fei, "Digitization and Logistical Reform," *Jiefangjun bao*, August 27, 1996, in FBIS-CHI-96-209, August 27, 1996.

training, spare parts, and maintenance systems for the weapon systems and other military equipment it purchases abroad. One reason for giving these areas short shrift is, of course, to save money. However, neglect in these areas seems also to reflect a deeply ingrained lack of recognition that these are key elements of a combat capability.

Put succinctly, the real question is whether there will be good reason, adequate will, and sufficient resources within the PLA to sustain such an integrated logistic system. The likely answer is that the system, if it evolves, will be riddled with "Chinese characteristics," raising the further question of whether, with such encumbrances, it can function at all. As with other aspects of the systems integration problem, it is far from a foregone conclusion that recognition of the problem followed by (probably token) efforts to effect a solution will lead to effective results. There is the strong prospect that attempts at developing an integrated logistic system at this stage of PLA modernization may become costly excursions into a nether world.

There are other serious underlying problems, arguably more fundamental than those described above. Chinese research and development is immature, isolated, fragmented and unfocused, all of which have stymied the gathering of needed momentum for the development of advanced military technologies and integrated weapon systems.[52] With respect to computer technology, China has until recent years emphasized hardware rather than software development, and currently domestic Chinese collaboration in software development remains deficient because of both technical and economic barriers. Contacts between developers and users, especially military users, are lacking. The capability to produce the complex and flexible programs needed for military applications is limited. Software development is proceeding apace elsewhere in the world, often leaving China behind, even with the recent attention there to software development.[53] Among the many reasons is the problem of adapting programs to the Chinese language, using Chinese-language processing, or training operators to use applications written for native speakers of other languages. This is further complicated by individual systems using different languages that must be integrated into a functional combat system. The language problems are significant and not restricted to the software area.

The author and others with whom he has spoken have seen on Chinese ships and in Chinese naval and military training facilities and simulators that manuals and equipment are frequently in English or another language other than Chinese. Chinese officers have remarked about misunderstandings, badly translated manuals and operating instructions, and decried the amount of training time that must be spent in learning or improving English comprehension to be able to maintain and operate these systems.

Other areas present formidable problems for China as well as for other countries. For example, the provision to sophisticated and delicate equipment of electrical

---

[52]Erik Baark, "Military Technology and Absorptive Capacity in China and India," p. 109.

[53]*Ibid.*, p. 103; also, extensive treatment of software problems in China is provided in Erik Baark, "China's Software Industry," *Information Technology for Development*, Vol. 5, No. 2, June 1990, pp. 117–136.

power of the needed stability, voltage, frequency, and phase is not a simple problem even for advanced militaries. It is certainly not one that the PLA can ignore. Of course, when one is considering combat applications there must be redundancy in many areas, including that of stable electrical power. The PLA has given little attention to redundancy and to other provisions so that its ships and other platforms could continue to fight after sustaining combat casualties or even "normal" malfunctions. The simple, if not sole, reason for this deficiency is that the PLA has not mastered keeping the first-line systems operating, much less worrying about redundancy, back-ups, work-arounds, and coping with combat damage or other casualties.

PROSPECTS FOR THE PLA: SOME CONCLUSIONS

Is the PLA gaining or losing in the race? Many observers of the PLA and analysts of PLA modernization assert that the PLA is anywhere from ten years to two generations behind modern armed forces in technological acquisition, assimilation, and systems integration. One should not be surprised at this variation in estimates. It is not reasonable to insist on any sort of precision or even accuracy in making generalizations of this sort. But there is no doubt that the PLA is not in the same league with truly modern armed forces, that it has a long way to go and is not getting there very quickly. An important question not often asked is whether the PLA is gaining or losing in this competition. Certainly there are elite PLA units that are making progress; some of these units may even achieve minor successes in systems integration. However, modern armed forces, especially those of the U.S. (which for various reasons the PLA uses as a standard for its progress), are developing at a rate that most of the world, including the Chinese, little appreciate.

Advanced technologies viewed as vulnerabilities. Some PLA writers hold out the prospect that the advent of armed forces dependent on the most advanced computer technologies and the complexities of systems integration will produce vulnerabilities for the more modern militaries and, concomitantly, opportunities for the PLA to exploit this inordinate dependence on exotic technology. The far more likely situation, albeit hardly assured, is that China will be unable or unwilling to devote the attention and resources required to develop the advanced technologies and systems integration capabilities needed for such exploitation and that this will continue to be a profound shortcoming of the PLA for the foreseeable future.

In a broader context, China is not likely to catch up with the U.S. or advanced countries in the region, like Japan and Australia. Taiwan is a different issue. The jury is out on whether Taiwan can, with U.S. assistance, achieve advances and systems integration in key areas, even if not across the board. On the other hand, with respect to the other countries of the region, China will be able to hold its own because the other countries are experiencing similar problems with technology and systems integration, although in some cases for other reasons.

For all the reasons described above, the increasing importance of extremely advanced technologies and the sweeping scope of systems integration are likely to produce an environment in which China, as the years pass, will be even more disadvantaged than at present. Put colorfully, the PLA may rely on its dream of leapfrogging through technology exploitation and yet awaken ten years into the next century to find itself still somewhere between ten years and two generations behind. To make its dream a reality, China would have to change much more than most consider feasible and would have to embrace concepts of change far more sweeping than Chinese leaders seem willing to risk.

The PLA will certainly attain some limited success in systems integration. Some of those areas may be significant, even troublesome, in the delicate balance of forces with Taiwan in certain warfare areas and the ability to cause consternation for U.S. forces in other areas. However, overall, the odds are very high that systems integration will prove, for the foreseeable future, to be yet another area where the PLA will suffer from the problems with which China as a whole has not been able to come to closure. The PLA is not likely to be able to take advantage of this or other sea changes in technology to overcome its shortcomings compared to truly modern armed forces. It will likely slip further behind in that regard. However, China's armed forces will likely apply a mix of indigenous and imported technologies to achieve a greater comparative advantage with respect to most other regional armed forces, with the notable exceptions of Russia, Japan, and Taiwan, including the crucial area of systems integration.

12. THE PLA AND THE TELECOMMUNICATIONS INDUSTRY IN CHINA

James Mulvenon and Thomas J. Bickford[1]

INTRODUCTION

The PLA has long suffered from an inadequate telecommunications infrastructure, characterized by outdated technology, limited capacity, and lack of secure communications. In the past, these weaknesses have severely limited the military's ability to transmit and process large amounts of information or coordinate activities between the various military regions, thereby reducing military effectiveness. For example, a number of observers believe inadequate communications were a major factor in the high level of losses suffered by the PLA during China's invasion of Vietnam in 1979.[2] In stark contrast, the PLA is very much aware of the critical role played by information-based C4I (command, control, communications, computers, and intelligence) technologies in the 1991 Gulf War, and the importance of these technologies in securing the eventual Allied victory against a force made up of largely Soviet and Chinese equipment.[3]

To overcome these deficits, the PLA has embarked on a well-financed effort to modernize its C4I infrastructure. An important goal of this modernization has been the acquisition of advanced telecommunications equipment from abroad, on the premise that the technologies of the information revolution provide China with the opportunity to "leapfrog" and vastly improve capabilities in areas related to C3I. The transfer of these technologies to China in general and the PLA in particular has been facilitated by two mutually supporting trends. First, there is enormous competition among Western telecommunications firms to get a share of the relatively backward but rapidly expanding Chinese telecommunications market, which is the largest market in the world. Naturally, the lure of potential billions has attracted every major player, including Lucent, Nokia, Ericsson, AT&T and countless others. From these

---

[1]Dr. Thomas Bickford received a Ph.D. in political science from the University of California, Berkeley in 1995. He has served as an assistant professor in political science at the University of Wisconsin-Oshkosh since 1995, specializing in China and International Relations.

[2]See James Mulvenon, "The Limits of Coercive Diplomacy: The 1979 Sino-Vietnamese Border War," *Journal of Northeast Asian Studies,* Fall 1995.

[3]For an example of PLA writings on this point, see Li Qingshan (ed.), *Xin junshi geming yu gaoshuji zhangzheng* (The New Military Revolution and High-Tech Warfare), Beijing: Military Science Press, 1995, especially pp. 122–125.

companies, China is buying between fifteen and twenty billion dollars worth of telecom equipment a year. Indeed, the statistics are staggering. China is reported to account for about 25% of the world's market for telecommunications equipment and is expanding rapidly. China's mobile phone network and paging market have averaged 100 percent annual growth for the last few years and show no signs of stopping. In real terms, the Chinese mobile market is growing by 25,000 new subscribers and 33,000 new handsets a day,[4] and the paging market is already the biggest in the world.[5] According to a report in the *South China Morning Post*, the PRC is installing about 15 million new fixed telephone lines a year,[6] a number that is exceeded only by sales in the United States.[7]

Much of this growth is achieved through sales by foreign telecommunications companies and by joint ventures with Chinese partners, which brings us to the second important trend. It is no exaggeration to say that the PLA is one of the key players in China's telecommunications modernization. For historical reasons, the PLA controls large sections of commercially exploitable broadcast bandwidth in China, most notably the 800-MhZ spectrum that is well-suited for cellular communications. Since 1978, the Chinese military's commercial enterprises have been free to marketize this infrastructure privilege, generating profits for military units at all levels of the system. More important, the commercial joint-venture relationships between the PLA and foreign companies have provided the military with access to advanced technology necessary for its C4I modernization. For example, PLA units provided the labor for the laying of most of China's fiber-optic networks. In return, the PLA received a percentage of the laid fiber for their own purposes, and was able to purchase additional equipment to lay dedicated landline networks between military region headquarters and other essential command and control nodes. While recent announcements regarding the divestiture of the military's business empire suggest that these types of commercial arrangements might be reduced or ended altogether, signals from Beijing suggest that the PLA's role in telecommunications, while perhaps forced to assume a much lower profile, will continue to be to actively acquire equipment from abroad for both military and civilian uses.

The purpose of this paper is to examine the relationship between the PLA's enterprises and the telecommunications market in China. The first section examines the historical reasons for the PLA's involvement in the economy in general, and the telecommunications sector in particular. The second outlines the types of cooperation and joint ventures that exist between PLA companies and foreign telecommunications companies, and assesses what degree of technology transfer might occur. The third section evaluates the long-term implications of these transfers

---

[4]Ibid. A different story by the *Financial Times* put the figure at 16,000 new subscribers a day, which would still mean that China currently accounts for more than 10% of all new subscribers for mobile phones. See Tim Burt, "Phones: Mobile Manufacturers Go With the Flow," *Financial Times*, July 31, 1998.

[5]Andrew Chetham and Mark O'Neill, "The Future Begins to Take Shape," *South China Morning Post Internet Edition*, China Business Review Special Report on Telecommunications, June 11, 1998.

[6]Ibid.

[7]"Silicon Valley, PRC," *The Economist*, June 27, 1998, pp. 64–65.

for the C4I modernization of the PLA and the export control policies of the United States Government.

THE PLA ENTERPRISE SYSTEM AND MILITARY TELECOMMUNICATIONS

For most of the history of the PRC, responsibility for telecommunications has been split between three organizations: the Ministry of Posts and Telecommunications, the defense industries (in particular, the Ministry of Electronics Industry), and the PLA. The responsibilities of these three institutions varied considerably, providing a nominal division of labor but also fostering significant bureaucratic friction.

Ministry of Posts and Telecommunications. The Ministry of Posts and Telecommunications (MPT) and its institutional predecessors oversaw the civilian telephone network and most other aspects of civilian communication (though responsibility for news and other information services was under the Radio Ministry). In the reform era, the MPT's primary commercial arm, China Telecom, has been a major actor in the development of telecommunications services in the Chinese economy. The MPT was generally believed to favor a more monopolistic approach to China's telecommunications modernization, in sharp contrast to the desire of other actors, primarily the consortium of the Ministry of Electronics Industry, Ministry of Railways and the PLA, who sought to offer a competitive alternative.

Ministry of Electronics Industry. During the Mao era, elements of the defense industry were responsible for the manufacture of telecommunications equipment and the construction of internal networks for the government and the military. Originally, the Fourth Ministry under the State Council was charged with these tasks, but eventually the relevant state-owned companies were reorganized under the Ministry of Electronic Industry (MEI). In the 1990s, the MEI was closely associated with China Unicom, which is currently the country's second largest telecommunications enterprise. The MEI was the largest shareholder in Unicom, though other ministries (Railways, Power) and some foreign partners held shares.

In March 1998, following the major reorganization of China's government after the 9th NPC, the MPT, MEI, and parts of the Ministry of Radio, Film and Television were combined into a new organization: the Ministry of Information Industry (MII). At least initially, however, it seems that the former MPT is the winner in this reorganization as the head of the new ministry, Wu Jichuan, is the same person who led the former MPT and many of the new MII's senior staff also appear to be former MPT people. Some reports at the time of this writing indicate that former MPT personnel may be put in charge of Unicom.[8] At the very least there appears to be a balance of power in favor of the old MPT within the MII leadership.[9] Some reports indicate that these former MPT officials plan to turn MII into a government

---

[8]Andrew Chetham, "Foreigners Await Outcome of Post-Merger Restructuring," *South China Morning Post Internet Edition*, China Business Review Special Report on Telecommunications, June 11, 1998.

[9]Kristie Lu Stout, "Liu Shuffle Shifts MII Power Balance," *South China Morning Post Internet Edition*, July 14, 1998.

regulatory organization, not unlike the FCC in the United States. Under this structure, the commercial elements of the telecommunications infrastructure would first be centralized, then consolidated, and finally privatized under MII regulation. China Telecom itself will likely be divided along functional rather than geographic lines, splitting into separate mobile, long-distance, and paging companies. Unicom, which has already been restricted to mobile communications, will likely be retained to provide an element of competition, but few believe that it will ever be allowed to challenge for supremacy in the market.

The Chinese People's Liberation Army. The third organization that traditionally has been a key player in telecommunications is the PLA. The role of the military in telecommunications stems from a combination of structural factors. The first factor in the PLA's connection with telecommunications services was the structure of the Chinese Leninist state. The military had its own dedicated infrastructure, including railway lines, ports, and airfields, as well as priority access to the civilian infrastructure. Rather than piggyback on the civilian telecommunications backbone, however, the PLA had its own separate telephone system, built by the predecessors of the Ministry of Electronics Industry.[10] In addition, the PLA was given control over large sections of the broadcast spectrum for reasons of national security.

The second structural factor is the PLA enterprise system, which provided the vehicle for exploitation of this telecommunications access. Before Liberation, the Chinese military developed an extensive system of farms and factories, with the goal of making the military self-sufficient. After the CCP came to power in 1949, the PLA was permitted to retain these enterprises, which were considered separate from both the state civilian sector and the national defense-industrial complex.[11] Over time, this enterprise system continued to grow in scale and importance, adding unit-level farms and factories for the dependents of officers. By 1978, the enterprise system was an indivisible feature of the Chinese military.

After 1978, however, Deng Xiaoping and the leadership sought to reduce the budgets of the Chinese military, in order to generate investment for economic reforms. The military's budget was slashed by almost 25% between 1978 and 1980. To make up for these lost funds, the PLA was given permission to gradually marketize its internal economy. Initially, these commercial forays were limited to agriculture, but eventually the PLA began to diversify its business interests, developing market positions in hotels, transportation services, and light industrial production.

One of the most logical sources of income was the PLA's aforementioned access to critical infrastructure. Just as the military was able to commercially exploit its transportation network, it also sought to marketize the unused portions of its

---

[10]Matthew Miller, "Early Starter Becomes a Late Developer," *South China Morning Post Internet Edition*, China Business Review Special Report on Telecommunications, June 11, 1998. According to the same source, Beijing had only 80,716 lines in 1978 and the crossbar exchange system adopted by the MPT in the 1970s actually represented a step backward in technology from the pre-1949 system. The source did not indicate if the numbers cited included the military telephone system or just the civilian apparatus.

[11]For more on the differences between national defense industry and PLA enterprises, see the chapter by John Frankenstein in this volume.

communications system. For example, the PLA's phone system had excess capacity that could be leased to provincial authorities.[12] Additionally, many of the bandwidths reserved for the military had been left unused. Rather than surrender control of these bandwidths to civilian authorities, the PLA made commercial use of them. Of the frequencies under PLA control, perhaps the one with the most potential is the 800-MHz band, which is well-suited for mobile cellular communications.

Soon, a variety of PLA telecommunications companies began appearing in China. Perhaps the most important is China Electric Systems Engineering Company (CESEC), which is operated by the Communications Subdepartment (4th Department) of the General Staff Department. CESEC is the key to PLA telecommunications, with interests ranging from mobile communications to encryption, microwaves, computer applications, and dedicated military C4I systems. The organizational chart for CESEC and its regional branches, "daughter" companies, affiliated research institutes, import-export companies, and wholly owned subsidiaries is displayed in Figure 1.

Other PLA enterprises that are involved in telecommunications are Poly Technologies, Kaili (Carrie) and others. In addition to these centrally based enterprises, there are many regionally based PLA enterprises at the military region and district levels that are involved in telecommunications, particularly in radio paging markets. One example is the Guangzhou Bayi Telecommunications Group, owned by the Guangzhou Military Region Air Force, which runs a commercial radio paging service that claimed 100,000 subscribers in 1994.[13]

THE PLA AND TELECOMMUNICATIONS: TWO CASE STUDIES AND A CAUTION

The structural advantages outlined above have given the PLA the necessary assets for potentially rapid and lucrative expansion in telecommunications services. On the whole, however, the PLA has been slow to take advantage of foreign investment and foreign joint ventures. This is true for most areas of PLA economic activity, not just telecommunications. While some deals were struck in the late 1980s and early 1990s (the PLA's first commercial venture in telecommunications was in 1988),[14] most of the activity has taken place over the last few years in the form of joint ventures with foreign companies. Motorola, Lucent, and AT&T are just some of the companies involved in projects with PLA participation.[15]

---

[12]Information Office of the State Council of the People's Republic of China, "China's National Defense," *China Daily*, July 28, 1998, pp. 4–6.

[13]Kathy Chen, "Soldiers of Fortune: Chinese Army Fashions Major Role for Itself as a Business Empire," *Wall Street Journal*, May 24, 1994, pp. A1, A9.

[14]Andrew Chetham, "PLA Muscling into Sector with its Separate Network," *South China Morning Post Internet Edition*, China Business Review Special Report on Telecommunications, June 11, 1998.

[15]Tai Ming Cheung, "Can the PLA be Tamed?" p. 52.

250

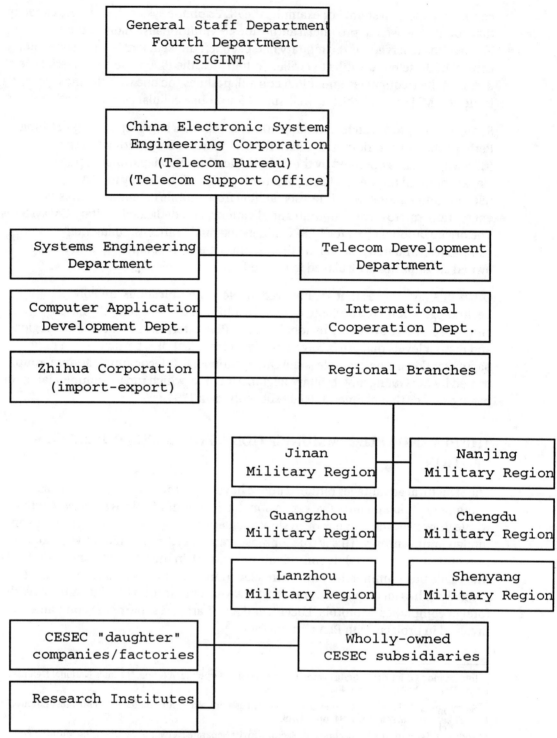

Figure 1—CESEC Organization Chart

These foreign investment and joint ventures are extremely important for the PLA, not just in terms of potential expansion of the PLA's commercial abilities, but also for what they may mean in terms of transfer of advanced telecommunications technologies for the military's C4I modernization. Joint ventures are a highly viable way to transfer technical skill and manufacturing capacity to China. The agreements with Motorola discussed below, for example, provide access to code division multiple access (CDMA) technology, which allows many users to use the same bandwidth without interfering with one another. Asynchronous Transfer Mode switching is another cutting edge technology that China now is able to produce itself and which, like CDMA, is useful for military communications.[16] Indeed, most aspects of civilian telecommunications technology have potential spin-offs that could help the PLA. Exactly how much military potential might result from such technology transfer is debatable. At a minimum, it means more reliable communications facilities with much improved capacity and, in the case of fiber-optics, much more secure communications. At a maximum, it could mean significant transfer of technical know-how and equipment that could accelerate the development of C4I capabilities well beyond what China could do with her own resources.

CESEC and China Great Wall CDMA

To generate profits from a 10-MHz section of the military's 800-MHz bandwidth that otherwise would have been left in unproductive static, a joint venture was forged in late 1995 between the MPT's China Telecom and the General Staff Department Communications Subdepartment's China Electronic System and Engineering Company. In less than three years, the resulting company, China Telecom-Great Wall Communications, assumed a dominant position in the rapidly expanding Chinese cellular market. The primary goal for Great Wall is to build a nationwide cellular network, based on CDMA (Code Division Multiplexing Access) technology. Foreign participation has been invited to bid for contracts at the provincial and local levels. All the big global equipment suppliers—Motorola, Northern Telecom, Ericsson, and Lucent—are competing to supply dozens of new provincial operators when the Great Wall system gets up and running.[17] Qualcomm of San Diego, CA—won the contract to supply US$350 million worth of handsets to Great Wall Telecom.[18] Motorola has the contract to supply ground stations to Great Wall Telecom in the Beijing market.[19] Samsung has established a network with Great Wall in Shanghai.[20]

One of the largest participants in this project is San Mateo, CA–based International Wireless Communications, which is involved in 11 cities of the network through a

---

[16]"Silicon Valley, PLA," *The Economist*, p. 65.

[17]Karl Schoenberger, "Joining Forces with China's PLA," *Fortune International*, February 3, 1997, pp. 49–53.

[18]David Welker, "The Chinese Military-Industrial Complex Goes Global," *Multinational Monitor*, June 1997, pp. 9–13.

[19]Ibid.

[20]Chetham, "PLA Muscling In."

joint venture with Hong Kong Star Telecom.[21] Nelson Wong, the patent-laden engineer who founded Star Telecom, got his first contract in 1990 to provide paging services to the Guangzhou Military Region.[22] Since then, he has expanded his technology-sharing partnership with the PLA to 13 local cell phone systems in seven provinces. Wong is a strong competitor for the Great Wall contracts, which allow outsiders to share revenues with Great Wall's provincial operating companies in exchange for supplying technology and capital. Wong's partner, IWC, invested US$20 million and committed another US$28 million for a 40% stake in a subsidiary of Hong Kong's Star Telecom called Star Digital that plans to help the PLA build the Great Wall cellular networks. They hope to raise an additional US$150–200 million in financing. A large portion of that will probably come from Indonesian tycoon Oei Hong Leong, who in December 1996 agreed to buy half of Star Telecom.

Another related joint venture involving CESEC is a deal approved in February 1998 with Dutch telecommunications concern KPN (PTT Nederland). Under the terms of the deal, which was signed during the visit of the Dutch Minister of Economic Affairs, a holding company will be set up in Nanjing. KPN will have an 80% stake in the holding company with the PLA having the remaining 20%. The holding company will invest in a range of activities related to the telecommunications sector, including satellites, mobile phones, paging, and Internet services among others. All of these activities will be operated by separate economic entities under the holding company. Under Chinese law, KPN will have less than 50% of the stake in these individual companies. The first project of the new holding company is most likely to be in the area of satellite technology.[23] Initial capitalization of the venture is quite small, a reported US$5 million, but the Dutch partners at least expect the venture to expand quickly once specific projects are finalized.[24] KPN hopes that this initial venture will be "a stepping stone to set up local, regional and national telecommunications projects with mainland partners."[25]

The PLA and Radio Paging

A second and equally profitable spin-off from the PLA bandwidth monopoly has been mobile radio paging. In Guangzhou, for example, three of the ten largest pager companies are owned by the PLA Air Force (PLAAF), the Guangzhou Military District, and a COSTIND subsidiary.[26] Similarly, the Guangzhou Bayi Telecommunications

---

[21]Andrew Tanzer, "The People's Liberation Army, Inc.," *Forbes*, March 24, 1997, pp. 44–46.

[22]Schoenberger, pp. 49–53.

[23]Foo Choy Peng, "KPN Forges Partnership with PLA Unit," *South China Morning Post Internet Edition*, February 27, 1998.

[24]Ibid.

[25]Andrew Chetham, "Dutch Giant Links up with PLA Arm," *South China Morning Post Internet Edition*, February 25, 1998. Also see *South China Morning Post Internet Edition*, June 11, 1998.

[26]Guangzhou Enterprise Evaluation Association, *Zhongguo Guangzhou daxing qiye paixu* (Ranking of China Guangzhou's Large-Sized Enterprises), Guangzhou: Zhongshan University Publishing House, 1994, pp. 115–117.

Group, owned by the Guangzhou Military Region Air Force, runs a commercial radio paging service that claimed 100,000 subscribers in 1994.[27]

Foreign participation has been critical to this sector. First, Hong Kong businessman Paul Kan and his company Champion Technology have partnered with PLA companies in Guangdong to develop radio paging.[28] Now these businesses have expanded to include paging franchises in dozens of Chinese cities, mainly through partnerships with firms controlled by local PLA units. Kan estimates that PLA units are involved in about 25% of China's 36 million paging subscribers, though the market is expanding by one million new customers a month, so the PLA's proportion of the market is likely to change considerably. Second, the PLA has laid at least 20,000 kilometers of fiber optic cable across China. [29] In one specific case, PLA units are laying an optic cable line supplied by AT&T and Lucent from Guangzhou to Wuhan.[30] Third, a US$30 million joint venture has been established between National Semiconductor Corp. (USA) and China Electronic Systems Engineering Company to produce mobile PBX systems.[31] How much profit the PLA has made through its telecommunications activities is far from clear. Available Chinese figures do not allow for a breakdown of PLA enterprises by sector nor do they give an accurate reflection of the amount of money generated. Recent estimates of total profits for PLA enterprises have run anywhere from 1–10 billion U.S. dollars.[32] The PLA's auditing system is at best minimal and underreporting of profits and transfers of money to offshore havens are rampant and major reasons contributing to the recent decision to shut down, or at least curb, PLA business activities. One aspect of PLA telecommunications enterprises is, however, quite clear. As this discussion shows, the PLA's telecommunications enterprises have been slow to move out of areas such as paging, where the PLA has the primary advantage. It remains to be seen whether conditions under the new MII will allow the PLA to continue to expand its telecommunications services or whether it will be largely limited to exploitation of its control over bandwidths. Current efforts to reduce the PLA's business activities should also strengthen the control of the old MPT over the telecommunications sector in coming months.

Some Cautions

In discussing present and future developments, however, it is important to keep in mind four potential problems for the PLA. One is the possible negative publicity that foreign companies might face in conducting business deals with PLA enterprises. Recent events in the United States (such as the investigations of the Cox Committee) underscore this fact. On the other hand, the lure of potential profits and the fear that

---

[27]Kathy Chen, "Soldiers of Fortune: Chinese Army Fashions Major Role for Itself as a Business Empire," *Wall Street Journal*, May 24, 1994, pp. A1, A9.

[28]Tanzer, pp. 44–46.

[29]"China's National Defense" op. cit.

[30]Cheung, "Can the PLA Inc. Be Tamed?" p. 52.

[31]Ibid.

[32]Tai Ming Cheung, *Far Eastern Economic Review*, August 6, 1997.

competitors might gain an edge provides considerable motivation for many companies to seek commercial agreements despite financial and political risks. One-tenth of Motorola's global sales are in China and other international telecom corporations are equally dependent on the China market.[33]

Second, the business environment in China and the global telecommunications market in general are unpredictable, to say the least. Despite the flurry of activity in the last few years and the obvious advantages of the PLA connection, for example, the Great Wall telecommunications project and related CESEC ventures remain high-risk propositions for two reasons. On the technological front, there are now serious doubts about the efficacy of the CDMA technology as a cellular standard, particularly given MII's clear commitment to the rival European GSM standard. Foreign financiers will have to invest money without being given an equity stake in the project, since China will not allow direct foreign ownership of telecommunication services or infrastructure. The KPN deal is a case in point. While it has a majority stake in the holding company, its ownership in all the entities under the holding company (and it is these entities that will be taking on the actual projects) is less than 50% and must remain so under current mainland law. Perhaps it is not surprising, therefore, that rumors have arisen that KPN has bought out the PLA's share of the deal.[34]

The third potential problem is bureaucratic. While the PLA is in a position to exploit its assets, the Ministry of Information Industry, which is dominated by former MPT personnel, is in a strong position to block both the PLA and the former MEI from undermining China Telecom's market dominance. The MPT and now the MII have certainly been successful in terms of limiting Unicom's success, limiting their participation to mobile communications. In addition, the MII have openly questioned the legality of the CCF (Chinese-Chinese-Foreign) business model, which has been the primary source of Unicom's funding.[35] There have even been reports that PLA telecommunications deals are being blocked by the MPT/MII.[36]

Four, as of the summer of 1998, there was a very strong probability that PLA enterprises would be shut down entirely, or at least the PLA would lose equity in most if not all its business operations. Since Jiang Zemin's announcement of PLA divestiture at an anti-smuggling conference in July, significant progress has been made in curtailing the commercial activities of the military. Between August and October, investigation and work teams were sent down to the units to compile lists of companies and assets. By 15 December, these companies were formally transferred to nonmilitary holding companies, though their exact fate and the precise amount of compensation have yet to be finalized. The fate of the PLA's telecommunications are less clear, however, as there are consistent reports that CESEC and other companies,

---

[33]"Telecom Aims to Go Global," *China Daily*, July 10, 1998, p. 1.

[34]Communication with informed source, December 1998.

[35]See, for example, Andrew Chetham, "China Telecom, Bureaucracy Stonewall Unicom," *South China Morning Post Internet Edition*, China Business Review Special Report on telecommunications, June 11, 1998.

[36]Chetham, "PLA Muscling In."

by virtue of their critical importance to the PLA's C4I modernization, have received "get-out-of-jail-free" cards from the military leadership. At best, they may be required to cut their ties with high profile projects or foreign partners, and perhaps even formally sever their ties to the PLA, but there is no evidence to suggest that they will not maintain close informal links to the military.

IMPLICATIONS PAST AND FUTURE

There are several implications deriving from the PLA's past and future role in the telecommunications sector. Before the divestiture, the lucrative potential of China's rapidly growing telecommunications market offered considerable opportunity (though no guarantees) for the PLA to significantly improve its earning capacity. This certainly helped individual units to supplement wages, food subsidies, and barracks subsidies. Far more important, however, was the potential of these telecommunications deals for dual-use technology transfer. Despite the fact that China has placed considerable effort into developing its own telecommunications technology, such as PLA investment in research facilities at Xidian and other universities, China continues to rely on foreign sources for equipment, parts and expertise.[37] Indeed, many of the telecommunications technologies involved in these deals can be used to improve the military's C4I infrastructure. For example, it is widely assumed that the B-ISDN (broadband integrated services digital network) and ATM (asynchronous transfer mode) technology involved in the controversial deal between SCM/Brooks (US) and Galaxy New Technology (PRC), whose primary shareholder is COSTIND, was presumably shared with the PLA.[38] Another example is the China Telecom-Great Wall CDMA project with Qualcomm. CDMA technology was originally developed for the U.S. military, which sought a system that could sustain a high volume of communications traffic in a small area.

An area of dual-use technology that is especially significant is fiber-optics. Units of the PLA have laid most of China's fiber, and in return the military reportedly receives a percentage of the fibers for their own use. In addition, the PLA has reportedly laid its own dedicated fiber-optic landline networks, connecting Beijing to military region headquarters and Second Artillery brigades. The advantage of fiber-optic cables is that they can carry considerably more communications traffic than older technologies that were available to the PLA in the past and they are far faster. The cables are able to transmit data at rates of 565 megabytes a second and higher. This represents an enormous jump over older copper-wire–based systems, thereby significantly improving the PLA's ability to transmit, receive, and process large amounts of information (including visual images) from the central government and the various military regions. Apart from speed, fiber optic cables are also less prone

---

[37]See *Jane's Defense Weekly*, April 23, 1998, p. 16; "PRC: High-Speed Optical Fiber Transmission System Developed," *Xinhua*, in FBIS-CHI-96-076, March 26, 1996; and "China: Mobile Telecommunications Industry Viewed," *Jingji ribao*, in FBIS-CHI-98-208, July 27, 1998.

[38]For more details on this point, see James Mulvenon, *Chinese Military Commerce and U.S. National Security*, p. 30. For more information, see Bruce Gilley, "Peace Dividend," *Far Eastern Economic Review*, January 11, 1996, pp. 14–16; and Bruce Gilley, "Not Over Yet: U.S.-China Technology Deal Raises Congress's Hackles," *Far Eastern Economic Review*, January 18, 1996, p. 15.

to corrosion and electromagnetic interference, making them more reliable. Their light weight and small size make them ideal for mobile battlefield command as well as fixed military headquarters. Most important of all, it is extremely difficult for American and other intelligence services to monitor military communications conducted over fiber-optic cable, particularly landline connections between Beijing and strategic command and control centers throughout the country. In the 1980s, some U.S. government agencies were opposed to the sale of fiber optics to the Soviet Union and other countries, including China, for this very reason.[39]

Moreover, the importance of fiber-optics is not limited to military communications. Fiber optics can be used as sensors in sonar arrays, as well as in perimeter defense systems and even biological weapons detection. Fiber optics can also be used for local area networks (LANs) in warships and in precision guided munitions.[40] Therefore, acquisition of fiber-optics means better weapons and C4I, in addition to the clear benefits to communications. For these reasons, some quarters in the U.S. government are opposed to sales of cable to China. However, not only does the U.S. sell fiber-optic cable to China, it is now manufactured in China through joint ventures. Shanghai Lucent Technologies Fiber Optics Co. is the largest fiber optics manufacturer in the PRC. Last year the company sold 680,000 kilometers of fiber optics, accounting for just under a quarter of the Chinese domestic market.[41] The joint venture's production capacity is expected to expand 1.5 million kilometers of cable a year in 1998.[42] How much of this goes to the PLA is unclear, but as already noted, Lucent is involved with China Telecom-Great Wall, which is half-owned by the PLA's CESEC. This should facilitate the PLA's progress towards its goal of connecting all of its military region, district, and group army headquarters with fiber-optics, as well as its application to battlefield communication and weapons systems.

To this point, we have skirted the ultimate question raised by this discussion of dual-use technology transfer; namely, will the PLA be able to acquire, integrate, and effectively employ these technologies in a combat environment? While the answer to this question is unknowable in advance, it can be reasonably asserted that this equipment will not allow the PLA to "leapfrog" the U.S. military, as some have asserted. Yes, the telecommunications industry occupies a special place in China's economic modernization as one of the few sectors unburdened by a bankrupt network of bloated "legacy" factories (aviation is a stellar example). As such, China can take advantage of Gershenkronian "late modernization" by laying fiber where copper wire or perhaps even nothing at all existed previously. In this respect, the Chinese have "leapfrogged," but bypassed only stages of development, not specific competitors. Thus, we conclude that while parts of China may gradually equal or even outpace elements of the U.S. telecom infrastructure, the Chinese military

---

[39]See Mastanduno, *Economic Containment*.

[40]For a good summary of the various new applications for fiber optics in the military, see Mark Hewish, "Penetrating the Fiber-Optic Fog: Embracing the Next Generation of Communications and Sensors," *Jane's International Defense Review*, April 1998, pp. 51–57.

[41]Zhang Yan, "Lucent Commits to China Market," *China Daily*, February 15, 1998, p. 2.

[42]Ibid.

cannot hope to match, much less exceed, the current or future capabilities of its American counterpart.

POSTSCRIPT: PLA DIVESTITURE

As of December 1998, there are clear indications that the conduit for the PLA's acquisition of advanced telecommunications equipment may be coming to an end. The divestiture of PLA, Inc. has achieved some remarkable successes, and in the end will have fundamentally changed the character of the military's participation in the economy. Minor PLA telecom concerns, such as the small radio paging companies attached to individual units, will probably be taken away from their owning unit and folded into the civilian telecommunications infrastructure. By contrast, the national companies like CESEC will most probably continue to operate, though they may be forced to divest themselves from high-profile projects like China Telecom-Great Wall CDMA and may even be required to publicly disassociate themselves from their former units (the General Staff Department Fourth Subdepartment in the case of CESEC). Sources in Beijing suggest, however, that CESEC is far too valuable to the PLA's C4I modernization effort to be abolished entirely. Instead, the company will likely persist in its acquisitions of advanced telecommunications equipment, albeit with less fanfare and publicity. Thus, we expect PLA involvement in the telecommunications arena to continue for the foreseeable future, as will the military's acquisition of advanced technology for its C4I modernization, though the future use and effectiveness of this equipment in warfighting situations cannot be predicted in advance.

13. A NEW PLA FORCE STRUCTURE

Dennis J. Blasko[1]

In 1995, the Military Commission of the CPC Central Committee had further decided that in building the army, it is necessary to pay more attention to quality instead of quantity and scale and attach greater importance to scientific and technological development instead of manpower. The formulation of the military strategic guideline in the new period and the decision to effect the "two transformations" in army building are explicit characteristics of the times. They reflect the objective need of army building in the new period . . . We have no existing examples to follow in ensuring quality army building.

> Chen Bingde
> Commander, Nanjing Military Region[2]

Our objectives are to develop the PLA into a revolutionary, modernized, and regular army with Chinese characteristics. We believe: A streamlined army of a reasonable size will be helpful to improving the international environment for arms control and disarmament, be conducive to enhancing mutual trust among countries, and be more beneficial for us to concentrate our energies to properly develop the economy.

> Defense Minister Chi Haotian
> Speech to the Japanese National
> Institute for Defense Studies,
> February 4, 1998[3]

As the modernization of the Chinese armed forces continues into the 21st century, changes in its force structure will be inevitable. Newer, more modern weapons and a new doctrine emphasizing joint operations necessitate that the size, organization, and command and control structure of the force be adapted to meet the new circumstances. A new force structure will seek to integrate the force's new

---

[1]Dennis J. Blasko served as an army attaché in Beijing and Hong Kong from 1992 to 1996. Previously, he had been assigned to infantry units in Germany, Italy, and Korea, and to Headquarters, Department of the Army and the Defense Intelligence Agency. He is now a senior analyst at the Washington-based consulting firm International Technology and Trade Associates.

[2]Chen Bingde, "Intensify Study of Military Theory to Ensure Quality Army Building; Learning From Thought and Practice of the Core of the Three Generations of Party Leadership in Studying Military Theory," *China Military Science*, No. 3, August 20, 1997, pp. 49–56, in FBIS-CHI-98-065, March 6, 1998.

[3]Le Shaoyan and Gang Ye, "China: Chi Haotian on PRC Defense Policy," *Xinhua Domestic Service*, February 4, 1998, in FBIS-CHI-98-035, February 4, 1998.

capabilities, maximize the performance of its new weapons, and effectively execute its new doctrine.

Because of domestic conditions and constraints that make China different from other nations, Nanjing Military Region Commander Chen Bingde correctly observes that there is no example for China to follow as it reshapes its forces. The Chinese military is constantly reminded of its role in society and its place among national modernization priorities. It is well aware that military modernization will be severely limited by funding constraints. In March 1998, President Jiang Zemin reiterated these realities in a speech to the military delegation at the Ninth National People's Congress:

> The level of China's productive forces is still not high, and our economy is not that strong. Therefore, we must concentrate our energies on economic development. Without a highly developed economy, it is also impossible to promote the modernization of national defense and the army. We must always insist on taking economic development as the central task while paying adequate attention to modernizing the national defense, and seek coordinated development for both the economy and national defense. We must blaze a trail of modernizing national defense and the army with Chinese characteristics.[4]

From this passage it is clear that "paying *adequate* attention to modernizing the national defense" is a condition that Chinese military planners will have to live with. Therefore, by structuring their military organization more efficiently, the Chinese may be able to put to better use the limited funding available. Force structure reform thus becomes an integral part of military modernization.

Overall, it is important to note that the ultimate objective of modifying the Chinese military force structure is to better organize itself to achieve China's national military objectives. These national military objectives may be summarized as:

- Protect the Party and Safeguard Stability

- Defend Sovereignty and Defeat Aggression

- Modernize the Military and Build the Nation.[5]

Although the senior Chinese civilian and military leadership has outlined the general trends and directions that changes in the force structure will take, a detailed blueprint has not been made public, if one has been fully developed. Such a plan would certainly be considered sensitive or classified information. However, based on Chinese writings and speeches, it is possible to speculate about what the Chinese military of the early 21st century will look like. But first, a brief description of the current force may be useful as a point of reference.

---

[4]"China: Jiang Zemin on Army Building at NPC," *Xinhua*, March 10, 1998, in FBIS-CHI-98-070, March 11, 1998.

[5]See David Finkelstein's chapter in this volume.

THE CURRENT FORCE STRUCTURE

Article 22 of the PRC Law on National Defense adopted on March 14, 1997 states "The armed forces of the People's Republic of China are composed of the active and reserve units of the Chinese People's Liberation Army (PLA), the Chinese People's Armed Police Force (PAP), and the people's militia."[6] The missions of this three-tiered force are defined as:

> The active units of the Chinese People's Liberation Army are a standing army, which is mainly charged with the defensive fighting mission. The standing army, when necessary, may assist in maintaining public order in accordance with the law. Reserve units shall take training according to regulations in peacetime, may assist in maintaining public order according to the law when necessary, and shall change to active units in wartime according to mobilization orders issued by the state. Under the leadership and command of the State Council and the Central Military Commission, the Chinese People's Armed Police force is charged by the state with the mission of safeguarding security and maintaining public order. Under the command of military organs, militia units shall perform combat-readiness duty, carry out defensive fighting tasks, and assist in maintaining the public order.[7]

The active duty PLA consists of ground forces (army and the Second Artillery, also known as the Strategic Rocket Forces), the navy (including marines and some aviation units), and the air force (including airborne forces and some antiaircraft artillery units). Reserve forces are mostly ground forces, although a limited number of navy and air force units reportedly have been formed. The above passage would suggest that the primary mission of the active duty force is external defense, while the PAP is tasked with internal or domestic security. As a secondary mission, the active duty and reserve PLA forces and militia may assist the PAP in maintaining domestic security.

Apart from mission, the force structure also reflects the three schools of military thought prevalent in the PLA today: People's War, Local War, and the Revolution in Military Affairs (the RMA school). These three schools are reflected in the PLA's doctrinal development, equipment, and scenario planning. The relationship of the three schools to one another and Chinese force structure can be visualized as a triangle or pyramid composed of three tiers.

The base of the pyramid consists of the People's War school—the vast majority of the PLA today. The military thought of Mao Zedong provides the theoretical foundation for this school. This doctrine has little utility beyond the borders of China, but a considerable portion of all Chinese military writing still must pay homage to the heritage of People's War. Probably about 80% of the PLA ground forces, navy, and air force is best suited to fight a People's War and is equipped with weapons designed in the 1950s and 1960s that would be museum pieces in many countries. This school relies upon the use of "existing weapons to defeat an enemy equipped with high

---

[6] "'Law of the People's Republic of China on National Defense,' Adopted at the Fifth Session of the Eight National People's Congress on March 14, 1997," *Xinhua Domestic Service*, in FBIS-CHI-97-055, March 14, 1997. All quotations of the law are from this source. Hereafter cited as "National Defense Law."

[7] Ibid.

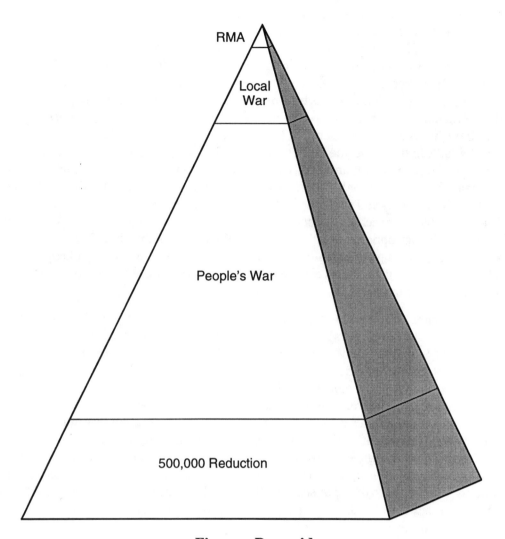

Figure—Pyramid

technology weaponry." These forces are trained to defend the mainland, its adjacent seas, and air space from invasion. They would fight along side the militia and swallow up an invader using concepts devised by Mao 60 years ago, now modified slightly to account for "modern conditions." The tactics these units practice are similar to those used in the War Against Japan, the War of Liberation, the Korean War, and the 1979 conflict with Vietnam.[8]

The second tier of the PLA pyramid is the Local War school—maybe 15% of all army, navy, and air force units. Deng Xiaoping provided the critical strategic direction for this school. Local War is understood to be a limited war on the periphery of China

---

[8]The campaign against Vietnam was the PLA's last major engagement against a foreign foe and its shortcomings provided the stimulus for military modernization efforts of the 1980s.

that will be short but intense, utilizing advanced technology weapons, with units fighting in joint and combined arms efforts. It envisions an element of force projection (i.e., the ability to transport combat forces beyond China's borders), but by definition is regional, not global, in nature. China usually regards Local War as its "next war"; the Persian Gulf War is often a point of reference for this school. China has no combat experience in this type of conflict. At this time, the development and dissemination of doctrine on how the PLA will fight such a war are in progress. The number of units actually prepared to live up to these modern standards is problematic. In the 1980s, the PLA began its current modernization program, focused on rapid reaction units and experimental forces. Some of these units, but by no means all, have received numerically limited imports of Russian hardware. Many units in this category still are equipped with outdated indigenous equipment and, like the People's War school, must devise ways to use their existing weapons to defeat a high technology opponent. This segment of the PLA probably does, however, receive more training opportunities than do units dedicated to fighting a People's War. This portion of the PLA is expected to grow in the future as the People's War segment shrinks in size.

The RMA school is at the top of the pyramid and is represented by only a very small portion of the PLA—thinkers in its premier academic institutions, a few officers in the General Staff Department and new General Equipment Department, some of the missile units in the Second Artillery, and a few other units equipped with modern cruise missiles. These elements are among the "pockets of excellence" described in the professional literature. The weapons that represent this school are also being incorporated into China's doctrine for Local War. The Chinese military and defense industries are investigating the entire scope of new technologies and theories applicable to RMA. Chinese defense industries are undertaking serious research efforts to identify areas upon which they should focus. No senior Chinese leader has lent his imprimatur to the RMA school. The lack of a focused, high-level vision of future war may slow the development of many of the concepts currently being explored by thinkers at lower ranks.

GENERAL TRENDS FOR A FUTURE PLA FORCE STRUCTURE

In recent years, the senior Chinese military leadership has outlined the general trends for the development of China's armed forces in the near- and mid-terms in professional writings and public speeches. Efforts have already begun in the following areas, and gradually the force structure will be modified to better implement these strategic directions:

1. Active duty PLA forces will become quantitatively smaller, with an emphasis on technological quality.

2. Reserves and the People's Armed Police will increase in size.

3. The PLA will retain many existing weapons and attempt to develop new tactics and techniques to defeat a high-technology enemy.

4. Only limited amounts of foreign weapons and equipment will be introduced into the forces; the indigenous Chinese defense industry will be the source of the majority of modern weapons.

5. Capabilities will emphasize rapid response and joint operations, focusing on precision attack, air operations, naval operations, information warfare, and space operations.

6. Command and control organizations will be reorganized to better manage the requirements of future warfare.

Like economic modernization, these elements of military modernization are considered long-term goals, which should be accomplished by the middle of the 21st century or 100 years after the founding of the People's Republic (2049). No specific milestones to achieve the different elements have been announced. However, by the year 2010 the general trends will have been in motion for over a decade and progress in these areas will be more apparent than they are at present. Chinese military capabilities will be improved but will still fall far behind many other contemporary modern forces. The remainder of the chapter will examine each of these six elements in more detail.

Smaller, but Better

In September 1997, President Jiang Zemin announced a 500,000-man reduction in the strength of the PLA, to be completed over the next three years. In reality, that reduction had begun a year earlier, as 14 ground force divisions in lower readiness categories were transferred to the PAP.[9] By the time of Jiang's announcement, the number of PLA personnel transferred to the PAP probably ranged from 110,000 to 150,000. As the reduction continues, more, but not all, of the PLA forces to be reduced may be transferred to the PAP. The forces subject to this reduction come from the bottom of the PLA force structure pyramid, those best suited to fight a People's War. Many of the units that have been or will be eliminated are likely to be those manned at less than 100% strength, some even below 50%. Therefore, it is possible that *more authorized slots will be reduced than actual personnel.* Unless the PLA changes its policy on openness to outsiders, the specific units and numbers of personnel reduced may never be announced officially and thus remain the subject of debate and disbelief among many outside of China.

The majority of the forces to be eliminated in the ongoing reduction will be ground forces. According to the July 1998 Defense White Paper, ground forces will be reduced by 19%, naval forces by 11.6%, and air force personnel by 11%.[10] These percentages amount to a reduction of about 418,000 ground forces, 31,000 naval

---

[9]Liu Hsiao-hua, "Armed Police Force: China's One Million Special Armed Troops," *Kuang chiao ching*, No. 307, April 16, 1998, pp. 42–47, in FBIS-CHI-98-126, May 14, 1998. See also Dennis J. Blasko and John F. Corbett, Jr., "No More Tiananmens: The People's Armed Police and Stability in China, 1997," paper presented at Staunton Hill Conference on the PLA, September 1997.

[10]Information Office of the State Council of the People's Republic of China, *China's National Defense*, July 1998.

personnel, and 52,000 air force personnel.[11] Of the 500,000 personnel to be reduced, the ground forces will account for nearly 84% of the total. In the year 2000, at the end of this reduction, the PLA will number approximately 2.5 million personnel, with ground forces (including the Second Artillery) comprising about 1.78 million, the navy 234,000, and the air force 418,000. During this period of troop cuts, the PLA will also experiment with organizational changes. Successful experiences during this period of experimentation will later be applied throughout the force.[12]

An important implication of the 500,000-man reduction underway is that the percentage of PLA ground forces within the total force structure will decrease as the percentages of naval and air forces increase. This condition parallels the increasing significance of the navy and air force in Chinese military strategy for the 21st century. In the past, the PLA was oriented to a continental defense strategy, which called for a large, dominant ground force. Now, as the PLA shifts its doctrine to local wars on the periphery of China, the navy and air force have risen in importance, receiving priority in PLA modernization efforts. They will naturally grow in proportion to the total force.

Presently, ground forces (including the Second Artillery) comprise 73% of the total force structure, with the navy and air force comprising only about 10% and 17%, respectively. If only for reasons of history, geography, and inertia, the PLA is likely to remain dominated by ground forces for several more decades, and many army units will still be best suited for the defense of mainland China using the People's War doctrine. However, as the PLA's focus shifts further from continental defense to a maritime orientation, naval and air capabilities will become more important and will better counterbalance the weight of the ground forces than they do today.

For a point of rough comparison, personnel numbers of the United States armed forces, which have global responsibilities that the PLA is not envisioned to assume, are much more heavily weighted toward naval forces (including marine forces) than they are to either the army or air force. Naval forces comprise 40% of U.S. forces, while the army and air force comprise 33% and 27%, respectively. For many reasons, the PLA is unlikely to select the U.S. force structure as a model for its modernization; however, over the years, the proportion of the PLA service arms will move further away from its nearly total dominance by the ground forces, as has been the case since the founding of the Red Army.

A force of 2.5 million will still be larger than needed for an adequate defense in the 21st century, especially as Chinese military doctrine stresses the use of high-technology weapons and equipment. The May 16, 1998 issue of the Hong Kong magazine *Wide Angle* predicts, "As the international environment relaxes and the national economy develops, further troop reduction may be required to ensure that

---

[11]These specific numbers are derived by multiplying the White Paper's percentages by figures of 2.2 million, 265,000, and 470,000, found in International Institute for Strategic Studies, *The Military Balance, 1996/97*, London: Oxford University Press, 1996, pp. 179–181.

[12]Kuan Cha-chia, "Commander Jiang Speeds Up Army Reform, Structure of Three Armed Services to be Adjusted," *Kuang chiao ching*, No. 305, February 16, 1998, pp. 14–18, in FBIS-CHI-98-065, March 6, 1998.

the troops are well-equipped and highly-mobile."[13] Another recent prediction in the Hong Kong press envisions additional cuts of 100,000 personnel per year through much of the next decade, ending with a total force of about 2 million by 2010.[14] Further reductions in personnel would be looked upon favorably by China's neighbors (see Defense Minister Chi's comments above) and benefit its own military modernization as its defense budget could be focused on fewer troops.

Therefore, it is likely that early in the 21st century Beijing will announce another significant reduction in the size of its standing forces. The bulk of future reductions beyond the current reduction can be expected to be felt again in the ground forces, and again in the units of a lower readiness category. The navy and air force may internally reorganize their forces and eliminate certain units (for example, many of the air force's antiquated fighters will be retired and many of its anti-aircraft artillery units may be transferred to the reserves as more air defense missile units are activated), but for the purpose of this analysis it is assumed that the number of personnel in these service arms will remain constant or perhaps even increase. For the sake of argument, we will assume that the next reduction will also number 500,000 personnel.[15]

A little recognized fact is that civilians in the PLA are included among the total numbers of China's active duty forces.[16] Known as *wenzhi ganbu*, these PLA civilians wear uniforms and can be given ranks if necessary. Most PLA civilians serve in technical and logistics capacities, such as doctors, instructors, computer specialists, headquarters personnel, or technical service personnel, and would not be considered combat personnel. Some *wenzhi ganbu* also work in PLA commercial activities. The exact number of PLA civilians is not known, but possibly constitute 20–25% of the total force.[17] After the 500,000-man reduction when the PLA numbers approximately 2.5 million, a conservative estimate of the number of PLA civilians in the force would be around 500,000.[18]

If this number seems high, the number of civilians who worked for the U.S. Department of Defense in 1997 was over 767,000, or an additional 53% added to the U.S. active duty strength of 1,443,000.[19] Because the PLA is currently less technically

[13]Liu Hsiao-hua, "Jiang Zemin Convenes Enlarged Meeting of Central Military Commission, Policy of Fewer But Better Troops Aims at Strengthening Reserve Service Units," *Kuang chiao ching*, No. 308, May 16, 1998, pp. 50–53, in FBIS-CHI-98-161, June 10, 1998.

[14]Willy Wo-Lap Lam, "New-Look PLA Plans More Cuts," *South China Morning Post*, June 30, 1998.

[15]This number was derived independently, but is consistent with those found in Willy Wo-Lap Lam, "New-Look PLA Plans More Cuts."

[16]Shaoguang Wang, "Estimating China's Defense Expenditure: Some Evidence From Chinese Sources," in *China Quarterly*, September 1996, p. 891. The author's conversations with PLA officers in 1995 and 1996 support this assertion. The 1998 Defense White Paper also contains the statement that "different from many other countries, China includes . . . civil cadre . . . in the overall strength of the PLA."

[17]Author's conversation with PLA civilian in September 1996.

[18]In early 1997, *Wide Angle* magazine estimated that about 500,000 people serve in "military enterprises and such departments as military research, medical care, literature and art, sports, and education." Yuen Lin, "China's Military Strength and Peripheral Military Situation," *Kuang chiao ching*, No. 293, February 16, 1997, pp. 62–67, in FBIS-CHI-97-059, February 16, 1997.

[19]*Defense 97 Almanac*, Issue 5, Washington: Department of Defense, 1997, p. 17.

complex than U.S. forces, it is not unreasonable that they would have a smaller percentage of civilians than does the United States. Most countries do not include civilians in the number of their active duty forces.

China would be able to *reduce significantly the size of its military without any impact on its capabilities,* if it were to declare openly the true numbers of its *wenzhi ganbu* and disaggregate them from the PLA active duty force statistics. Such a decision would conform to generally accepted international standards and would make comparisons between the size of the PLA and other militaries more accurate and illuminating. However, such an announcement would probably be interpreted by some critics as a disingenuous attempt to deceive the world about Chinese military strength. Nevertheless, for the purpose of this analysis, 20% of the PLA year 2000 end-strength will be subtracted from each component of the force to represent the approximate number of *wenzhi ganbu,* to more accurately portray the size of PLA forces.

The PLA could be further streamlined by removing from the active force the officers and enlisted troops devoted to commercial activities. The Hong Kong newspaper *Ming Pao* reported in May 1998 that President Jiang "clearly demanded that all army-run enterprises be separated from the army in three years" at an unpublicized meeting of senior military officers.[20] A recently announced policy has prohibited noncombat, as well as combat, units at Group Army level and below from engaging in commercial activities.[21] In mid-July 1998, President Jiang announced that the army and the PAP must not engage in commercial enterprises.[22] This edict was pronounced during an anti-smuggling meeting and probably applies mainly to commercial enterprises, such as major hotels, restaurants, real estate ventures, trade and investment operations, and other ventures in which crime and corruption are rampant. Traditional agricultural and light-industrial sideline production at the unit level and the PLA's system of numbered factories, which produce nonlethal material and logistics supplies, will probably be affected only minimally, if at all. Most of the commercial management personnel about which Jiang spoke are found at higher headquarters, where they would not be considered deployable combat personnel. These personnel may be redesignated as *wenzhi ganbu* or non-military-related civilians and dropped from the active duty rolls once their enterprises are separated from the PLA.

Some higher echelon engineer and transportation units have for years been dedicated to military and civilian construction and commercial projects. Since these units probably do minimal training for their wartime missions, they too could be transformed into organizations manned by *wenzhi ganbu* to support the PLA. Any attempt to quantify the number of PLA officers and enlisted currently performing

---

[20] "Jiang Zemin Criticizes Slow Progress in Army Reform; Reducing the Army by 500,000 Not Proceeding on Schedule," *Ming Pao,* May 19, 1998, p. A12, in FBIS-CHI-98-140, May 20, 1998.

[21] *Guangming ribao,* April 11, 1998; and "China: Noncombat Units Should Not Engage in Business Production," *Xinhua,* April 10, 1998, in FBIS-CHI-98-100, April 13, 1998. This prohibition does not include sideline production found in unit farms.

[22] Seth Faison, "China's Chief Tells Army to Give Up Its Commerce," *The New York Times,* July 23, 1998.

commercial activities would be a guess. Therefore, this option will simply be mentioned for consideration, and no personnel subtracted from the active duty strength in the following projection.[23]

If, over the next decade, the PLA does not include its civilians in active duty personnel numbers and reduces another 500,000 from the ground forces, by about the year 2010 it will have a total manpower strength of approximately 1.465 million personnel.[24] The ground forces will comprise about 64% of that number, a drop of 9% from its current proportion of the forces. Assuming that the navy and air force are not subjected to major personnel reductions, but rather redistribute personnel among units, the proportions of these two service arms will grow to 14% and 23%, respectively. (See Table 1.)

Table 1

Comparison of U.S. and Chinese Active Duty Forces

Component	U.S. 1998	PLA 1998	PLA 2010(5)
Army (1)	483,000/33%	2,090,000/73%(4)	932,000/64%
Navy (2)	578,000/40%	280,000/10%	199,000/14%
Air Force (3)	383,000/27%	470,000/17%	334,000/23%
Total	1,443,000/100%	2,840,000/100%	1,465,000/101%

Sources: *Defense 97 Almanac* for U.S. forces and *The Military Balance, 1997/98* for the PLA.

Notes: 1. PLA Army figures include 90,000–125,000 Second Artillery personnel.

2. Navy figures include marine forces in both countries.

3. PLA Air Force includes all airborne and some antiaircraft artillery personnel.

4. The number of PLA ground forces (Army) in 1998 evidently reflects the impact of the initial phase of the 500,000-man reduction announced in 1997. This number is 110,000 smaller than the 2.2 million listed in previous years.

5. Percentages do not add to 100 because of rounding.

The proportion of naval forces could be further expanded if Beijing decides to increase the size of the existing 5,000-man marine force by changing the uniforms and mission of several ground force infantry units stationed near the coast. If five infantry divisions, approximately 60,000 men, were converted to marines, the percentage of naval forces would grow to about 18% of the total force, while the ground forces would drop to 60%.[25] Such a decision would be politically sensitive internationally and probably be considered threatening by Taiwan, Japan, and countries having territorial disputes with China in the South China Sea. However, it would provide the PLA greater flexibility in its protection of its maritime claims.

Significantly, *a reduction of one million* from the 2.2 million-strong ground forces (as of 1996, prior to the current round of reductions) conducted over the next 12 years

---

[23]If *Wide Angle* is correct (see footnote 13), many of the personnel in these units may already be included in the 500,000 personnel the author has assumed to be considered civilians working for the PLA.

[24]This number would be consistent with Willy Wo-Lap Lam's earlier estimate.

[25]According to *Jane's Defence Weekly*, "Rapid Deployment Key to PLA Modernization," April 15, 1998, the 31st Group Army in Nanjing Military Region has three infantry divisions capable of amphibious operations.

would have no adverse impact on the PLA's ability to project force beyond their borders. Currently, the PLA can move only a few tens of thousands of troops beyond its borders using air and sealift. With fewer forces, the military budget will be able to stretch farther than it can now. In an important long-term investment for the PLA, more funds could be made available for aircraft and ships suitable for transporting and supplying airborne troops, ground forces, or marines. Remaining troops will be able to undergo more training and receive more modern equipment than they currently do.

The size of PLA combat units will become smaller as newer, more capable weapons and communications and mobility equipment enter the force. There are too many factors, too many types of units, and too many unknowns as to exactly when and what new weapons will be incorporated into the inventory to speculate about the specific size of any tactical unit. However, it is well understood that the basic form of many units will change. As Li Xueyong of the Army Command Academy said at a 1998 "Theoretical Symposium on Characteristics and Laws of Hi-Tech War":

> combat forces are bound to become smaller in size but stronger in combat effectiveness. As a result, smaller units are likely to become "comprehensively composed" and capable of fighting bigger battles.[26]

Though Professor Li was referring to ground force units, the principle he outlines is applicable to other services as well. His reference to "comprehensively composed" units would translate into combined arms units, which organically integrate various service arms so the capabilities of each individual arm complement and enhance the others.

As Chinese military modernization proceeds beyond the first decade of the 21st century, the proportions of naval and air forces can be expected to continue to grow as more resources are shifted away from the ground forces. This trend will reflect a major transformation in the culture of the PLA. No longer will China's security be oriented toward army-dominated continental defense, but rather the PLA will turn its focus outward to its maritime periphery using naval, air, and missile forces.

More Reserves and PAP

Defense of the Chinese mainland from land invasion cannot and will not be ignored by PLA planners. Neither will the PLA's role in domestic stability be forgotten. However, large active duty ground forces may not be the most cost-effective way to perform those missions in the 21st century. For the defense of the mainland from land invasion, a larger reserve force may prove more suitable than a large standing active duty force. For domestic security, PAP forces have been tasked officially by the National Defense Law to safeguard security and maintain the public order.[27]

---

[26]Huang Youfu, Zhang Song, and Zhao Guifu, "Take Note of Trial of Strength on Network, Greet Battlefield Changes—Summary of Theoretical Symposium on Characteristics and Laws of Hi-Tech War," *Jiefangjun bao*, May 5, 1998, in FBIS-CHI-98-154, June 3, 1998.

[27]"National Defense Law."

A land invasion of China is unlikely to be a lightening strike or bolt from the blue. Rather, PLA planners can assume a reasonable warning period during which they could mobilize reserve forces to augment the standing army. Even with active duty ground forces numbering less than a million, many units will be located near traditional "avenues of attack" into China and will be able to act in concert with the local reserve and militia forces to trade space for time, utilizing People's War tactics. New smaller, more mobile ground forces will be able to be shifted from one part of the country to another to reinforce units in an area under attack. (Reserve and militia units will have an important role in supporting active duty forces from other regions once they arrive from their home bases.) Moreover, a more modern air force and mobile missile forces will be able to support the defense of a land attack against the mainland.

According to *Wide Angle*, an April 1998 meeting of the Central Military Commission emphasized the need to expand the reserve forces. After the meeting, the Military Districts were ordered to step up the implementation of plans to build reserve units.[28] At present, Chinese reserve forces are estimated to number 1.2 million.[29] Much of the equipment and many of the personnel affected by reductions in the ground forces (who do not go to the PAP) in the next decade can be expected to find their way into the reserves. In addition to army reserves, more naval and air force reserve units will be formed as older PLA equipment is retired and their units disbanded. A new form of reserves, similar to the U.S. Individual Ready Reserve, in which officers are centrally managed but not assigned to specific units may also have been instituted in the PLA.[30] These soldiers often are specialists used to augment headquarters elements at higher echelons.

Maintaining reserve forces is less expensive than active duty forces—according to *Wide Angle*, one-tenth the cost of an active army division.[31] Eventually, the reserves could outnumber the total of PLA active duty forces, perhaps up to a total of 2 million if the PLA undergoes another 500,000-man reduction. A larger number of reserves than active duty forces would not be unique to the PLA. In 1997, total numbers of U.S. reserve forces (including National Guard units) were more than 1,449,000, slightly larger than the 1,443,000 on active duty.[32]

A larger reserve force also would be able to assist many of the disaster relief and community service missions that the PLA, PAP, and militia are often called to perform. These missions will continue to be an essential role for the armed forces of China no matter what their size and composition. Such missions test the organization and command and control structure of the forces, as well as contribute to the national military objective of "building the nation."

---

[28]Liu Hsiao-hua, "Jiang Zemin Convenes Enlarged Meeting of Central Military Commission."

[29]*The Military Balance, 1997/98*, p. 176.

[30]Thanks to Dr. David M. Finkelstein for providing information on this new type of PLA reserve officer.

[31]Liu Hsiao-hua, "Jiang Zemin Convenes Enlarged Meeting of Central Military Commission."

[32]*Defense 97 Almanac*, Issue 5, p. 23.

As the reserve force grows in size, the requirement for maintaining a large militia force will probably be reevaluated. Much of the existing militia strength would be of questionable military value in a modern conflict, and as more reserve units are established, some militia forces may be eliminated. However, because of the difference in wartime missions between the reserves and militia, the reserves are not envisioned to totally replace the militia. To formally disband much of the militia would appear to be a rational act (to many Western observers), but to do away with the entire militia would be difficult to justify as long as the PLA continues to hold Mao's military thought as the basis for all military strategy. Therefore, in the first decade of the 21st century, the Chinese militia will probably gradually be reduced to a smaller force than exists today, but not eliminated completely.

As the reserves expand, so too will the PAP. Currently the PAP strength is approximately 800,000,[33] and is probably on its way to about one million as the PLA continues its reduction through the year 2000. Once they get rid of their heavy weapons, the PLA's lower readiness disbanded light infantry and artillery units will be well organized and equipped to handle the internal security mission. The units will need specialized training and some specialized equipment in their newly assigned role, but the transition should not be too difficult. Many will likely become rapidly deployable, mobile reaction units.

Strengthening the PAP will make intervention by the active duty PLA less necessary, and therefore less likely, in a future domestic crisis (though always an alternative). Both the PAP and PLA will be able to focus on and train to perform their respective primary missions, rather than spending undue amounts of time on secondary missions. As the PLA becomes more technically advanced and complex, it will become less suitable for domestic security missions and will require specific, intensive training to maintain its proficiency in its mission to defend China from external foes.

Use Existing Weapons to Defeat a High-Technology Enemy

At a size of 3 million, the entire PLA could not be equipped adequately with modern equipment. Even at half that size, equipping the force with weapons of the late 20th century would be a daunting and expensive task. Beijing's decision in the 1980s to selectively equip only a portion of the force with the most modern equipment continues to make sense. The gradual introduction of modern equipment into the force allows for experimenting with how the PLA may best put the new equipment to work, as well as allowing time for doctrine to be developed and disseminated. At the same time, the education and sophistication level of the soldiers, sailors, and airmen has risen and the general mind-set of the PLA has been modified to accept the need for high-technology equipment. This is not a trivial transformation for a military that proudly continues to trace its roots back to a technologically inferior guerrilla force.

---

[33] *The Military Balance, 1997/98,* p. 179; and Liu Hsiao-hua, "Armed Police Force: China's 1 Million Special Armed Troops."

It also prepares the way for the mental shift necessary for naval and air force operations, not land warfare, to be the centerpiece of most future PLA operations.

As new equipment is introduced into the force, the PLA will still retain large numbers of older, lower technology weapons. Excerpts from the 1995 RAND study *China's Air Force Enters the 21st Century* illustrate this fact.

Though new information may revise some of the numbers slightly (see the following paragraph and Table 3), the trend is obvious. The majority of PLA Air Force fighters will be second-generation F-6 and F-7s well into the first decade of the 21st century. Though they may be upgraded with more advanced avionics, engines, and weapons systems, the survivability of these aircraft against the fourth-generation fighters of many potential foes is highly questionable. It will take many years before the proportion of truly modern aircraft outnumbers the older fighters in the inventory.

Table 2

Chinese Fighter Force Projection (excerpt)

Aircraft	Number in 1994	Number in 2005
F-6	2,824	544
F-7	586	919
F-8	205	466
Su-27	26	70

Source: Kenneth W. Allen, Glenn Krumel, and Jonathan D. Pollack, *China's Air Force Enters the 21st Century*, Santa Monica: RAND, MR-580-AF, 1995, p. 163.

Willy Wo-Lap Lam has reported that the fighter force will be reduced to about 1,000 aircraft in the next decade.[34] This number is supported by a recent projection by Ken Allen, who estimates that by the year 2010 the numbers of *relatively modern fighters* in the force will be less than 1,000. Table 3 estimates the composition of the "modern" Chinese fighter aircraft force in 2010.

Allen acknowledges that these total numbers may be on the high side. Significantly, more than half of this total figure will be the F-7-III, a modification of the MiG-21, an aircraft first designed in the 1950s. Army, Navy, and Second Artillery units all face similar challenges with the majority of equipment in their inventories.

All estimates of this type are based on imperfect information and are likely to be proven inaccurate in many details over time. However, the general trend indicated above cannot be denied—unless a drastic political decision is made by Beijing to change the priority for funding PLA modernization, the Chinese armed forces will continue to be equipped with older, but modified, equipment well into the 21st century.

Falling back on their Red Army heritage, the senior Chinese military leadership has emphasized that they will have to learn to make do with what they have got by

---

[34]Willy Wo-Lap Lam, "More Cuts Planned for Lean, Mean PLA," *South China Morning Post*, June 25, 1998.

creating new tactics and techniques that will optimally employ their existing weapons to defeat an enemy with high technology weapons. Chen Bingde joins the chorus as he repeats this mantra:

> We must focus on defeating a strong and superior force with a weak and backward force . . . our Army still generally must rely on inferior weapons and equipment to defeat enemies with superior weapons and equipment . . . Comrade Jiang Zemin pointed out that we must study strategies and tactics to defeat the enemy with our Army's existing weapons and equipment, especially the strategic concept of fighting a people's war under conditions of high technology.[35]

Table 3

Chinese Fighter Aircraft in the Year 2010

Type	Number
F-7-III	480
F-8-II	240
Su-27	128
F-10	30
Total	928

Source: Kenneth W. Allen, "PLAAF Modernization: An Assessment," in James R. Lilley and Chuck Downs (eds.), *Crisis in the Taiwan Strait*, Washington DC: National Defense University Press in cooperation with the American Enterprise Institute, 1997, p. 244.

Units all over the army are investigating ways to implement this directive. A Group Army in the Shenyang Military region attacked the problem with vigor:

> They mobilized the masses in launching the activity in which everybody assiduously studied and thought out "a few methods by which the inferior can defeat the superior." Over the past 15 months, everywhere in the barracks there have been fiery scenes of "I offer a stratagem or a method for 'winning a hi-tech war'." From the armies and divisions down to the companies, more than 320 teams, formed to tackle key problems, have been active on the training grounds, staging contest platforms at every level. All people, be they generals or soldiers, have got into action and racked their brains to think up methods for "winning a hi-tech war" and defeating the enemy.[36]

Their efforts were successful in that they:

> have attained more 320 achievements, such as the "mechanized army group's wartime ammunition supply system" capable of raising work efficiency by 108-fold, the "rocket mortar ground-wind allowance automatic measurement and calculation equipment" capable of raising shooting accuracy by 10-fold, and the "tank rapid-warmup system" which raises the capability of mechanized units to set out quickly under bitter cold conditions. They have "grafted" their hi-tech achievements to the

---

[35]Chen Bingde, "Intensify Study of Military Theory To Ensure Quality Army Building."

[36]Zhang Haiping and Liu Shiren, "Army Group Under Shenyang Military Region Starts Mass Training in Science and Technology," *Jiefangjun bao*, May 21, 1998, p. 1, in FBIS-CHI-98-155, June 4, 1998.

existing equipment, inventing more than 120 methods for countering hi-tech weapons of the powerful enemies, such as thermal imaging surveillance and electronic jamming. In the whole army, they are the first to realize a leap from surface to armored cars in terms of "field command automation system." Given their success in attaining 800-plus achievements and solving 280-plus difficult problems of "winning a hi-tech war," the mighty mechanized troops can move more quickly and become even stronger in the hi-tech battlefields.[37]

A significant aspect of this report is that it indicates one element of "using existing weapons to defeat a high tech enemy" simply involves improving the performance of equipment that has been in the inventory for decades. For example, the "rocket mortar ground-wind allowance automatic measurement and calculation equipment" and the "tank rapid-warmup system" probably do not involve great technological innovations. This implies that for many years the PLA's training on this equipment was performed at less than maximum capability under less than realistic modern battlefield conditions. Had these units actually been training consistently under realistic conditions, they would have confronted and been forced to solve many of these problems much earlier. That such an effort to develop methods to operate their weapons and equipment systems at maximum effectiveness was undertaken only in 1997 says much about the previous state of training in the PLA. On the other hand, the seriousness with which they have applied themselves to overcoming this problem indicates a step up on the ladder of military professionalism.

The same spirit is also being applied in theory to information warfare of the 21st century. It is evident that the PLA has studied assiduously the 1991 Persian Gulf campaign. Nearly all the writings about future battle plans begin with attack on enemy command and control and air defense units:

> we should learn to fight an information battle by relying upon existing equipment. After an information battle starts, we should immediately launch and all-round attack on the enemy's C3I system by relying upon artillery, airmen, and campaign strategic missile units, and dispatch special units to an enemy's rear . . .[38]

The untested question is whether such intentions can be successfully executed. So far, most of the techniques and tactics the PLA has developed are the result of academic studies of conflicts involving foreign militaries. There is little indication that the PLA has tested any new techniques they have developed *against actual high technology weapons*. They simply do not have access to the kind of weapons and systems they are seeking to defeat for them to test the effectiveness of their innovations. While they can quantitatively evaluate whether they have improved the effectiveness of their weapons, the PLA cannot be confident that the theoretical methods they have developed to defeat high technology weapons will be successful on a modern battlefield.

Perhaps, the large numbers of existing weapons will best fit into camouflage, concealment, and deception (CC&D) schemes. The vast majority of existing

---

[37]Ibid.

[38]Zhang Deyong, Zhang Minghua, and Xu Kejian, "Information Attack," *Jiefangjun bao*, March 24, 1998, p. 6, in FBIS-CHI-98-104, April 14, 1998.

weapons in the PLA inventory, even when their capabilities are maximized by equipment modification or employment techniques, simply do not have the range to be used in an offensive manner against many modern high technology weapons systems with long-range target acquisition, stand-off, and precision strike capabilities. As PLA leaders have the opportunity to observe personally modern military capabilities as part of their foreign diplomacy efforts, a telling indicator of their understanding of modern warfare will be if they continue to believe that existing weapons are capable of defeating a high technology foe.

Foreign Imports vs. Local Production

In the 1997 book entitled *The Third-Generation Leadership Group of the Party and the Building of the Quality of Armed Forces*, published by the Chinese Commission of Military Sciences and the Academy of Military Sciences, Chengdu Military Region Commander Liao Xilong states that:

> Jiang Zemin has emphasized time and time again that self-reliance should be the key word in strengthening our Army's modernization. Judging by this, in developing its arsenal for cross-century purposes, the PLA will continue to adhere to the principle of mainly relying on self-reliance and drawing on foreign experience to a limited extent. As far as some leading-edge weapons are concerned, in particular, domestic production will be the top priority.[39]

The balance between self-reliance and foreign import has long been a matter of debate, but appears to have been resolved with the emphasis on self-reliance. Speaking at the macro-planning level, Cao Gangchuan, currently director of the General Equipment Department, is quoted in the book mentioned above when he was Minister of the Commission of Science, Technology, and Industry for National Defense (COSTIND):

> Recently, Jiang Zemin pointed out that at present and for some time to come, it would be impossible to improve all the weapons and equipment of the PLA. It is imperative to identify priorities and find out what needs to be done and what can be left aside for the time being. In particular, we must make up our minds to concentrate financial resources, materials, and research resources on the research and development of critical technologies and critical weapons, in order to achieve breakthroughs and innovations . . . On the one hand, we should focus on achieving a breakthrough in key technologies . . . we should set our eyes on the leading edge of science and technology world-wide . . . On the other hand, we should focus on the development of new-generation weapons and equipment. In the scheduling of defense research programs, substantive measures should be taken to strike an overall balance between demand and possibility; as far as financial resources allow, make up our mind to cut non-key projects . . .[40]

These words indicate that hard choices in priorities must be made. The acquisition of limited numbers of a few types of foreign equipment has been approved.

---

[39] Kuan Cha-chia, "Military Regional Commanders Express Support for Jiang Zemin, Military Works Out Development Plans for the 21st Century," *Kuang chiao ching*, No. 300, September 16, 1997, pp. 12–17, in FBIS-CHI-97-288, October 15, 1997.

[40] Ibid.

However, the PLA leadership would prefer that most of the new equipment entering the force be of Chinese origin. And, for the most part, that means the majority of the PLA's equipment will still lag behind world standards. Therefore, the PLA leadership has resigned itself to a mix of old and new equipment for the foreseeable future. As Chen Bingde says, the PLA must "energetically explore new methods of operations to make use of the combination of high-, medium-, and low-grade weapons in combat."[41]

The ability of the Chinese defense industries to produce advanced weapons and deliver them in large numbers to the forces is debatable. Shenyang Aircraft Corporation is reported to have begun production on the first Su-27 to be assembled from knockdown kits supplied by Russia.[42] The annual production target is 10–15 aircraft, which will not be achieved for several years. Annual production of the F-8 series fighter is estimated to be about 24 per year and F-7 about 50.[43] The F-10 reportedly has recently made its initial test flight; flight-testing could go on for up to two years before it goes into production.[44] If the F-10 goes into production early in the 21st century, it will probably replace F-7 production at Chengdu. The Allen, Krumel, and Pollack RAND study referenced earlier suggests that China cannot afford more than one full-scale primary fighter development program at any one time.[45] Thus, it is likely that F-8 or Su-27 production will suffer. Once the F-10 reaches full-scale production, it *could* reach 75 aircraft per year and become the mainstay of the early 21st century PLA Air Force.[46] However, based on the aviation industry's past experience, a production figure of 75 aircraft a year is a highly optimistic goal, and unlikely to be attained within the first decade of F-10 serial production. In any case, as demonstrated in Table 3 above, it will take many years before the F-10 outnumbers the older F-7s in the PLA Air Force's inventory.

Unless military procurement budgets are drastically increased, total fighter production will be about 100 aircraft per year after the turn of the century. Allen et al. predict a 45% drop in the numbers of the fighter force if existing production rates are continued.[47] For a point of reference, the Soviet Union at the end of the 1980s produced 575–625 fighters and fighter-bombers, mostly of the fourth-generation represented by the Su-27 and MiG-29.[48] Thus, a policy of self-reliance in military equipment production will result in a significantly smaller, if technologically improved, force. Production at such a pace can hardly be characterized as "rapid military modernization."

---

[41]Chen Bingde, "Intensify Study of Military Theory."

[42]"Beijing Builds Su-27 Fighters from Russian Kits," *Jane's Defence Weekly,* June 10, 1998, p. 12.

[43]John Frankenstein and Bates Gill, "Current and Future Challenges Facing Chinese Defense Industries," *The China Quarterly Special Issue: China's Military in Transition,* No. 146, June 1996, p. 413.

[44]"First Flight for F-10 Paves Way for Production," *Jane's Defence Weekly,* May 27, 1998, p. 17.

[45]Allen, Krumel, and Pollack, p. 165.

[46]Frankenstein and Gill, p. 415.

[47]Allen, Krumel, and Pollack, p. 164.

[48]*Military Forces in Transition 1991,* Washington: Government Printing Office, 1991, p. 23.

A similar situation exists in all the defense industries. Modern equipment is likely to continue to be introduced only gradually to selective units in all services over the next decade or more. Contrary to the desires of the PLA leadership, most truly modern military equipment introduced into the force, with a few exceptions, will be of foreign origin well into the next decade.

The singular important exception to this condition may be strategic, fixed communications. The PLA has benefited, like the rest of China, by the opportunity to skip a generation of hard-wire telephony by moving quickly into optical fiber, mobile, and satellite communications systems. (This sector may be the best, and only(?), example of a real leapfrog in technology.) These advancements will enhance national strategic command and control, but will only improve battlefield communications on the margin. Most of these new systems have yet to be transformed into reliable, survivable, mobile, tactical communications equipment available to the lowest unit level. As the U.S. Army has discovered in its attempts to digitize its tactical operations centers (TOC), the common computer equipment that works well in an office environment requires "huge quantities of power cables and computer connector cables" to operate in the field.[49] These cables and their electric generators make the U.S. TOCs difficult to move, and unless they are mobile, they are unlikely to survive on a modern battlefield.

Some communications equipment, like beepers, mobile telephones, and hand-held commercial radios, are currently in use in the city and in administrative environments. However, not all of them are applicable for use in the field where conditions are much more harsh and a supporting infrastructure does not exist (such as relays for beepers and cell phones). Some communications systems, such as the Iridium satellite communications system, overcome these obstacles (and will be used by U.S. forces). But these foreign systems are very expensive now, and therefore will probably be only in experimental use in the PLA for the near- to mid-future.

Surprisingly, the PLA leadership appears to be ready to accept this state of affairs. A slow introduction of modern equipment allows for personnel to be trained to operate and maintain it, whenever it arrives. According to *Wide Angle*, Jiang Zemin has set the requirement, particularly for the navy, air force, and Second Artillery, that "we should let qualified personnel wait for the arrival of equipment rather than let equipment wait for qualified personnel to operate it."[50]

One final point related to self-reliance is the PLA's fascination with "secret weapons." The Chinese military literature is replete with references to developing "'secret weapons' that can effectively have the enemy by the throat," as Chief of the General Staff Fu Quanyou wrote in March 1998.[51] These weapons may include methods of attacking information and electronic systems, advanced physics weapons, or low-

---

[49] Colin Clark, "Key Force XXI Systems Fail Grade," *Defense Week*, June 15, 1998, p. 1.

[50] Kuan Cha-chia, "Military Authorities Define Reform Plan; Military Academies To Be Reduced by 30 Percent," *Kuang chiao ching*, No. 306, March 16, 1998, pp. 8–9, in FBIS-CHI-98-084, March 25, 1998.

[51] Fu Quanyou, "Make Active Explorations, Deepen Reform, Advance Military Work in An All-Round Way," *Qiushi*, No. 6, March 16, 1998, pp. 2–6, in FBIS-CHI-98-093, April 3, 1998.

yield tactical nuclear weapons. One problem the PLA may face with this type of weapon is keeping them secret while testing their effectiveness and perfecting methods of employment. Without testing and doctrine for employment, the final military utility of "secret weapons" is problematic.

MODERN CAPABILITIES

Of course, many in the Chinese military are not completely happy with such a strategy of equipment modernization. Instead of setting priorities, some call to do it all at the same time, especially when it comes to weapons and equipment needed for information warfare. Passages such as this in the *Liberation Army Daily* newspaper are not uncommon:

> If we take the matter lightly and let the opportunity slip past, we will once again be discarded by history when developed countries have completed their work of building an information army by the middle of the 21st century. The opportunity created by the new military revolution is a chance of a lifetime. Our army enjoys many favorable conditions for informationization.[52]

The author then goes on to say that "it is quite obvious" that in the reform of the structure of military organizations, equal attention should be paid to firepower, mobility, and the rapid flow of information.[53] A big order, but one that seems to cover the priorities in military capabilities the PLA has set for itself. The Chinese military has identified selected systems with the following capabilities as a focus of its equipment modernization program:

- Long-rang precision attack
- Air operations
- Naval operations
- Information warfare
- Space operations.

Long-Range Precision Attack

The weapon that first comes to mind with the capability for long-range precision attack is the cruise missile. China has several types of air- and sea-launched cruise missiles, but none is capable of attacking land targets. A land-attack cruise missile must be a high priority for development or acquisition. These weapons will give the navy and air force capabilities needed for several local war scenarios. Over the next decade, new cruise missile-equipped units can be expected to be added to the PLA force structure as existing ones are upgraded with more accurate and powerful versions of weapons in the inventory.

---

[52]Wang Baocun, "Talk on Deepening Reform."
[53]Ibid.

In the ground forces, precision guided munitions (PGM) can be expected to be distributed to existing artillery and tank units. PGMs will enhance the capabilities of the artillery by their ability to hit discrete targets. Most forms of artillery-delivered PGMs require that the target be designated by a device, such as a laser aimed by personnel on the ground or in the air. Thus, secure, reliable, and rapid tactical communications links between forward observers and firing units are essential. Precision-guided anti-tank rounds can be fired by ground troops, artillery, tanks, or helicopters. Again, communication is as important as the weapons themselves. PGMs will probably first be imported in small quantities, with the eventual goal of mass-production by the indigenous Chinese ordnance industry. Their introduction into the force will require minor structural changes to ensure that the targeting and communications requirements can be achieved.

PLA ground forces are likely to put priority on building helicopter units. Currently only extremely limited numbers of helicopters are found in the force. However, the PLA's command and control, reconnaissance, mobility, and attack capabilities could all be greatly enhanced by additional helicopter formations at lower echelons of the ground forces. A major investment here could prove to be one of the army's most important decisions in shaping the force for the 21st century.

The PLA historically has looked at its strategic missile force as an extension of its conventional artillery, hence the name Second Artillery. Battlefield and strategic missiles are incorporated routinely into battle plans. Given the Congressional investigations that began in the spring 1998 concerning alleged U.S. technology transfers which may have led to improvements in the Chinese missile force, it is unnecessary to mention that the PLA seeks to improve the accuracy of these weapons, both tactical and strategic.

Until the PLA can build a more modern and effective conventional force, the role of cruise and ballistic missiles will become increasingly more important. These two weapons are the PLA's most visible *modern, high technology* weapons and their psychological value will continue to be emphasized for deterrent purposes. It is likely that they, and possibly China's nuclear forces, will increase in numbers gradually in the first decade of the 21st century.

Alastair Iain Johnston writes that some Chinese military strategists may have determined that China ought to upgrade its nuclear force from its current minimal deterrent capability to one capable of "limited deterrence."[54] To the Chinese, their existing minimal deterrence force requires only the ability to carry out a simple, undifferentiated countervalue second strike. Very few warheads are needed to accomplish this task, and the small number of weapons leaves the force vulnerable to an opponent's first strike.[55] These analysts advocate that China should instead build a limited deterrence force, capable of limited counterforce warfighting. One Chinese study determined that such a force would be required to:

---

[54] Alastair Iain Johnston, "China's New 'Old Thinking'," *International Security*, Vol. 20, No. 3, Winter 1995/96, p. 5.

[55] Ibid. p. 18.

- Strike enemy strategic missile bases and weapons stockpiles, major naval and air bases, heavy troop concentrations, and strategic reserve forces, thus destroying the enemy's strategic attack capabilities;

- Strike at the enemy's theater through strategic political and military command centers and communications hubs, thereby weakening its administrative and command capabilities;

- Strike at the enemy's strategic warning and defense systems;

- Strike the enemy's rail hubs, bridges, and other important targets in its transportation networks;

- Strike basic industrial and military industrial targets;

- Strike selectively at several political and economic centers so as to create social chaos; and

- Launch warning strikes in order to undermine the enemy's will to launch nuclear strikes, and thereby contain nuclear escalation.[56]

A limited deterrence force would be able to respond to any level of attack—from tactical to strategic—with an option appropriate to the scope of the initial attack. One set of Chinese strategists argues that such a force would require:

- A greater number of smaller, more accurate, survivable and penetrable ICBMs;

- SLBMs as countervalue retaliatory forces;

- Tactical and theater nuclear weapons to hit battlefield and theater military targets and to suppress escalation;

- Ballistic missile defenses to improve the survivability of the limited deterrent;

- Space-based early warning and command and control systems; and

- Anti-satellite weapons to hit enemy military satellites.[57]

Johnston concludes that China does not now have the operational capabilities to implement this vision. Rather, this proposal appears to establish a wish list of capabilities from which Beijing must choose within the economic, technological, and arms control constraints the nuclear modernization program faces.[58] If Beijing made the decision to do so, Johnston assesses *China has the technical capacity to increase the size of its nuclear forces by about two to three times* and to improve its operational flexibility to be better able to execute a doctrine of limited deterrence.[59]

---

[56]Ibid. p. 20.

[57]Ibid.

[58]Ibid. p. 6.

[59]Alastair Iain Johnston, "Prospects for Chinese Nuclear Force Modernization: Limited Deterrence Versus Multilateral Arms Control," *The China Quarterly,* No. 146, June 1996, p. 548.

However, there is no authoritative evidence to confirm that the senior Chinese leadership has made the political decision to adopt a limited deterrence doctrine or that such a doctrine is being translated into military plans.[60] The Chinese writings cited above may only be that part of the debate accessible to outsiders. Foreigners simply do not know which theorists have the greatest influence on Chinese decisionmakers and what nuclear doctrine or force structure has been adopted by Chinese warfighters for the 21st century. The move to a limited deterrence force would emphasize the need for a wide array of precision attack weapons and the command, control, communications, computer, and intelligence (C4I) systems necessary to acquire and target long-range weapons.

Air Operations

The primary trend in the development of the PLA Air Force force structure was summed up by its commander Liu Shunyao as a "switchover of the air force from air defense to combined offensive and defense."[61] The principal role of the vast majority of the aircraft in the forces has historically been local defense of the Chinese mainland. That is the mission most pilots have trained for, and command and control systems have been designed to support. However, as can be seen from the purchase of Su-27s and the efforts to develop in-flight refueling and airborne command and control capabilities, the emphasis in the past decade has switched to acquiring an offensive-oriented force projection capability. Newer air-to-air missiles and air-launched cruise missiles will be an essential element of this aspect of modernization. The ground-based, logistics support for newer, more offensive-oriented units will grow as weapons systems become more sophisticated. More civilian technicians will probably be needed to keep the modern systems operational. The trend toward larger logistics units also will be found in the other services as the number of high technology systems increase throughout the PLA.

The number of long-range transport units in the force is also scheduled to increase as units dedicated solely to air defense decrease. A larger, long-range air transportation capability is essential as the PLA seeks to improve its strategic mobility. Long-range transport will be necessary to support not only the air force's operations in various parts of the country, but also ground and naval operations. Because of their cost, these larger aircraft will probably be added to the force incrementally. However, as China's strategic airlift expands, Beijing must be prepared to explain this, and other modifications in the force structure, to its regional neighbors or risk the inference that these developments threaten China's neighbors and are destabilizing to the region.

---

[60]Personal correspondence with Dr. Johnston, May 1998.

[61]Kuan Cha-chia, "Military Regional Commanders Express Support for Jiang Zemin."

Naval Operations

The $64,000 question is when will the PLA Navy deploy an aircraft carrier? According to the June 3, 1998 issue of *Jane's Defence Weekly*, the answer is "China is prepared to wait until 2020 to have a fully functioning aircraft carrier at sea."[62] China believes it needs a carrier to complete its naval modernization plans. Currently, it appears that the decision has been made to build one in China rather than buy one from abroad. *Jane's* reports that the Central Military Commission is prepared to wait until the year 2000 to begin a two-year feasibility study on the project, which is then estimated to take 18 years to complete construction, fitting out, sea trials, and training. Funding of $500 million for the program has not yet been secured.[63]

If this is the case, then the PLA Navy has more time to incorporate the capabilities that will allow an aircraft carrier at sea to survive beyond the first seconds of a high-intensity exchange. The PLA Navy's shortfalls in air defense and anti-submarine capabilities are well documented, not to mention its shortcomings in logistical support at sea. A decision to delay the introduction of a carrier will allow the PLA Navy time to build the capabilities, train the personnel, and form a battle group to protect a carrier. As a result, additional modern destroyers, frigates, logistics support ships, and submarines are likely to be added to the force before the one high-value, high-profile carrier becomes a reality.

Another PLA Navy deficiency, modern amphibious ships and craft, also can be expected to be a focus of acquisition efforts. These vessels will be particularly important if the decision is made to expand the size of the marine force.

Information Warfare

The precise manner in which information warfare (IW) will affect the PLA force structure is difficult to predict. As *China Electronic News* points out, information warfare is "a style of warfare; it is not a category of war . . . IW has to do with the substance of warfare . . ."[64] The article divides IW weapons into three types:

1. Weapons that destroy information infrastructure, such as telecommunications systems, electrical power systems, transportation systems, etc.

2. Weapons that use procedures to induce powerful psychological reactions in personnel and control their actions.

3. Weapons that use wireless suppression methods to defeat the enemy's electrical, sonar, or infrared equipment.[65]

---

[62]Paul Beaver, "China Will Delay Aircraft Carrier," *Jane's Defence Weekly*, June 3, 1998, p. 26.

[63]Ibid.

[64]Liang Zhenxing, "China: New Military Revolution, Information Warfare," *Zhongguo dianzi bao*, October 24, 1997, p. 8, in FBIS-CHI-98-012, January 12, 1998.

[65]Ibid.

The Chinese military literature is full of discussions about IW and this article was chosen as only an example. However, it implies that IW capabilities will be added onto, and incorporated into, existing and future forces. It is unlikely that large organizations will be designated specifically as IW units; however, most, if not all, units will have IW missions.

The Chinese often look to "secret weapons" under development by their defense industries to be applicable to IW. According to Chinese theory, these future weapons expand the three dimensional concept of military operations of air, land, and sea to include the additional operational dimensions of electromagnetism and space. A significant portion of research and development efforts has been focused on what are known as "advanced physics weapons," some of which may have nuclear components. An article in *Contemporary Military Affairs* noted that:

> the weapons systems produced by the third military revolution mainly use sound, electromagnetism, radiation, and other destructive means. Operational actions in which armed forces use radiation-damaging energy to strike at the enemy's electronic equipment, weapons systems, military equipment and personnel, and other military targets are called "radiation combat." The main radiation weapons are laser weapons, microwave weapons, particle beam weapons, and subsonic wave weapons; they possess enormous military potential.[66]

Significantly, but left unstated in the Chinese article, some weapons used to conduct "radiation combat" may have a nuclear device as an integral component of the weapon. The enhanced radiation warhead (i.e., "neutron bomb") is the most obvious example. Other weapons, such as electromagnetic pulse (EMP) weapons, may use small nuclear reactions to initiate a powerful secondary effect, such as the local disabling of electromagnetic systems like computers. If such a weapon were to be used, the threshold for the employment of other, more traditional nuclear weapons would become less distinct than it is today. The lowering of the nuclear threshold may be an important unintended consequence of the pursuit of advanced weapons to conduct information warfare in the future.

Space Operations

The Chinese have accepted that space will be an integral dimension of warfare in the 21st century. Generally, operations in space fall into two categories: 1) weapons, including missiles traveling through space or space-based systems that can be used against missiles, satellites, or targets on the surface of the earth, or 2) support to operations, such as communications, global positioning, intelligence collection, and weather systems. Though the criticality of space systems is not disputed, the cost, technical feasibility, and suitability of the whole array of space systems are a matter of debate.

---

[66]Chen Huan, "The Third Military Revolution," in Michael Pillsbury (ed.), *Chinese Views of Future Warfare*, p. 396.

Officially, Chinese policy advocates a complete ban on weapons of any kind in outer space.[67] Based on this policy, the relatively low national priority given to military modernization, and the limited resources likely to be available to the PLA and defense industries, it appears that the most likely course to be pursued will be one concentrating on space-based support operations. Two officers from COSTIND's Command Technical Academy writing in *China Military Science* concluded:

> Economically, the development of a space force consisting mostly of information support might is now most economical . . . in the current stage, the technology is advanced and mature enough for building a space force consisting mostly of information support might . . . But building a large-scale space attack force would be very risky technically, as well as exceeding the economic limits of national might.[68]

Space-based systems, as well as the other capabilities discussed above, will all contribute to the PLA's ability to deploy rapidly and conduct joint and combined arms operations. The Chinese military literature and developing doctrine have fully embraced these concepts. To implement them, however, changes will have to be made in existing basing arrangements and command and control structures.

ADJUSTMENTS IN THE COMMAND AND CONTROL STRUCTURE

As long as the PLA is "subject to the leadership of the Communist Party of China,"[69] one of its military objectives will be to protect the Party. To do so, there will be a need to continue the political commissar system that parallels the operational chain-of-command. At the top of the hierarchical order will be the Central Military Commission (CMC). There is no indication that any major changes to the existing political structure are being contemplated, though some of its manpower may be reduced slightly.

On the other hand, there has been discussion about strengthening the office of the Ministry of National Defense (MND) so that the Defense Minister has institutional power, in addition to his personal power and influence derived mostly from his position on the CMC. One controversial way for this to occur would be to appoint a civilian to be Defense Minister.[70] Such a decision could be interpreted positively by the international community as a further separation of the PLA from involvement in national politics. It would also resemble the civilian command of the military found throughout much of the world. A civilian Defense Minister would probably focus mostly on the larger issues of national military-political strategy. To do so would require a staff, some of which may be civilian deputies or assistants. The establishment of such a system could assist the Chinese military's foreign diplomacy by creating Chinese counterparts for civilians often found in other nations' defense

---

[67] *China's National Defense,* July 1998.

[68] Ping Fan and Li Qi, "A Theoretical Discussion of Several Matters Involved in the Development of Space Forces," *Zhongguo junshi kexue,* May 20, 1997, pp. 127–131, in FBIS-CHI-97-302, October 29, 1997.

[69] Article 19, National Defense Law.

[70] See Willy Wo-Lap Lam, "Proposal to Give Civilian Top PLA Job," *South China Morning Post,* June 1, 1998, Internet Edition.

establishments. The initial difficulty of this proposal would be finding candidates with appropriate experience with defense issues. For the purpose of this analysis, the civilianization of the Defense Ministry will be assumed as well as inserting the MND in the chain-of-command between the CMC and the General Departments.[71]

However, the most important changes in the PLA's structure that affect its operational capabilities will be found at the levels of command one and two levels below the General Departments and within the forces themselves. As Chen Bingde has written:

> Our Army's command system is still far from meeting the requirements to organize and direct local warfare under the conditions of high technology . . . Our methods in conducting military operations are relatively backward, and our command system remains inefficient. Our reconnaissance, early warning, command and control, and electronics countermeasures capabilities are still relatively weak.[72]

One method of improving the command system that apparently has been proposed is the reduction of Military Regions from seven to five.[73] There appears to be debate about what the five new headquarters will be called—possibly "theaters," "war zones," or no change in name. Five "theaters" (used for lack of a better term) could be drawn to logically divide the major strategic directions China must defend:

- The Northeast, oriented toward Russia, Korea, and Japan;

- The Northwest, oriented toward Central Asia and Russia;

- The East, oriented toward Taiwan and Japan;

- The South, oriented toward the South China Sea, Indochina and India; and

- The Central Reserve and Capital Region, primarily used as a holding area from which additional troops can be dispatched to China's four corners, as well as protection of Beijing.

These headquarters are likely to be smaller in size than existing Military Region headquarters, with some current local functions assumed by Beijing. Nan Li writes that these new joint commands will give prominence to the command departments and the battlefield functions of intelligence, decision control, communications and electronic warfare, and fire control and coordination.[74] As the PLA gets smaller, it may actually be easier to exercise central control and standardization than it has been with larger forces. A true indicator of the PLA's commitment to joint operations would be for the commander of the Eastern or Southern Theaters to be a naval officer or the Central Reserve/Capital Region commander to be an air force officer.

---

[71] However, there was not one reference to the Ministry of National Defense or the Minister of Defense in the 1998 Defense White Paper. Therefore, this assumption may prove to be premature.

[72] Chen Bingde, "Intensify Study of Military Theory."

[73] Hsiao Peng, "Seven Major Military Regions To Be Changed Into Five Major Theaters—A Great Change in PLA Commanding System Is Under Deliberation," *Sing tao jih pao*, April 15, 1998, p. A4, in FBIS-CHI-98-105, April 15, 1998; and Willy Wo-Lap Lam, "PLA Faces Streamlining of Regions," *South China Morning Post*, April 16, 1998.

[74] See Nan Li's chapter in this volume.

One way to give the ground force more clout as it suffers the bulk of manpower reductions would be to form an "Army Headquarters" subordinate to the General Staff Department, equal in status to the headquarters of the navy and air force. Army headquarters would be responsible for training, manning, doctrine, and equipment policy for the entire ground force. The General Departments would then be responsible for coordinating the policies and efforts of all the services. If such decisions were taken, then the PLA would appear to be adopting a U.S.-style Joint Staff system. However, the Chief of the General Staff would probably be first among equals at the General Department level, and no position of "Chairman" created. Those chairman duties would be reserved for the CMC.

As mentioned earlier, the PLA during the 500,000-man reduction will experiment with organizational structures. Some headquarters elements will be eliminated. Reportedly, up to 30% of PLA academies may be disbanded.[75] Within the ground forces, two things will probably occur to accommodate the reduction: 1) Many group armies will lose a division, usually an infantry division, as units are demobilized or transferred en masse to the PAP; and 2) several group armies, perhaps five or six in all, will be eliminated.[76] The group army headquarters to be disbanded will most likely come from the Shenyang, Beijing, and Jinan Military Regions, where there is a higher density of ground forces than in other parts of the country. In the process of the 500,000-man reduction, some units that survive the elimination of their higher headquarters will be reassigned to other remaining headquarters.

In the past, though group armies have appeared to be under the command of the Military Regions in which they are stationed, local commanders in reality have had only extremely limited authority to move troops. Any unit movement larger than a battalion or any movement outside a regional boundary has to be ordered by the CMC working through the General Staff Department.[77] Even as ground force units become fewer and more mobile, this rigid control system is unlikely to change.

An important indicator to watch will be whether larger ground force units consolidate their subordinate organic elements closer together to facilitate rapid deployment and combined arms training as units reorganize themselves. Presently, many group armies and divisions are spread over wide areas, with individual regiments often in isolated locations, which slows the time it takes for these units to marshal for deployment and makes routine combined arms training that much harder because of the distance units must travel to operate together. Therefore, it would seem logical that some of the smaller, reorganized combat units will consolidate at railheads or near airfields to improve their rapid response capabilities.

As the number of existing Military Regions is reduced, it will be necessary for naval and air force headquarters to follow suit. Thus, the air force could ultimately end up with five regional air forces corresponding to the "theaters" and the navy possibly could eliminate one fleet headquarters, probably the Northern Fleet, as it basically

---

[75]Kuan Cha-chia, "Military Authorities Define Reform Plan."

[76]"Rapid Deployment Key to PLA Modernization," *Jane's Defence Weekly*, April 15, 1998, p. 32.

[77]Swaine, *The Military & Political Succession in China*, pp. 122–124.

shifts to an eastward and southern orientation. Elimination of these headquarters will free up personnel slots for reallocation to technical support roles needed to sustain the new equipment entering the forces.

For new capabilities to be properly allotted throughout the forces, contrary to the general trend to reduce headquarters, two new smaller national-level headquarters may be formed: Space Forces and Special Operations Forces. As the PLA's capabilities in these two very specialized functions expand, their operations may become too complex for simple inclusion in existing headquarters. Moreover, these two functions will be involved in any future military scenario, so it seems reasonable for them to be controlled by central headquarters.

The Space Force Command would probably have the status of the other services and rank behind the Second Artillery in order of precedence. It would serve as a centralized location for the integration of communications and intelligence systems that will be essential for the conduct of any military operation. All theater commands will have access to the capabilities of this organization in routine planning and in times of emergency. It will be able to augment the theater headquarters as required.

The Special Operations Forces will be relatively small, composed primarily of ground troops, and could reasonably be subordinated to the new Army Headquarters as long as this headquarters retained the ability to go to the air force and navy for direct support as necessary. These troops can be expected to be the best of the ground forces, tasked with *strategic* long-range reconnaissance and surveillance missions, as well as precision strike at important enemy targets. These units will be separate from, but related to, tactical reconnaissance units found at lower organizational levels. Because of their strategic orientation, political sensitivity, and specialized training requirements, special operations units would best be consolidated at a

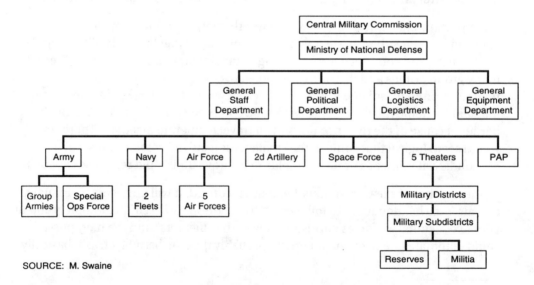

SOURCE: M. Swaine

Figure 2—Postulated PLA Command Structure

national level headquarters. The concept of dedicated special operations forces in the PLA is still in its infancy and its development will take time and significant resources.

SUMMARY AND CONCLUSIONS

The Chinese force structure developments speculated in this essay have been based on a straight-line projection of international and domestic security conditions. Major changes in the international or China's internal security situations could result in unforeseen modifications to the PLA to cope better with the new reality at hand.

In summary, these are the major trends foreseen:

- The PLA will be reduced in size to perhaps 1.5 million strong.
  - The percentages of naval and air forces will increase as ground forces decrease.
  - PLA civilians and business operations may be stripped from the active duty rolls.
- The numbers of reserves and People's Armed Police will increase.
- For the foreseeable future, units will have a mix of high-, medium-, and low-technology weapons and equipment and will strive to find ways to maximize the use of their existing equipment to defeat a high-technology enemy.
- The numbers and types of logistics and technical units will increase throughout the force to maintain and support the PLA's modern equipment.
- The Chinese defense industries will be able to produce limited amounts of modern weapons for the PLA, but most truly advanced weapons will be of foreign origin and relatively few in number.
- Rapid deployment of conventional forces will be enhanced through acquisition of transport ships and aircraft as well as by unit consolidation near points of embarkation.
- Naval and air forces will acquire more offensive capabilities and the ability to operate farther from the Chinese land mass, but an operational aircraft carrier capability will not enter the force until at least the end of the second decade of the 21st century.
- Cruise missile, ballistic missile, and nuclear forces will be improved gradually and incrementally and will remain the key to China's deterrent force.
- Changes in the command and control structure will contribute to better integration of forces and capabilities.
  - Several regional headquarters will be eliminated, resulting in five "theater-like" headquarters.
  - A few smaller headquarters will be formed for the Army, Special Operations Forces, and Space Forces.

- Tactical units will be restructured during a period of experimentation.

During the period of reorganization, it is likely that some units will suffer a decrease in effectiveness until all the kinks of the new structure are worked out. Eventually, as modern systems are linked together, the PLA will realize an improvement in overall capabilities. However, they will not be transformed into a force capable of long-range, sustained force projection for several decades to come. Integrating the pockets of modernity into integrated systems will probably be the PLA's biggest challenge. Force structure changes will not solve these problems by themselves. Training, doctrine, and attitudes are the key to systems integration.

Beijing must also contend with problem of how to explain its military modernization to its neighbors and the rest of the world. There are already many misperceptions about the pace and scope of China's military modernization that Beijing has not adequately addressed. For China to achieve its goals by the mid-21st century, it must find a way to inform the world in a credible manner about its national intentions. Any visible improvement in Chinese military capabilities will raise questions, particularly among China's Asian neighbors. The "Defense White Paper" is only part of the answer. Greater Chinese willingness to allow foreigners to observe and understand their forces is essential. When asked, the Chinese must be willing to answer questions, not simply respond with a dismissal of uncomfortable inquiries. Certainly, many details need remain secret, but a greater openness that can be verified through observation would contribute significantly to the perceptions of China's neighbors and other concerned observers.

No nation knows what its military will look like in 2010. This paper is based on no "inside" information and is, at best, only partially informed speculation. Changes in the international or domestic security situations could have major impacts on the future force structure. For example, China may feel compelled to more rapidly expand and modernize its strategic missile force if India builds a nuclear arsenal or if a Theater Missile Defense system is deployed in Japan or near Taiwan. Likewise, extended domestic unrest caused by economic and social change underway in China could force the PLA to reemphasize its secondary mission of ensuring domestic stability. Given these uncertainties, it is essential to monitor the trends identified above to determine how the PLA interprets its environment and translates its perceptions into a concrete force structure capable of achieving its national military objectives in the 21st century.